UNIT OPERATIONS IN RESOURCE RECOVERY ENGINEERING

UNIT OPERATIONS IN RESOURCE RECOVERY ENGINEERING

P. AARNE VESILIND
ALAN E. RIMER

Duke University
Durham, North Carolina

PRENTICE-HALL, INC.
Englewood Cliffs, N.J. 07632

Library of Congress Cataloging in Publication Data

VESILIND, AARNE P
 Unit operations in resource recovery engineering.
 (Environmental engineering series)
 Includes bibliographical references and index.
 1. Recycling (Waste, etc.) 2. Refuse and refuse
disposal. 3. Separation (Technology) 4. Refuse as
fuel. I. Rimer, Alan E., joint author. II. Title.
III. Series
TD794.5.V47 1980   628'.445   79-25757
ISBN 0-13-937953-3

© 1981 by Prentice-Hall, Inc.
Englewood Cliffs, New Jersey 07632

All rights reserved. No part of this book
may be reproduced in any form or by any means
without permission in writing from the publisher.

Printed in the United States of America
10 9 8 7 6 5 4 3 2 1

Editorial Production/Supervision by Theodore Pastrick
Manufacturing Buyer: Gordon Osbourne

PRENTICE-HALL INTERNATIONAL, INC., *London*
PRENTICE-HALL OF AUSTRALIA PTY. LIMITED, *Sydney*
PRENTICE-HALL OF CANADA, LTD., *Toronto*
PRENTICE-HALL OF INDIA PRIVATE LIMITED, *New Delhi*
PRENTICE-HALL OF JAPAN, INC., *Tokyo*
PRENTICE-HALL OF SOUTHEAST ASIA PTE. LTD., *Singapore*
WHITEHALL BOOKS LIMITED, *Wellington, New Zealand*

CONTENTS

1. SOLID WASTE CHARACTERISTICS, PRODUCTION, AND POTENTIAL FOR RESOURCE RECOVERY   1

Historical Background   1
Materials Flow in Society   3
Solid Waste Quantities and Characteristics   9
Potentials for Reclamation of Useful Materials
    and Energy from Solid Waste   36
Obstacles to Resource Recovery   37
Parting Shots   39

2. COLLECTION OF SOLID WASTE   43

Solid Waste Collection Systems   43
Effectiveness of Solid Waste Collection   64
Collection of Source-Separated Materials   69
Litter   73
Parting Shots   78

3. STORAGE, HANDLING AND SHREDDING OF SOLID WASTE   83

Storage and Conveying   83
Compaction   91
Shredding   93
Pulping   125
Roll Crushing   126
Parting Shots   129

4. MECHANICAL SEPARATION   134

General Expressions for Materials
    Separation   135
Hand Sorting   140
Screens   141
Air Classifiers   154
Jigs   177
Stoners   183
Sink/Float Separators   184
Inclined Tables   189
Shaking Tables   191
Flotation   194
Color Sorting   198
Parting Shots   200

5. MAGNETIC AND ELECTROMECHANICAL SEPARATION PROCESSES   205

Magnetic Separation   205
Eddy Current Separation   214
Electrostatic Separation   231
Magnetic Fluids   236
Parting Shots   241

6. BIOCHEMICAL PROCESSES   246

Methane Generation by Anaerobic Digestion   247
Methane Generation in Landfills   253
Composting   263
Glucose Production by Acid and Enzymatic Hydrolysis
    270
Other Biochemical Processes   275
Parting Shots   275

7. COMBUSTION — 280

Historical Background   280
Principles of Combustion   281
Thermal Balance in Combustion   287
Materials Balance in Combustion   289
Process Variables   289
Parting Shots   291

8. INCINERATION AND ENERGY RECOVERY — 294

Introduction   294
Incinerator Design Concepts   295
Other Incineration Processes   322
Waste Heat Recovery   331
Air Pollution Control   340
Parting Shots   352

9. PYROLYSIS — 357

Process History   358
Process Characterization   359
Process Control Variables   362
Reactors   366
Additional Pyrolysis Products   383
Environmental Concerns   387
Parting Shots   387

10. ULTIMATE DISPOSAL — 391

Options in Ultimate Disposal of Residues   391
Landfills   392
Hazardous Substances   417
Parting Shots   420

APPENDICES 426

Particle Size 426
Sieve Sizes 432
Analytical Techniques for Solid Wastes 435
Conversion Factors 442

PREFACE

Accelerated extraction and continuing scarcity of raw materials, coupled with environmental and ethical constraints on the disposal of wastes, has resulted in a steadily increasing interest in the recovery of materials and energy from solid waste—especially mixed municipal refuse. The estimates of the rapid development of this field range from phenomenal to merely impressive.

The problems associated with the extraction of energy and materials from a heterogeneous and time-variable waste such as municipal refuse has spawned a discipline known popularly as resource recovery engineering. The foundations for this profession lie mostly in mining, civil, chemical, and mechanical engineering, but also draw on the knowledge of biologists, chemists, and social scientists. In the sense that knowledge from many and diverse fields must be borrowed in order to accomplish the recovery of resources from wastes, it clearly is a "dirty" profession with few externalities.

The objective of this book is to borrow the necessary knowledge from these diverse sources so as to develop a comprehensive text on resource recovery engineering. We wanted this book to be more than a review of existing descriptive information on resource recovery facilities, and thus introduce the various materials separation and energy recovery operations from a fundamental viewpoint. Accordingly, this book provides the basic information for a rigorous analysis of the unit operations which can then be designed to develop complete resource recovery processes.

This book is not intended primarily to be a basic text on solid waste engineering, although the subjects of collection, landfills, and other standard solid waste topics are covered. This coverage is, however, more analytical than descriptive, and instructors wishing to use the book as a first text would probably find supplemental descriptive materials such as published by the Environmental Protection Agency to be useful.

The initial impetus for this book originated while the authors were working on a National Science Foundation grant for the development of course materials in resource recovery engineering. Much of the text by the senior author was prepared while he was on sabbatical leave as a Fulbright Senior Lecturer at the University of Waikato, in Hamilton, New Zealand. The cooperation and enthusiastic support of the University and especially Mr. Tom Fookes, the executive director of the Environmental Studies Unit, is greatly appreciated.

The initial impetus for this book originated while the authors were working on a National Science Foundation grant for the development of course materials in resource recovery engineering. Much of the text by the senior author was prepared while he was on sabbatical leave as a Fulbright Senior Lecturer at the University of Waikato, in Hamilton, New Zealand. The cooperation and enthusiastic support of the University and especially Mr. Tom Fookes, the executive director of the Environmental Studies Unit, is greatly appreciated.

Portions of the manuscript were reviewed and criticised by Ernst Schloemann, Charles O. Velzy, Jerry L. Jones, Raymond Ragan and L. G. Austin. In addition, many of our students at Duke contributed constructive suggestions, and their perspective was of significant value.

From the onset, the authors received valuable guidance and editorial assistance from their colleague, George W. Pearsall. His comments on and wholesale rewrites of some sections of the book were of immeasurable value, and his participation in this effort is gratefully acknowledged. Much of the typing and general organization of the manuscript was done by Ms. Judy Edwards, to whom we extend our appreciation.

P. A. Vesilind
A. E. Rimer

1
SOLID WASTE CHARACTERISTICS, PRODUCTION, AND POTENTIAL FOR RESOURCE RECOVERY

HISTORICAL BACKGROUND

The emergence of the industrial age fostered the science of economics, and prompted many leading thinkers to attempt to bring rational order to the seemingly chaotic world around them. The rationalism that resulted led to the common belief that trends could be understood—and decisions made—best on the basis of numbers. This substitution of the quantitative for the qualitative still pervades modern society and influences our entire set of attitudes toward resources and how they should be distributed. Adam Smith, through his concept of "the invisible hand," introduced an element of positive faith and optimism. However, his efforts were overshadowed by a number of pessimists—analysts who predicted continuing misery, poverty, exploitation, and class discrimination. Ricardo, with his "iron law of wages," held that wages for the working people would always remain at the poverty level, since any increase in wages would result in a commensurate increase in population, and this would once again drive wages down.

Equally pessimistic was the view held by Thomas Malthus, who in 1798 reasoned that since population growth is geometric and the increased production of food is arithmetic, a famine must result. This "law of population" was part of the "laissez-faire" school of economic liberalism, and was in great part responsible for the earned reputation of economics as a "dismal science." Malthus held that overpopulation can be prevented only by two types of checks: positive and preventive. Numbered among the former are wars, plagues, and similar disasters. Preventative checks include abstention from marriage, limitations on the number of children, and the like. Although the latter is clearly preferable, Malthus had little hope for the world, and insisted that the poor were "authors of their own poverty," simply because they failed to use the preventive checks on population growth.

This thesis was widely believed for many years and held as basic economic dogma. But as populations grew, widespread famine and deprivation was avoided, and Malthus's writings fell from favor. Economists began to think of Malthus as an economic anachronism—to be studied, but only in the historical context. Technology, the new God, was able to preserve order, avert disaster, and lead us into the promised land.

This optimism was widely shared during the nineteenth and well into the twentieth century, with only a few disquieting voices. Thoreau's distrust of things technical was tolerated with bemusement as the ramblings of an ungrateful crackpot.

In the later 1950s and 1960s, a few more voices in the wilderness became audible. Paul Erlich, with his grand overstatements and predictions of doomsday, seemed strangely reminiscent of Malthus. Barry Commoner became the first *public* ecologist, and helped promote the feeling of disquiet. Slowly, through the 1960s, the public became convinced that there may indeed be something to this "doomsday" talk.

The most respected and well-publicized voice of pessimism became an interdisciplinary group of scientists at MIT. Funded by the Club of Rome, a group of concerned industrialists, this group of talented scientists and engineers developed a computer model of the world, based on projections of pollution, agricultural production, availability of natural resources, industrial production, and population. Their ambitious undertaking, led by Dennis Meadows and Jay Forrester, resulted in the publication of the final report, which indicated that even our most *optimistic* projections will eventually lead to the onset of famine, wars, and the destruction of our economic system [1]. It was, in short, a dismal outlook. Malthus would have been pleased.

The Meadows report has been criticized for inaccuracies and misinterpretations, and some of these accusations appear to be valid. Indeed, a revised model has shown an increased chance for world survival [2] and more accurate data would seem to reduce the level of pessimism.

Nevertheless, the dismal outlook of Malthus is reaffirmed by Meadows, and we are beginning to realize that our planet is finite—that it has only limited resources and living space. The scarcity of land and nonrenewable resources could indeed have the ultimate devastating effect envisioned by Malthus and now once again suggested by Meadows. At the very least, the concern is real, and we should begin to seek alternative life systems in order to have more assurance that these disasters can be avoided.

One (of many) possible potentially beneficial alternatives toward global stability is to eliminate the solid wastes generated by our materialistic society which are now deposited on increasingly scarce land. The recovery of these resources from solid waste would be a positive step toward establishing a balanced world system where society is no longer dependent on extraction of scarce natural ores and fuels. It seems quite clear that society has to adapt, using less technology in some instances, more in others, in order to achieve this balance. The technology and philosophy necessary for the implementation of resource recovery is the topic of this text.

MATERIALS FLOW IN SOCIETY

Reasons for Resource Recovery

The flow of materials in our society may be illustrated by the schematic diagram shown in Fig. 1-1. This diagram emphasizes the fact that we do not "consume" materials; we merely use them and ultimately return them, often in an altered state, to the environment. The production of useful goods for eventual use by those people called "consumers" requires an input of materials. These materials originate from one of three sources: raw materials, which are gleaned from the face of the earth and used for the manufacture of products; scrap materials produced in the manufacturing operation; and materials recovered after the product has been used. The industrial operations are not totally efficient, and thus produce some waste which must be disposed of. The resulting processed goods are sold to the users of the products, who, in turn, have three options after use: to dispose of this material; to collect the material in sufficient quantities to either use it for energy production or to recycle it back into the industrial sector; or to reuse the material for the same or a different purpose without remanufacture.

It is instructive to note that this is a closed system, with only one input and one output, emphasizing again the finite nature of our world. At steady state, the materials injected into the process must equal the materials disposal back into the environment. This process applies to the sum of all materials as well as to certain specific materials. For example, the manufacture of aluminum beverage containers involves the use of raw material—bauxite ore—which is refined to produce aluminum. The finished product—the cans—are sold to the consumers. Some of these cans are defective or for other reasons unfit for consumer use and are recycled as industrial scrap. The consumer uses the cans, and the empty

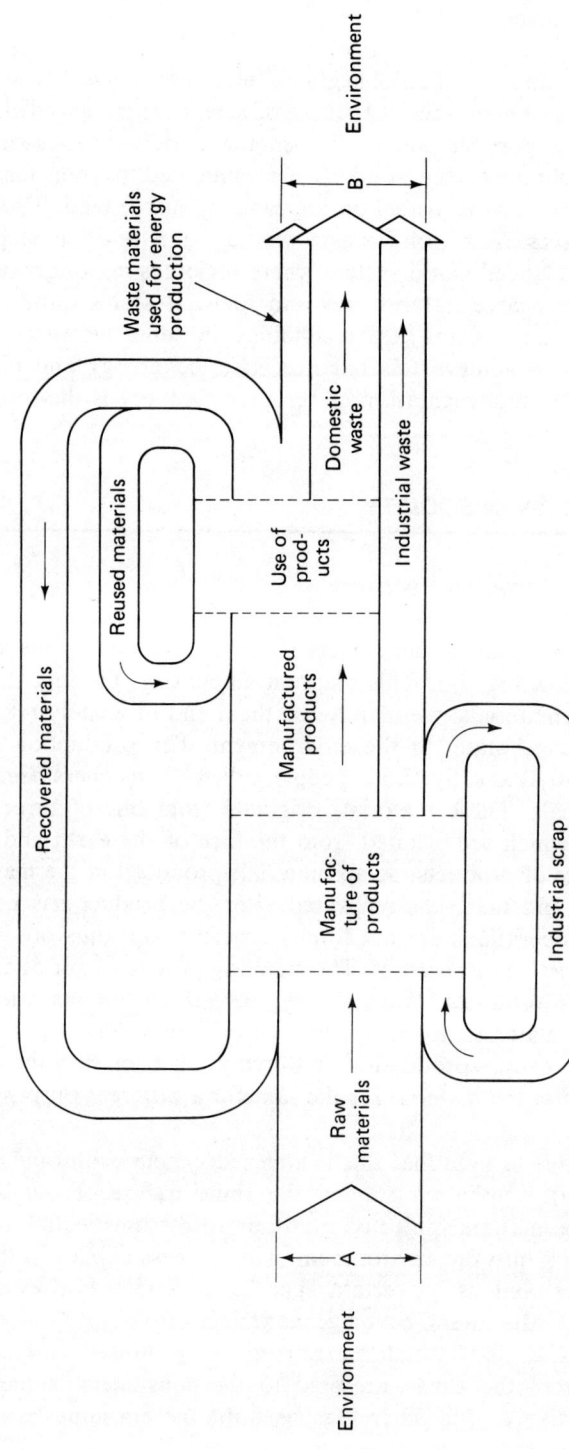

Figure 1-1. Flow of materials in society.

containers or other products are disposed of in the usual manner. Some of this aluminum is returned to the industrial sector (for remanufacture) and some of it might be used for other purposes in the home. The aluminum that is recovered and returned to the manufacturing process gets there only by a conscious effort by the community or other organizations which collect and recycle the material through the system. For many of the materials, this is often at a financial loss.

The interaction of the materials flow with the "environment" is at the input of raw materials and the deposition of wastes. In Fig. 1-1 these two interfaces are denoted by the letter "A" for raw materials and by the letter "B" for the materials returned to the environment.

It can be argued that both A and B should be as large as possible since there are many benefits to be gained by increasing these values. For example, a large quantity of raw materials injected into the manufacturing process represents a high rate of employment in the raw materials industry, which can have a residual effect of creating cheaper raw materials and thus reducing the cost of manufacturing.

A large B component is also beneficial in the sense that the waste disposal industry (which includes people as diverse as the local trash collector and the president of a large firm that manufactures heavy equipment for landfills) has a key interest in the quantity of materials that people dispose of. Thus a large B component would mean more jobs in this industry.

However, large A and B components also have detrimental effects. In the first place, a large raw material input means that great quantities of nonreplenishable raw materials are extracted (often using something less than environmentally sensitive methods, as exemplified by the present method of strip mining). Similarly, large quantities of waste can have a significant detrimental effect, in that wide land areas are being used for disposal of the waste, or that burning the waste in incinerators can result in serious air pollution problems in local situations.

A high rate of raw material extraction can eventually lead to a problem in the depletion of natural resources. At the present time in the United States we have already exhausted our domestic supplies of some nonreplenishable materials, such as copper, zinc, and tin, and are importing a substantial fraction of these materials [3]. It is obvious that if the rest of the world were to attain the standard of living that the developed nations have at the present, the raw materials supply would not be adequate to meet the demand. Our present life-style is based on obtaining these materials from concentrated sources (ores), and in using them we are distributing the products over a wide land area. Such a distribution obviously makes recovery and reuse difficult.

Finally, the question of national security for each country is predicated on the nation's ability to obtain reliable supplies of raw materials. We already have seen the problems that can be created by relying on other countries for such necessities as oil. There is little doubt that cartels will be developed by nations that have large deposits of other nonreplenishable materials, and that in the future the cost of such products as aluminum, tin, and rubber will increase substantially.

There is thus ample justification for reducing the wastes disposed of into the environment to the smallest quantities practical and we should clearly try to redesign our economic system to achieve this end.

Methods of Decreasing Raw Material Use and the Production of Waste Quantities

It is clear from Fig. 1-1 that if the system is in steady state, the input must equal the output. Hence a reduction of either A or B necessarily results in a concomitant reduction in the other. In other words, it is possible to attack the problem in two ways.

Looking first at the A component, a reduction in raw materials demand could be achieved by increasing the amount of industrial scrap reprocessed, by decreasing the amount of manufactured goods, or by increasing the amount of recovered materials from the postconsumer waste stream. Increasing industrial scrap would involve increasing either "home scrap" (waste material reused within an industrial plant) or "prompt industrial scrap" (clean, segregated industrial waste material used immediately by another company). But scrap represents inefficiency, and an ultimate goal of industry is to produce as little scrap as possible. Clearly, decreasing the demand for raw materials will require one of the other two approaches.

The second possibility for achieving a low use of raw materials is to decrease the amount of manufactured goods. This will necessitate a redesign of products in such a way as to use less material and less energy. The quantity of material used for manufactured goods might be reduced as a result of increased raw material cost, or it can be mandated by the government. Under the name "waste reduction," the federal government has evaluated several methods of legislating a lower rate of material use. Taxes on excessive packaging, a package charge (e.g., 1 cent/lb), mandatory longer life of manufactured products, and other options have been considered.

The third possibility is to increase the recovery of materials. If this is accomplished, the total amount of consumer goods produced and the amount of goods manufactured by industry need not be reduced, but it would be possible to reduce the raw materials input to this system and concurrently to reduce the amount of materials destined for disposal. This strategy seems not only feasible but economically, politically, and practically attractive.

Looking at the other end of Fig. 1-1, the disposal fraction, it is clear that the only two methods of reducing the quantity of materials to be disposed of is to increase the recovery and/or the reuse component, or to increase the use of waste for the production of energy. As defined in the figure, a reused product is one that the consumer can put to some other or to repeated use without the product going back to the manufacturer. On the other hand, the recovery of a material involves the remanufacture or processing of that material by industry. Both increased reuse and recovery will result in decreased raw material use. Increased use of waste for energy production will, on the other hand, only reduce the disposal quantities and will only indirectly affect raw material extraction.

In summary, it thus seems reasonable that the feasible options for achieving reduced material use and waste generation is by
1. Waste reduction.
2. Increased recovery.
3. Increased reuse.

The following paragraphs are devoted to a discussion of each of these potential methods of achieving reduction of solid waste and the use of natural resources.

Waste reduction

The savings in material use due to waste reduction programs could be significant. For example, an 80% shift to refillable beer and soft drink containers, which would have 18 trips to the bottling plant and back, is not unreasonable under the Oregon-type bottle legislation. Better automobile tires, which would last 100,000 to 130,000 km (60,000 to 80,000 mi), instead of the present 30,000 km (20,000 mi), certainly seem to be in the future. In almost all cases, products can be redesigned to produce a 10 to 15% increase in their life, at very minimal cost.

If just those three goals were achieved (refillable beverage containers, better tires, and longer product life), a reduction of 18 million tonnes* (20 million tons) of postconsumer waste, or a 15% reduction in the 1985 projected waste generation figures, is possible [4].

Waste reduction can be achieved in two basic ways: (1) reduction in the amount of material used per product without sacrificing the utility of that product and/or (2) increasing the lifetime of a product.

The reduction in material use per product can probably be achieved most readily by redesigning some of the packaging that is presently used in the marketing operation. For example, a drawn and ironed steel can result in a savings of about 25 to 30% in materials over the common seamed tin can. Redesign of the automobile to reduce by 5% the steel presently used

---

*"Tons" in this text means 2000 lb, and "tonnes" means 1000 kg.

would result in about 315,000 tonnes (350,000 tons) of steel saved annually.

The car can also be used as an example of what would occur if longer life is achieved. The average life of a passenger car in the United States is about 10 years (much less than in other countries). The average weight of a car is about 1800 kg (4000 lb) and increasing the life by only 2 years, to an expected life of about 12 years by the year 1990, will achieve a savings of about 5.4 million tonnes (6 million tons) of steel, 135,000 tonnes (150,000 tons) of aluminum, and 135,000 tonnes (150,000 tons) of zinc.

It should be reemphasized that such reductions and changes in product materials and design will undoubtedly produce some economic ramifications. Such rules and regulations, if enacted, must be drawn up with the full knowledge that economic repercussions will result.

Reuse

At the present time, many of our products are reused in the home without much thought being given to ethical considerations. These products simply have utility and value for more than one purpose. For example, paper bags obtained in the supermarket are often used to pack refuse for transport from the house to the trash can. Newspapers are rolled up to make fireplace logs, and coffee cans are used to hold nails. All of these are examples of reuse. Unfortunately, none of these secondary uses has much economic impact on the total quantities of raw material used by our society.

By contrast, the use of refillable beverage containers would constitute a major form of reuse. At the present time in the United States, there are about 60 billion beer and soft drink containers sold annually. This translates into about 8 million tonnes (9 million tons) of solid waste, or about 8% of the solid waste stream. More important perhaps is the fact that these products account for a large fraction of our visible litter.

The advisability of an Oregon-type bottle law is still hotly debated. The bottling industries are vehemently against it, because such a law would force changes in their bottling and distributing strategies. Most environmental groups, as well as the EPA, are in favor of such legislation.

Materials recovery

Many of the components of municipal solid waste can be recovered and recycled for subsequent use, the most important being paper, steel, aluminum, and glass.

About 54 million tonnes (60 million tons) of paper enter the solid waste stream annually, and only about 15% of this is recovered. It is estimated that about 27 million tonnes (30 million tons) per year could be recovered economically from the solid waste stream without the use of new and advanced technology.

In 1973, about 3.6 million tonnes (4 million tons) of steel cans were generated in urban areas of the United States, and only about 2% of those cans were recovered. Even with the new basic oxygen furnaces which require rather pure charges and can tolerate only limited amounts of recovered steel, the steel industry could still absorb about 3 million tons of scrap steel per year.

Aluminum is another material that is ideally suited for recycling. About 0.9 million tonnes (1 million tons) of aluminum enters the solid waste stream annually, 50% of this being cans and 30% being foil. Only about 4% of our aluminum is presently being recovered, although some aluminum can recycling operations have achieved 15 to 20% recovery in certain urban areas.

Glass is potentially an ideal material for recycling because it is clean and can be reprocessed many times without a loss in its structural strength or other attributes. However, purity specifications and color standards are stringent, and reprocessed glass requires approximately the same amount of energy as glass made from raw materials. Of the 13 million tons of glass that are discarded annually, only about 3% is presently recovered.

Energy recovery

The potential for energy recovery from solid waste is significant. For example, in 1975, 122 million tonnes (135 million tons) of waste was generated, 70% of which was combustible, yielding a heat value equal to about 500,000 barrels of oil per day. Recognizing that much of the waste generated in rural areas is not economically recoverable, the combustible materials from Standard Metropolitan Statistical Areas would still result in over 400,000 barrels of oil per day equivalent. This translates to about 10% of all the coal used by the utilities and 5% of all the coal used in the United States.*

SOLID WASTE QUANTITIES AND CHARACTERISTICS

Definitions

A great deal of confusion exists in the definition of "solid waste," and this leads directly to disagreements on the estimated quantities and composition of solid waste. There are both gross and subtle differences in the

---

*While it is convenient to translate Btus into barrels of oil, this is not realistic. The production of the Btu from refuse requires a certain amount of effort and expenditure of energy. Second, the refuse cannot be used in all of the many ways that oil can, and thus it cannot be a one-to-one substitution. This is discussed later in the text.

types and sources of such material, and indeed a question as to what is and is not "solid."

This text is devoted mainly to the recovery of resources from *postconsumer solid waste*, defined as the waste generated in private households, office buildings, and commercial and service establishments. Excluded are the wastes disposed of either into the sewerage system or directly into the atmosphere. This definition also eliminates many other types of solid wastes, including those resulting from mining, agricultural, and industrial processing. In these cases, not only is it often difficult to define what is and is not a "waste" (e.g., manure in a pasture vs. feedlot), but it is often meaningless to attempt to define "solid" (e.g., drums of waste pesticide or lubricating oil). Thus, although it is readily conceded that these wastes represent very real problems, we have chosen in this text to emphasize the waste materials typically collected by municipalities from homes and commercial establishments.

When solid waste generation sampling and composition studies are conducted and reported, it is also commonly assumed that the data do not include junked automobiles, street sweepings (including leaves), and sewage sludges, although all of these could well be included by the foregoing definition. In the vernacular, we are most interested in "what the garbage truck takes away."

The postconsumer solid waste thus defined is commonly referred to as *municipal solid waste*, or MSW. Municipal solid waste is comprised of a number of distinct categories of materials which have professionally accepted definitions:

Refuse—all MSW as transported from a home or commercial establishment, and comprising *garbage*—food waste (exclusively); *rubbish*—paper, cans, bottles, and so on; and *ash*—residue of coal burning.

Trash—tree limbs, leaves, and bulky items (refrigerators, large boxes, etc.).

Often some part of the trash is collected with the refuse, while in other communities the trash is either transported by the homeowners or is removed by special pickup. Whether some fraction of the trash is counted in the overall sampling survey has a large bearing on the outcome. For example, a solid waste sampling study for Honolulu yielded a surprisingly high organic fraction of over 85%. In that instance, all of the garden waste (mangoes, palm leaves, etc.) are picked up by the refuse truck, and this fraction can be as high as 30% by weight of the total load. The high organic fraction in Honolulu is thus readily understandable. In some communities, garden wastes may be counted as part of the refuse during one study, and not counted in another, thus yielding wide discrepancies of what is produced as MSW.

In this text, we define MSW as containing garden waste which is normally placed in the refuse can and collected with the remainder of the refuse. MSW as defined here does not, however, contain tree limbs, white goods and other bulky items collected separately.

Solid waste quantities should always be expressed in terms of weight, not volume, since the latter varies with compaction.

Solid Waste Generation

The quantities of MSW generated can be estimated in one of three ways:
Input analysis.
Secondary data.
Output analysis.

Input analysis is based on data from published industry production statistics. For example, it is estimated by the Glass Containers Manufacturers Institute [5] that the annual production of glass containers is about 9,000,000 tonnes (10,000,000 tons). Of this about 270,000 tonnes (300,000 tons) was recovered and did not enter the waste stream [6], resulting in about 9,350,000 tonnes (9,700,000 tons) of waste glass containers. Other glass products entering the waste stream are estimated at 997,000 tonnes (1,100,000 tons) per year [7]. The total tonnage of waste glass is thus approximately 9,800,000 tonnes (10,800,000 tons) per year.

Similar analyses yield national tonnages for other materials. The 1975 values recently published by EPA are shown in Table 1-1. The last line in the table indicates an annual national solid waste production of 115 million tonnes (128 million tons) or on a personal level, a contribution of about 1.6 kg/capita/day (3.5 lb/capita/day). The table has two total columns, "as generated" and "as disposed." The difference between the two columns is moisture transfer during storage and transport of the mixed refuse. Newsprint, for example, has in its original condition about 7% moisture, but absorbs considerable moisture from garbage and other sources, yielding an average moisture content of 23% after the moisture transfer has taken place.

The input method of estimating solid waste generation is applicable to national data (or for isolated areas), where the input figures can be readily obtained from highly specialized agencies which routinely collect and publish industry-wide data. This system also allows for regular updates of waste generation estimates due to low-cost data gathering. Further, since the data collected by the same institutions include future projections, it is possible to estimate future solid waste generation. Using such a method, the EPA has projected the growth in solid waste generation as shown in Table 1-2.

TABLE 1-1
Generation of Municipal Solid Waste in the United States,[a] (1973)

(millions of tons, as-generated wet weight)[b]

Material category	Product category						Totals			
	Newspapers, books, magazines	Containers, packaging	Major household appliances	Furniture, furnishings	Clothing, footwear	Food products	Other products	As-generated wet weight[b]		As-disposed wet weight[c]	
								Millions of tons	Percent	Millions of tons	Percent
Paper	9.8	19.1	tr.	tr.	—	—	8.3	37.2	29.0	44.9	34.9
Glass	—	12.2	0.1	tr.	—	—	1.0	13.3	10.4	13.5	10.5
Metals	—	5.9	2.1	0.1	—	—	4.0	12.1	9.6	12.6	9.8
Ferrous	—	(5.2)	(1.8)	(0.1)	—	—	(3.7)	(10.8)	(8.6)		
Aluminum	—	(0.7)	(0.1)	tr.	—	—	(0.1)	(0.9)	(0.7)		
Other nonferrous	—	—	(0.2)	tr.	—	—	(0.2)	(0.4)	(0.3)		
Plastics	—	2.7	0.1	0.1	tr.	—	1.5	4.4	3.4	4.9	3.8
Rubber and leather	—	tr.	tr.	tr.	0.7	—	2.6	3.3	2.6	3.4	2.6
Textiles	—	0.1	tr.	0.6	0.5	—	0.9	2.1	1.6	2.2	1.7
Wood	—	1.8	tr.	2.6	—	—	0.5	4.9	3.8	4.9	3.8
Total nonfood product waste	9.8	41.7	2.3	3.4	1.2	—	18.9	77.5	60.5	86.5	67.3
Food waste	—	—	—	—	—	22.8	—	22.8	17.8	19.1	14.9
Total product waste	9.8	41.7	2.3	3.4	1.2	22.8	18.9	100.3	78.3	105.6	82.2
Yard waste								26.0	20.2	20.9	16.3
Miscellaneous inorganics								1.9	1.5	2.0	1.6
Grand total								128.2	100.0	128.5	100.0

[a] Net solid waste disposal defined as net residual material after accounting for recycled materials diverted from waste stream.
[b] "As-generated" weight basis refers to an assumed normal moisture content of material in its final use prior to discard, for example: paper at an "air-dry" 7% moisture; glass and metals at 0%. Total waste, including food and yard categories, estimated at 26% moisture.
[c] "As-disposed" basis assumes moisture transfer among materials in collection and storage, but no net addition or loss of moisture for the aggregate of materials.

Source: Ref. 8.

TABLE 1-2
U.S. Solid Waste Generation Projections, Based on 1975 Data

Year	Million tons per year	Pounds per capita per day
1975	136	3.40
1980	175	4.28
1985	201	4.67
1990	225	5.00

Source: Ref. 8.

National averages are of limited value, however, because they can rarely be used with any degree of precision for local or regional purposes.

The second method of estimating solid waste generation is to use secondary information such as income, products sold in stores, and so on, and by means of a stepwise linear regression analysis, calculate MSW production. Some models are based on product availability (and hence disposal) information [9], whereas others rely on variables related to the collection routes.

For example, in one study [10] it was found that the tons of MSW generated in a route or subdistrict (W) could be estimated as

$$W = 0.0179(S) - 0.00376(F) - 0.00322(D) + 0.0071(P) - 0.0002(I) + 44.7$$

where W = waste generated, tons
  S = number of stops made by the truck
  F = number of families served
  D = number of single-family dwellings
  P = population
  I = adjusted gross income per dwelling unit

A similar model developed for a town in New England, yielded the following results [11]:

$$Y = 619(X_1) + 655(X_2) - 35(X_3) + 37(X_4) + 8.5(X_5) + 626$$

where Y = gallons of uncompacted solid waste produced per week
  X_1 = blockfaces with low income
  X_2 = blockfaces with middle income
  X_3 = single-unit dwellings
  X_4 = dwellings containing two to four living units
  X_5 = dwellings with five or more living units

In addition to the variables listed above, the frequency of collection affects the generation of solid waste [12]. In Portland, Maine, the amount

TABLE 1-3
Average Monthly Variation in Municipal Solid Waste
Collection in New Orleans

Month	Percent of Averages
January	94
February	88
March	101
April	101
May	103
June	106
July	108
August	110
September	103
October	95
November	91
December	100

Source: Ref. 15.

of solid waste collected increased by 20% when the collection frequency was increased from once to twice per week [13]. The obvious interpretation is that if the frequency of service is not sufficient, citizens will find other, perhaps less desirable means of solid waste disposal.

The trouble and expense of obtaining the necessary statistical information, coupled with unpredictable human behavior and the uncertainty inherent in applying such models, makes use of these models questionable [14]. Accordingly, it often becomes necessary to use the last of the three alternatives, the output-analysis technique, for estimating the local solid waste generation.

Output analyses by weighing the MSW as delivered to disposal have shown that refuse generation is not easily predictable. For example, refuse generation varies with time—seasonally, monthly, weekly, and daily. Seasonally, the high collection periods seem to be the spring and summer months (spring cleaning, yard wastes). Typical data [15] for the City of New Orleans are shown in Table 1-3. Solid waste generation is also greater during December, because of the holiday season.

Weekly variation may result from variable weather, holidays, and the season. These variations are shown graphically in Fig. 1-2, which was developed by the EPA from overall national data [16]. It is important to note that 95% of the year (49 of 52 weeks) the quantity of refuse does not exceed 15% of the average annual mean. For three weeks, however the quantities can be as high as 25% greater than the mean, and could strain the capability of collection systems.

The production of solid waste also varies with population density. Urban life contributes to larger waste generation, while rural people either

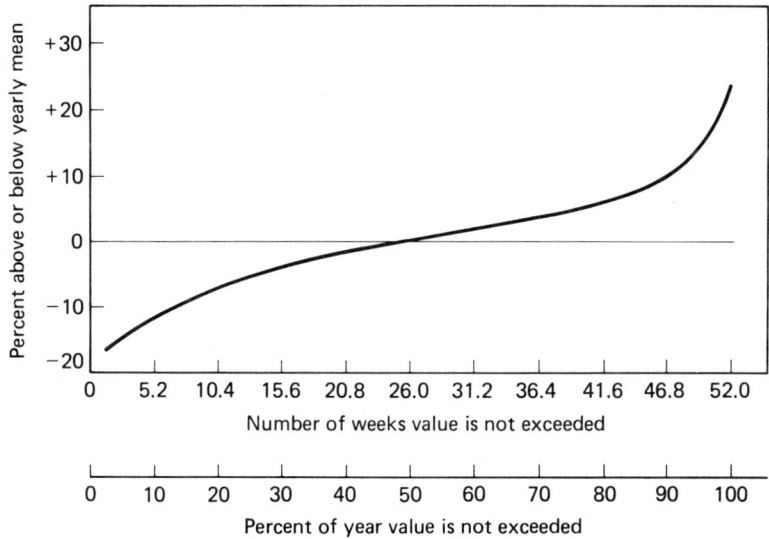

Figure 1-2. Weekly variations in refuse generation. (From Ref. 18.)

make do with less and discard proportionately smaller quantities, or reuse more of the products [17]. Figure 1-3 shows the effect of population density on solid waste generation.

A number of output studies of solid waste generation have shown that there is a significant difference in the amount generated, depending on one's life-style, age, and socioeconomic level. In one study [19], the average refuse quantities were found to be significantly influenced by socioeconomic level, as shown in Table 1-4. One indicator of socioeconomic level is the number of people per dwelling. Figure 1-4 shows the solid waste generation in low-income neighborhoods as a function of the number of people in a dwelling [20].

Although industries strive to minimize waste, substantial amounts of material are discarded annually. Table 1-5 is a partial listing of the solid

TABLE 1-4

Effect of Socioeconomic Level on Domestic Refuse Generation

	Socioeconomic Level		
	Low	Middle	High
Total annual generation as pounds per capita	1.86	1.98	2.07

Source: Ref. 19.

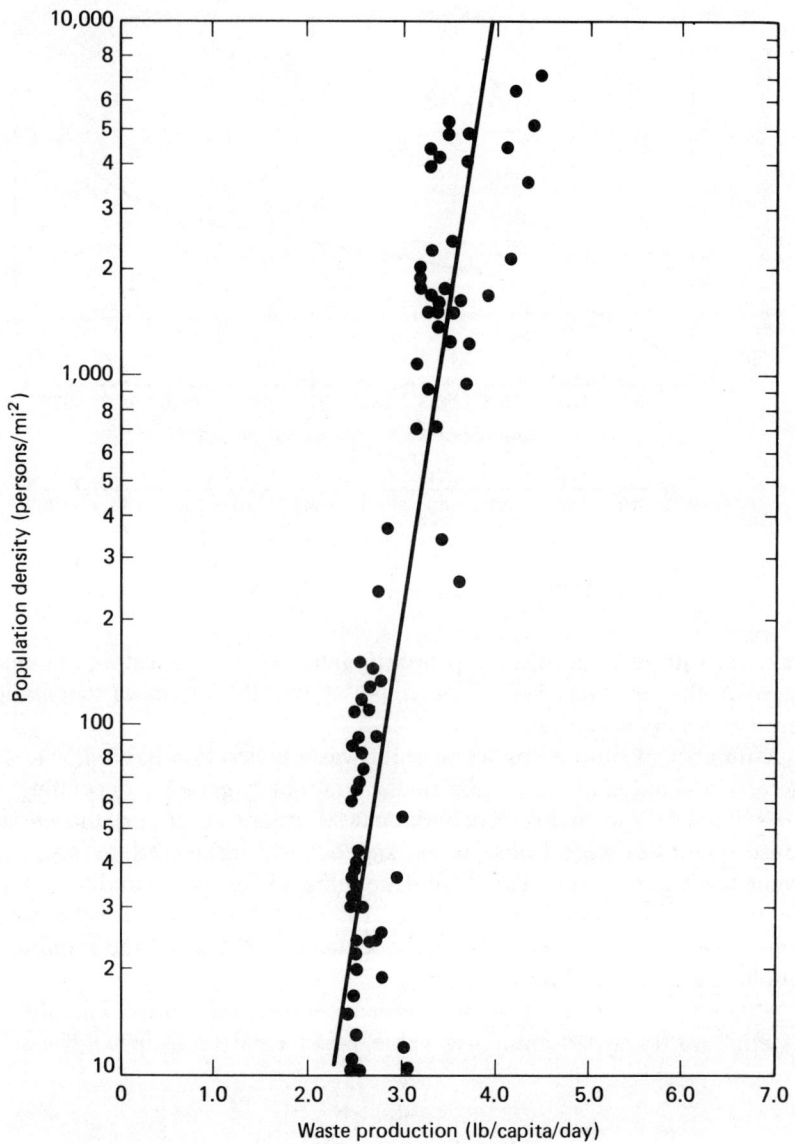

Figure 1-3. Solid waste generation as the function of population density. (From Ref. 18.)

TABLE 1-5
Industrial Waste Quantities and Composition

SIC Code	Industry	Paper	Wood	Rubber	Plastics	Metals	Glass	Textiles	Food	Bulk Density (lb/yd³)	Tons per Employee per Year
20	Food	52.3	7.7	—	0.9	8.2	4.9	—	16.7	1240[a] 886[b]	7.95
22	Textiles	45.5	—	—	4.7	—	—	26.8	—	?	2.16
23	Apparel	55.9	—	—	—	—	—	36.5	1.35	159	2.19
24	Lumber	16.7	71.6	—	—	—	—	—	—	894	8.53
25	Furniture	24.7	42.1	—	—	—	—	—	—	464	2.78
26	Paper	56.3	11.3	—	—	9.4	—	—	—	557	3.99
27	Printing	84.9	5.5	—	—	—	—	—	—	1671	5.83
28	Chemicals	55.0	4.5	—	9.3	7.3	2.2	—	—	895	8.86
29	Petroleum	72.1	6.8	—	15.3	4.4	—	—	—	?	1.59
30	Plastics and rubber	56.3	5.2	9.2	13.5	—	—	—	—	148	9.84
31	Leather	6.0	3.9	—	—	13.5	—	—	—	?	8.99
32	Stone and glass	33.8	4.3	—	—	8.1	12.8	—	—	2601	6.41
33	Primary metal	41.0	11.6	—	5.4	5.5	2.0	—	—	?	3.18
34	Fabricated metal	44.6	10.3	—	—	23.2	—	—	—	785	6.83
35	Machinery	43.1	11.4	—	2.5	23.7	—	—	—	650	3.19
36	Electrical machinery	73.3	8.3	—	3.5	2.3	—	—	1.2	756	2.94
37	Transportation	50.9	9.4	1.4	2.1	—	—	—	—	290	2.56
38	Professional and scientific	44.8	2.3	—	6.0	8.4	—	—	—	562	1.77
39	Miscellaneous	54.6	13.0	—	11.9	5.0	—	—	—	475	1.60

[a] For canning and preserving.
[b] For all other food processing.
Source: Ref. 21.

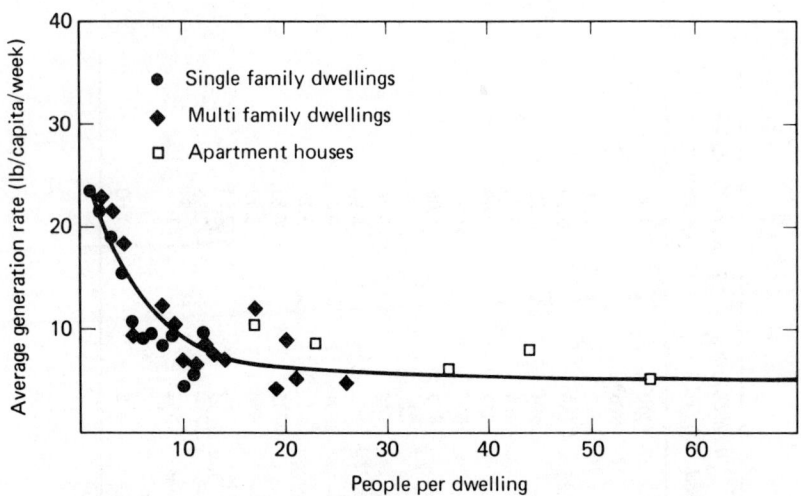

Figure 1-4. Solid waste generation as a function of the number of people per dwelling in low income areas. (From Ref. 20.)

TABLE 1-6

Composition and Quantities of Solid Waste from Office Buildings[a]

	Pounds/employee/day			
	Bank/Insurance Co.		General Office	
Material	Average	Percent	Average	Percent
Paper				
Computer tab cards	0.39	17	0.05	3
Computer printout	0.70	30	0.11	7
White ledger	0.70	30	0.51	33
Subtotal (high grades)	1.79	77	0.67	43
Colored ledger	0.12	5	0.09	6
Newspaper	0.07	3	0.25	16
Corrugated	0.05	2	0.14	9
Other[b,c]	0.14	6	0.17	11
Subtotal (paper)	2.17	93	1.32	85
Nonpaper[c]	0.14	7	0.23	15
Total	2.31	100	1.55	100

[a] Based on representative solid waste sampling conducted at six buildings studied by EPA; does not include cafeteria waste.
[b] Generally nonrecyclable paper: carbon paper, wax-coated or impregnated paper products, etc.
[c] Small quantities of garbage, metal, plastic, glass, textiles, wood, and other materials.
Source: Ref. 22.

wastes generated by the manufacturing and processing industries (Standard Industrial Code 20's and 30's). Much of this waste could be recovered with regional waste exchange programs.

The predominant waste in office buildings is paper, much of it high-grade white ledger paper which is readily recycled. A recent EPA study [22] estimates paper production in office buildings at over 2 lb/employee/day (Table 1-6).

Characteristics of Refuse

As long as the MSW is to be disposed of by landfill, there is little need to analyze the waste much further than to establish the tons of waste generated and perhaps consider the problems of special (hazardous) materials. When resource recovery is the objective, however, it becomes necessary to have a more complete picture of the MSW. Some of the characteristics of interest are:

Moisture content.

Particle size.

Composition by materials (steel, paper, etc.).

Chemical composition (carbon, hydrogen, etc.).

Density.

Mechanical properties.

Moisture content

A transfer of moisture takes place in the garbage can and truck, and thus the moisture content of various components changes with time. As described earlier, newsprint has about 7% moisture by weight as it is deposited into the receptacle, but the average moisture content of newsprint coming from a refuse truck often exceeds 20%. The moisture content becomes important when the refuse is processed into fuel or when it is fired directly. The usual expression for calculating moisture content is

$$M(\%) = \frac{w - d}{w} \times 100$$

where M = moisture content, %
w = initial (wet) weight of sample
d = final (dry) weight of sample

Drying is usually done in an oven at 77°C (170°F) for 24 h to assure complete dehydration and yet avoid undue vaporization of volatile material.

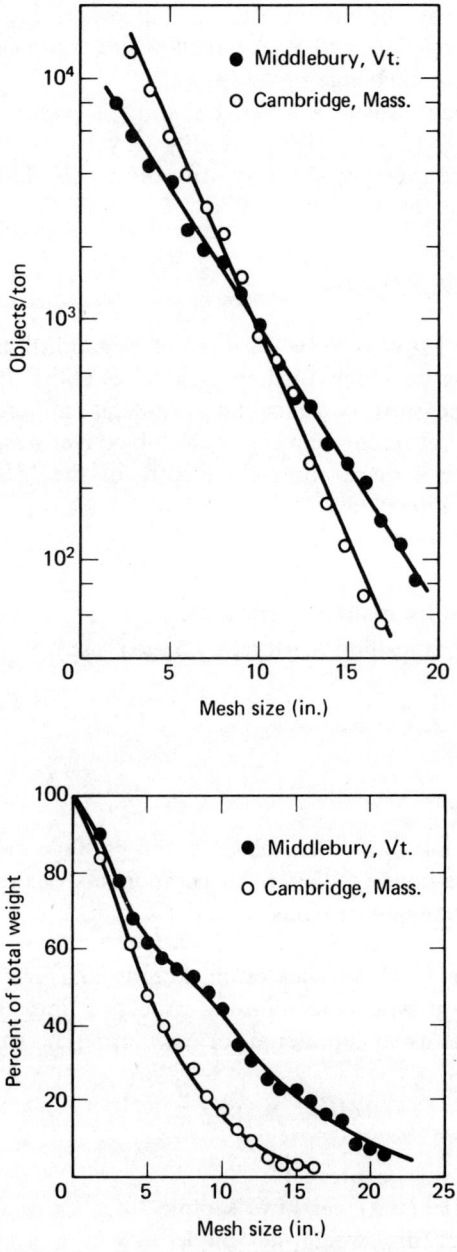

Figure 1-5. Particle size of mixed solid waste, as objects passing through a mesh of given size. (From Ref. 23.)

Particle size

The results of a study performed in two New England towns, illustrated in Fig. 1-5, shows a particle-size distribution by longest dimension and the ability to pass a sieve [23]. This parameter, vitally important in resource recovery, is discussed further in Chapters 3 and 4.

Composition by materials

Even more important for resource recovery than particle size is the materials makeup of the refuse. A serious handicap in this regard, however, is that considerable variation seems to exist in the methods used to measure refuse composition. This problem has surfaced in a number of cases where results of several studies performed in the same city show wide disagreement. Data from several sampling programs are shown in Table 1-7, and the large variation is immediately obvious.

The composition of refuse is also dependent on life-style, income, climate, and other factors. The data in Table 1-7 also show how the composition of refuse (dry basis) can vary considerably in various parts of the United States.

Although it would be difficult to specify "typical U.S. refuse," national standard of living and life-style certainly have significant effects on the composition of refuse. Table 1-8 shows average urban refuse composition in Australia, New Zealand, and several European countries. It is clear that as the standard of living increases, the fraction of garbage and organics decreases, and the metal and glass fraction increases. Sweden, which has a

TABLE 1-7
Composition of Municipal Solid Waste According to Various Surveys

Location	Percent by weight				
	Garbage	Glass	Metal	Paper	Yard
Oceanside, N.Y.	9.6	9.7	8.0	32.8	33.3
	10.2	9.5	8.2	39.8	19.0
	16.7	11.9	10.6	53.3	0.3
Flint, Mich.	29.1	12.7	14.5	13.0	26.7
	36.0	23.2	14.5	21.1	0.3
New Orleans, La.	18.9	16.2	12.2	39.4	9.2
San Diego, Ca.	0.8	8.3	7.6	46.2	21.1
Johnson City, Tenn.	21.1	7.0	7.5	59.8	0.9
	34.6	9.0	10.4	34.9	2.3
Purdue University, Lafayette, Ind.	12.0	6.0	8.0	42.0	12.0
Berkeley, Ca.	12.5	11.3	8.7	44.6	12.5

Source: Ref. 24.

TABLE 1-8
Composition of Municipal Refuse Overseas[a]

	Paper	Organics	Garbage and Metal	Glass	Plastics	Other	Reference
United Kingdom (av.)	28	13	7	8	—	54	25
Stevenage, U.K.	33	14	7	10	4	—	26
France (av.)	30	24	4	4	—	38	25
Holland (av.)	23	50	3	13	5	—	26
West Germany (av.)	20	21	5	10	—	44	25
West Germany (av.)	28	15	7	9	3	—	26
Sweden (av.)	55	12	6	15	—	12	25
Stockholm, Sweden	45	17	6	7	9	—	26
Spain (av.)	21	45	3	4	—	27	25
Rome, Italy	18	50	3	4	4	—	26
Poland (av.)	6	40	1	2	—	51	25
Auckland, New Zealand	28	48	6	7	—	11	27
Australia (av.)	23	10	10	16	—	—	28
Sydney, Australia	38	13	11	18	—	—	28
Vienna, Austria	35	24	10	9	6	—	26
Prague, Czechoslovakia	13	42	6	7	4	—	26
Sofia, Bulgaria	10	54	2	2	2	—	26

[a]Some of the figures have been rounded for clarity.

very high living standard has a total of 21% metal and glass, whereas Spain and Poland have only 7% and 3%, respectively.

Measuring the composition of a totally heterogeneous material such as mixed municipal refuse is not a simple task, but some determination of its components is necessary if the various fractions are to be separated and recovered. Some authorities suggest that since a major effort is required to establish the composition with reasonable accuracy, it is often not worth the trouble and expense, and a national average can be used. For example, if beverages are not sold in aluminum cans in the area, it is unlikely that enough aluminum will be found in the refuse to warrant recovery, and thus it makes little difference if the exact percentage is 1% or 2%.

In some cases, however, it is necessary to conduct a study of refuse composition. Municipalities seem loath to entertain resource recovery possibilities without such hard data—accurate or otherwise. Two analytical methods are available for conducting such studies: the *manual classification* method and the *photogrammetric* technique.

Refuse composition by manual sampling

Sampling studies for characterizing refuse must be designed so as to produce the most useful and accurate data for the least cost and effort. The two variables of importance in designing such a study are sample size and method of characterizing the refuse.

Through a series of studies using samples ranging from 54 to 770 kg (125 to 1700 lb) each, it was demonstrated that sorting and analyzing more than 90 kg (200 lb) would have little statistical advantage [29]. A 90-kg sample should be selected by dumping the contents of a "typical" truck and quartering the contents until one sample is about 90 kg.

Separation of the mixed sample into components requires the establishment of categories. In the past, EPA sponsored studies have divided refuse into nine categories:
1. Food waste (garbage).
2. Garden waste.
3. Paper products.
4. Plastics, rubber, and leather.
5. Textiles.
6. Wood.
7. Metals products.
8. Glass and ceramics.
9. Ash, rocks, and dirt.

It is obvious that this classification system is inadequate for studies where resource recovery is contemplated. For example, metal products could readily be divided into three separate components:
1. Ferrous.
2. Aluminum.
3. Other nonferrous.

Glass could similarly be further classified into:
1. Clear or green glass.
2. All other glass and ceramics.

In some cases, these classifications could be subdivided even further. It is prudent to identify the potential marketable products of resource recovery before the sampling study is undertaken.

Finally, it should be recognized that even careful sampling studies yield imprecise information because of the nature of the refuse. Not all items can be readily categorized into the desired components. For example, a tin can with an aluminum top and paper wrapper has four components: steel, tin, aluminum, and paper. Regardless of the final classification of this item, inaccuracies are introduced into the final values.

Refuse composition by photogrammetry

The foregoing comments for manual sampling apply equally for the photogrammetric technique for characterizing the components of MSW. The latter method involves photographing a representative portion of refuse and analyzing the photograph.

TABLE 1-9
Bulk Densities[a] of Some Refuse Components

Material	g/cm^3	lb/ft^3
Light ferrous	0.100	6.36
Aluminum	0.038	2.36
Glass	0.295	18.45
Miscellaneous paper	0.061	3.81
Newspaper	0.099	6.19
Plastics	0.037	2.37
Cardboard	0.030	1.87
Food	0.368	23.04
Garden waste	0.071	4.45
Rubber	0.238	14.9

[a]Including void space.
Source: Ref. 30.

The photograph should be taken directly at the refuse (90° angle) with a wide-angle lens, using 35-mm color film, with electronic flash fill to eliminate shadows. The 2 × 2 slide is then projected on a screen that has been divided into about 10 × 10 grid blocks. This grid can either be drawn on a piece of poster board, or a negative transparency projected from an overhead projector. The components in each grid intersection are then identified and tabulated. Using predetermined bulk densities (which include interior space, e.g., in a beverage can), the fraction by weight is then calculated. Some of the bulk densities are listed in Table 1-9.

Example 1-1.

Suppose that a projected picture of refuse on a grid yielded the following:

Grid intersections on:	Number	Percent
Newspaper	8	50
Tin cans	4	25
Aluminum cans	4	25
	16	100

Using the bulk densities from Table 1-9, the distribution by weight is calculated as follows:

Fraction Intersections		Bulk Density (g/cm^3)	Proportional Weight	Percent by Weight
0.50	×	0.099	0.0495	58.9
0.25	×	0.100	0.0250	29.8
0.25	×	0.038	0.0095	11.3
			0.0840	100.0

The percent by weight is calculated by dividing each proportional weight by the sum of the proportional weights and multiplying by 100.

Solid Waste Quantities and Characteristics

The photogrammetric technique suffers from two disadvantages. First, its accuracy is dependent on bulk density figures that must be fine-tuned. However, some practitioners found this technique to be quite accurate [31]. The second disadvantage is that the time required to analyze one picture is substantial, and it may be faster to simply classify manually and weigh. The technique does have one important advantage—the refuse need not be touched or smelled.

Number of samples required

With either technique, however, the non-homogenous nature of the refuse introduces a sampling error. The analysis of a single 90 kg (200 lb) sample may be quite accurate, but there is no guarantee that this sample in fact reflects the actual average composition of the waste. Using a number of 90 kg (200 lb) samples, Woodyard and Klee have developed curves for selecting the proper sample size for a desired level of precision for a number of components [37]. Fig. 1-6 shows their data, in simplified form, and indicates the approximate number of samples necessary in order to be 90% confident of being within the desired level of precision. They also note that these samples should be taken over the space of a single week, taking into account the variation in daily and hourly volume.

Obviously, any one of the 90 kg (200 lb) samples taken from a truck should be selected judiciously. Klee and Carruth [29] drive this point home

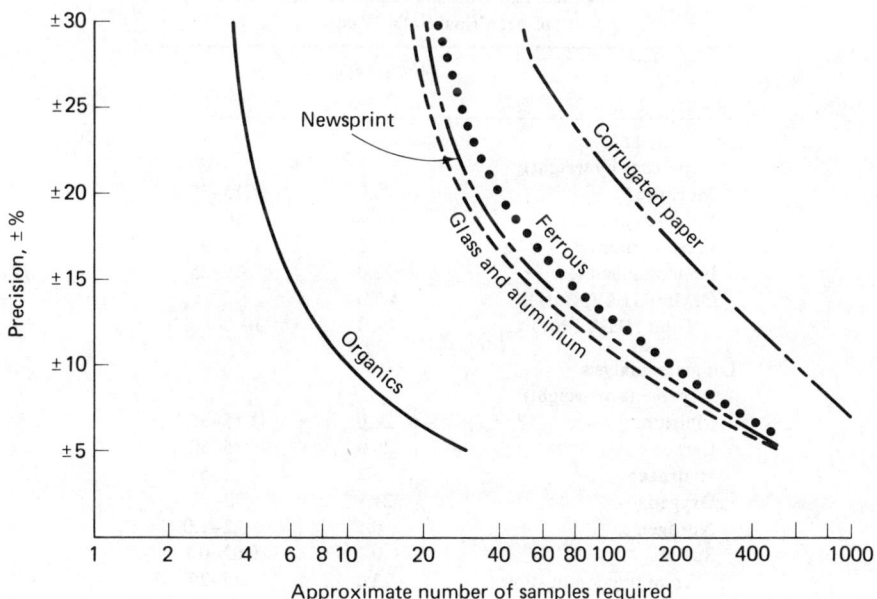

Figure 1-6. Approximate number of 90 kg (200 lb) samples necessary to achieve desired precision. (From Ref. 37.)

by their observation that

> when one attempts to obtain random samples from a truck ... the pragmatic man will reject the random sample that contains a dead horse if he perceives that the rest of the material does not contain dead horses. This is not only common sense, it is good decision statistics as well.

Chemical composition

The economic recovery of materials and/or energy often depends on the chemical composition of the refuse—the individual chemicals as well as the heating value. Both proximate and ultimate analyses for a specific refuse (Hempstead, N.Y.) and some general ranges published by EPA are tabulated in Table 1-10. These data are of limited value for design purposes because once again the heterogeneous nature of refuse and its variability with geography and with time result in rather wide ranges. Accurate information for a specific refuse can be attained only by concerted sampling and analysis.

An alternative to chemical testing is to use the tabulated results of chemical analyses for specific materials. Table 1-11 is a compilation of some ultimate analysis data, and further information can be found in reference 15.

TABLE 1-10

Proximate and Ultimate Chemical Analysis of Municipal Solid Waste

	Hempstead, N.Y.	EPA
Proximate analysis (percent by weight)		
Moisture	28.0	15–35
Volatile matter	43.4	50–60
Fixed carbon	6.6	3–9
Noncombustibles	22.0	15–25
Organic HHV, Btu/lb	4400	
Total HHV, Btu/lb	4500	3000–6000
Ultimate analysis (percent by weight)		
Moisture	28.0	15–35
Carbon	25.0	15–30
Hydrogen	3.3	2–5
Oxygen	21.1	12–24
Nitrogen	0.5	0.2–1.0
Sulfur	0.1	0.02–0.1
Total noncombustibles	22.0	15–25

Source: Hempstead data from Ref. 15; EPA data from Ref. 16.

TABLE 1-11
Ultimate Analysis of Typical Municipal Refuse Components

Refuse component	C (%)	H$_2$ (%)	O$_2$ (%)	N$_2$ (%)	S (%)	Inerts[a]	Btu/lb	Percent moisture	Percent as delivered
Newspapers	49.14	6.10	43.03	0.05	0.16	1.43	7,974	5.97	10.33
						1.52	8,480		
Brown paper	48.36	6.13	42.30	0.14	0.11	2.96	8,266	5.83	6.12
	44.90	6.08	47.84	0	0.11	1.01	7,256		
						1.07	7,706		
Magazine paper	32.91	4.95	38.55	0.07	0.09	22.47	5,254	4.11	7.48
Corrugated boxes	43.73	5.70	44.93	0.09	0.21	5.06	7,043	5.20	25.68
						5.34	7,429		
Plastic coated paper	45.30	6.17	45.50	0.18	0.08	2.64	7,341	4.71	0.84
						2.77	7,703		
Waxed milk cartons	59.18	9.25	30.13	0.12	0.10	1.17	11,327	3.45	0.84
						1.22	11,732		
Paper food cartons	44.74	6.10	41.92	0.15	0.16	6.50	7,258	6.11	2.27
						6.93	7,730		
Junk mail	37.87	5.41	42.74	0.17	0.09	13.09	6,088	4.56	3.03
Tissue paper	43.9	6.1	49.0			0.93	6,999	7.00	2.18
Cardboard	45.52	6.08	44.53	0.16	0.14	3.57	7,841		
Miscellaneous paper	44.00	6.15	41.65	0.43	0.12	7.65	7,793		
Vegetable and food wastes	49.06	6.62	37.55	1.68	0.20	1.06	1,795	78.29	2.52
Citrus rinds, seeds	47.96	5.68	41.67	1.11	0.12	0.74	1,707	78.70	1.68
Meat scraps, cooked	59.59	9.47	24.65	1.02	0.19	3.11	7,623	38.74	2.52
Fried fats	73.14	11.54	14.82	0.43	0.07	0.	16,466	0.	2.52
Garbage	41.72	5.75	27.62	2.79	0.25	21.87	7,246		
Leather	42.01	5.32	22.83	5.98	1.00	21.16	7,243	7.46	0.42
Rubber composition, heel, sole catch	53.22	7.09	7.76	0.50	1.34	29.74	10,899	1.15	0.42
Plastics									
Average	78.0	9.0	13.0				15,910		0.84
High	90.0	10.0					19,303		
Low	55.8	7.0	37.2				9,580		

TABLE 1-11. Continued

Refuse component	C (%)	H₂ (%)	O₂ (%)	N₂ (%)	S (%)	Inerts[a]	Btu/lb	Percent moisture	Percent as delivered
Polyethylene	85.6	14.4					19,950		
Vinyl	47.1	5.9	18.6 (chlorine = 28.4%)				8,830		
Plastic film	67.21	9.72	15.82	0.46	0.07	6.72	13,846		
Mixed, from municipal refuse, contaminated with food waste									
Other plastics, rubber, leather	47.70	6.04	24.06	1.93	0.55	19.72	9,049		
Paints, oils	52.1	13.1	34.8			0.	12,780		0.84
Vacuum cleaner	35.69	4.73	20.38	6.26	1.15	30.34	6,386	5.47	0.84
Evergreen trimmings	48.51	6.54	40.44	1.71	0.19	0.81	2,708	69.00	1.68
Flower, garden plants	46.65	6.61	40.18	1.21	0.26	2.34	3,697	53.94	1.68
Lawn grass, green	46.18	5.96	36.43	4.46	0.42	1.62	2,058	75.24	1.68
Ripe tree leaves	52.15	6.11	30.34	6.99	0.16	3.82	7,984	9.97	2.52
Softwood, pine	52.55	6.08	40.90	0.25	0.10	0.12	9,150		
Hardwood, oak	49.49	6.62	43.39	0.25	0.10	0.15	8,682		
Wood	49.00	6.0	42.00			2.28	6,840	24.00	2.52
Grass and dirt	48.30	5.97	42.44	0.29	0.11	2.89	8,236		
Rags	36.20	4.75	26.61	2.10	0.26	30.08	6,284		
Textiles	43.9	6.1	49.0			0.93	6,999	7.00	0.84
Dirt	46.19	6.41	41.85	2.18	0.20	3.17	8,036		1.68
Glass bottles	0.52	0.07	0.36	0.03		99.02	84		
Btu in labels, coatings, and remains of contents									
Glass, ash, ceramics						100.000			8.50
Glass, stones, ceramics			(same as above, glass bottles)						
Metal cans	4.54	0.63	4.28	0.05	0.01	90.49	742		
Btu in labels, coatings, remains of contents									
Metals						100.00	2,660		7.53

[a]Inerts—ash, glass, metal, stone, ceramics. Source: Compiled in Ref. 15.

Solid Waste Quantities and Characteristics

TABLE 1-12
Properties of Refuse Derived Fuel (RDF)

Fuel	Fuel Value		Composition (wt %)				
	(J/g)	(Btu/lb)	S	Cl	C	N	Ash
St. Louis MSW[a]	10,846	4675	0.1	0.3	—	—	20.0
RDF[b]	15,962	6880	0.2	0.5	37.1	0.8	22.6
RDF, $+\frac{3}{16}$ in.[c]	18,223	7855	0.1	0.5	43.6	0.7	11.3
Paper	24,900	7500	0.1	—	45.4	0.3	6.0
Coal	15,000–30,000	6500–13,000	1–5	0.05	—	—	8–12

[a]Shredded, but not air-classified; magnetics removed; not dried.
[b]Air-classified by NCRR, not dried.
[c]Same as RDF, but the oversize from a $\frac{3}{16}$-in. screen.
Source: Ref. 32.

The heating values of refuse are of some importance in resource recovery. Some published values for several fuels are shown in Table 1-12 to illustrate the variability of the fuels according to how they are derived. The data are from a compilation by the National Center for Resource Recovery. The production of RDF (refuse derived fuel) is described further in Chapter 8.

The energy value of RDF can also be calculated using the modified Dulong formula [33] as

$$\text{Btu/lb} = 14{,}096\, C_{\text{org}} + 60{,}214\left(H - \frac{O}{8}\right) + 1040 N + 3982 S$$

$$+ 8929\left[\frac{H - \frac{O}{8}}{2}\right] + 4274\left(\frac{O}{2}\right) - 6382\, C_{\text{inorg}}$$

where C_{org}, H, O, N, S, and C_{inorg} are the decimal fractions of organic carbon, hydrogen, oxygen, nitrogen, sulfur, and inorganic carbon found on ultimate analysis.

There has been considerable confusion over the energy equivalents of various fuels. For example, a comparison of the heats of combustion of coal and RDF in Table 1-12 suggests implicitly that RDF can be substituted directly for coal, by using amounts inversely proportional to the two heats of combustion. In this case, the heat of combustion of coal is about 24,000 J/g (10,000 Btu/lb), and that of RDF is about 17,000 J/g (7000 Btu/lb), so one might conclude (erroneously) that 10 tons of RDF represents the same energy value as 7 tons of coal. Many claims have been made, using this reasoning, of the equivalent tons of coal (or barrels of oil) represented by the combustible fraction of the nation's waste stream.

The fallacy of the fuel-equivalency reasoning is that energy "costs" have not been taken into account. Identifying RDF's heat of combustion with its energy value is a bit like arguing that coal or oil still in the ground has an energy value equal to its heat of combustion—neglecting the energies of mining, drilling, pumping, and so on. In the case of MSW, the principal energy costs associated with its preparation are shredding and air classification.

One attempt to clarify the difference between heat of combustion and energy actually available has resulted in a new vocabulary of energy equivalents [34]. According to this vocabulary, the energy equivalence that does *not* take processing energy costs into account is called an *arithmetic equivalence*, which is of theoretical interest only.

Recognizing that energy is required to produce RDF, it is possible to calculate the heat of combustion of RDF minus the energy cost per mass of RDF to produce it. If this energy value is compared to other fuels, the comparison is called the *conversion equivalence*: the mass of oil (or coal) per day that RDF can replace, taking account of the energy required to process the RDF. According to the second law of thermodynamics, it costs energy to obtain energy in a useful form.

Finally, it should be noted that if the new fuel is to be used as a supplement in conventional processes, the new fuel may operate at a different efficiency from that of the old fuel. For example, injecting shredded RDF in a coal-fired boiler usually results in greater quantities of bottom ash and may result in less complete burning, reducing the boiler's overall efficiency. Further, not all of the heat value as measured by calorimeters is available at the operating temperatures of furnaces. Martello [38] has shown that the heat produced by the oxidation of highly exothermic aluminum, for example, is measured in a bomb calorimeter but does not contribute to the heat derived from RDF in conventional combustion chambers. The net amount of old fuel that can be used to replace conventional fuel in a specific application without a reduction in overall net energy production is called its *substitution equivalence*.

The laws of thermodynamics dictate that

$$\begin{bmatrix} \text{arithmetic} \\ \text{equivalence} \end{bmatrix} \geq \begin{bmatrix} \text{conversion} \\ \text{equivalence} \end{bmatrix}$$

and practical considerations suggest that in most cases

$$\begin{bmatrix} \text{conversion} \\ \text{equivalence} \end{bmatrix} \geq \begin{bmatrix} \text{substitution} \\ \text{equivalence} \end{bmatrix}$$

When new fuels, such as RDF, are considered as components in total energy management and policy planning, their value should be calculated in terms of substitution equivalence, not arithmetic equivalence.

Density

Material densities, either as mixed refuse or as individual materials, can affect disposal as well as resource recovery operations. Some typical densities of mixed refuse are shown in Table 1-13, and Table 1-14 is a compilation of densities of specific materials commonly found in refuse.

The density of municipal refuse can be increased by mechanical compaction (Chapter 2). The energy requirements for achieving dense MSW are shown in Fig. 1-7, and Fig. 1-8 shows the energy required to densify individual components of MSW. These curves indicate small energy

TABLE 1-13

Refuse Densities

	Density (lb/yd^3)
Loose refuse, no processing	100–200
Refuse from a compactor truck, after dumping	350–400
Refuse in a compactor truck	500–700
Shredded refuse	600–900
Refuse baled in a paper baler	800–1200
Refuse in a landfill	500–900

^{a}To obtain kg/m^3, multiply by 0.59.

TABLE 1-14

Densities of Refuse Components

	Specific Gravity	lb/ft^3
Aluminum	2.70	168
Steel	7.70	480
Glass	2.50	156
Paper	0.7–1.15	44–72
Cardboard	0.69	43
Wood	0.60	37
Plastics		
Polyethylene	0.94	59
ABS	1.03	64
Acrylic	1.18	64
Polypropylene	0.90	56
Polystyrene	1.05	65
PVC	1.25	78

^{a}To obtain kg/m^3, multiply lb/ft^3 by 16.
Source: Ref. 15.

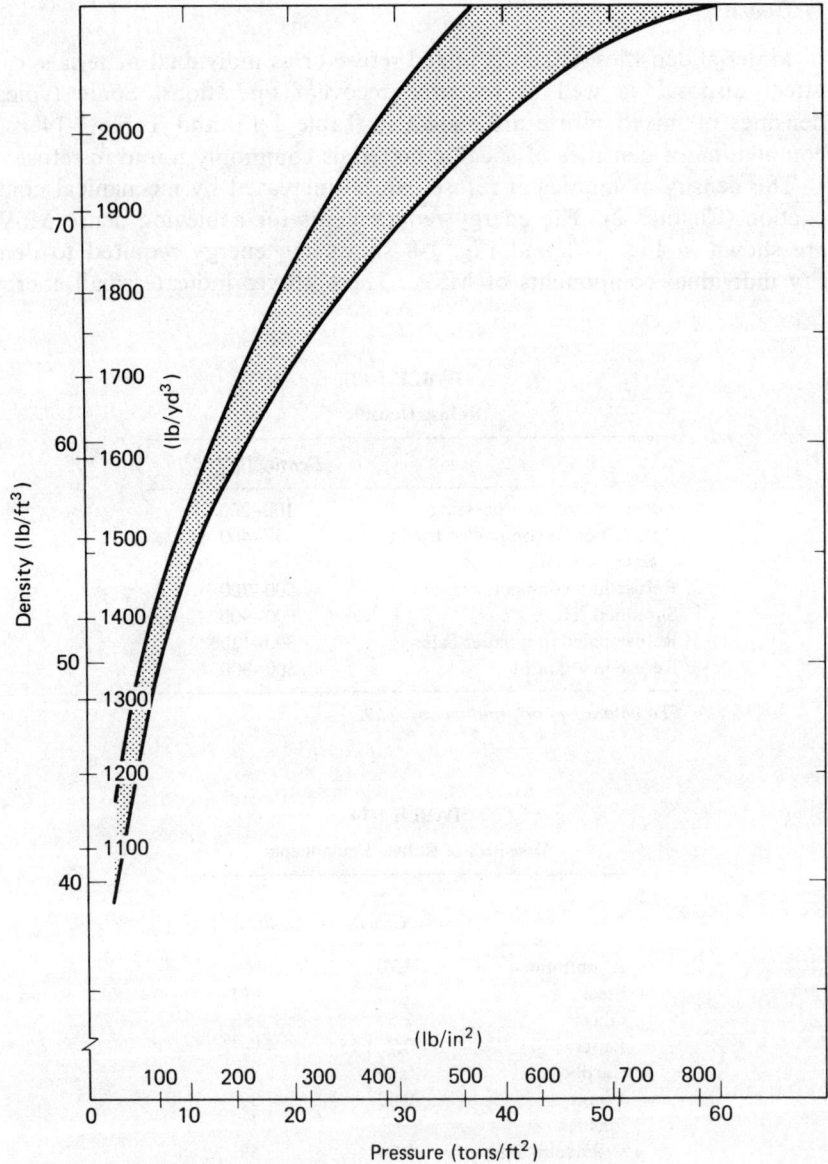

Figure 1-7. Compression of MSW.

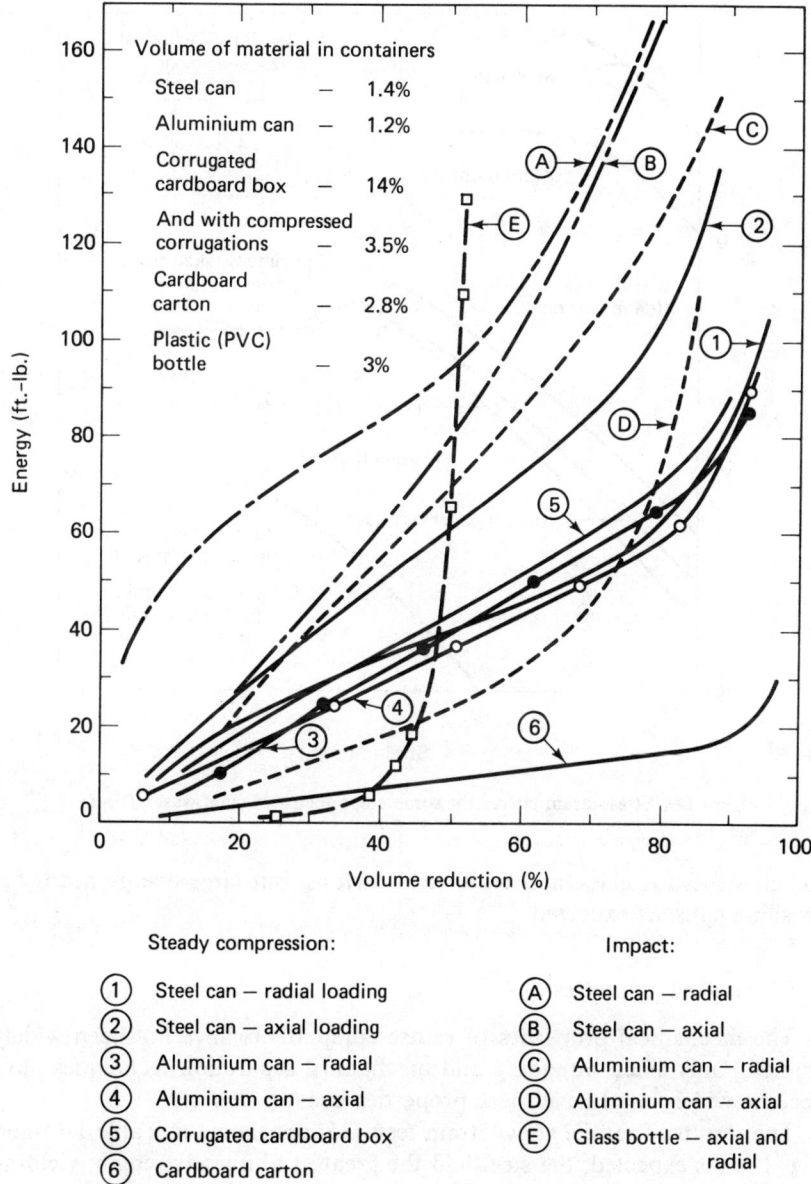

Figure 1-8. Compressive characteristics of some components of solid waste. (From Ref. 35.)

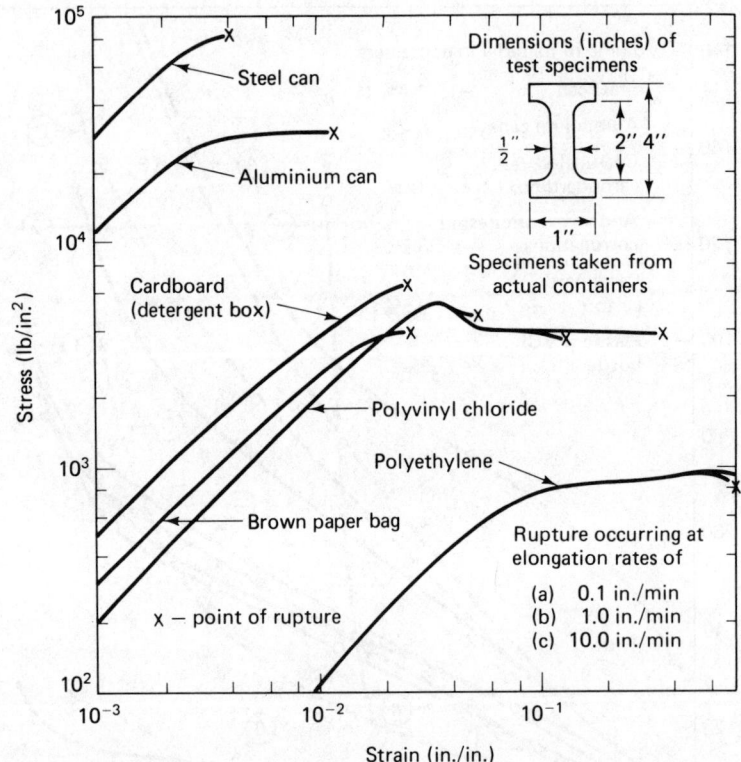

Figure 1-9. Stress-strain curves for some components of solid waste. (From Ref. 35.)

requirements for glass, and cardboard cartons, but large energy needs for crushing cans, as expected.

Mechanical properties

The mechanical properties of refuse components have not been widely studied. Only since shredding and mechanical separation techniques have become widely used have these properties been of concern.

The results of tensile stress-strain tests [35] are shown in Table 1-15 and Fig. 1-9. As expected, the steel had the greatest ultimate strength, yielding a computed modulus of elasticity (E) of 83.8×10^6 kg/cm² (28.5×10^6 lb/in.²), which compares favorably with the usual value of E of 85 to 88×10^6 kg/cm² (29 to 30×10^6 lb/in.²) for carbon and low-alloy steels. The E for aluminum was computed as 30×10^6 kg/cm² (10×10^6 lb/in.²), while PVC had an E of only 0.6×10^6 kg/cm² (0.2×10^6 lb/in.²), as expected.

TABLE 1-15
Stress-Strain Data for Components of Refuse

Material	Type of Container	Container Shape and Specimen Locations	Specimen Thickness (in.)	Ultimate Strength (lb/in.^{2})	Ultimate Strain (in./in.)	Rupture Energy (ft-lb/in.^{3})
Steel	12-oz can, beverage	Cylinder—spec. cut from side, axially and circumferentially	0.007	82,000	0.005	9.4
Aluminum	12-oz can, beverage	Same as above	0.006	31,000	0.013	26.5
Cardboard	Box, laundry detergent	Rectangular box—spec. cut from front and back panels	0.025	6,400	0.025	8.3
Paper	Bag, brown paper	"Grocery"-type bag—spec. cut at various locations	0.009	4,000	0.025	5.1
Plastic Polyvinylchloride	Bottle, liquid soap	Sculpted molding—spec. cut from front and back panels	0.19–0.026	4,000–5,000	0.360 for $\dot{e}$ = 0.1[a] 0.130 for $\dot{e}$ = 1.0 0.061 for $\dot{e}$ = 1.0	111 for $\dot{e}$ = 0.1 44 for $\dot{e}$ = 1.0 19 for $\dot{e}$ = 1.0
Polyethylene	Bottle, shampoo	Cylinder—spec. cut as in cans above	0.028–0.036	1,000	0.80 for $\dot{e}$ = 0.1 0.84 for $\dot{e}$ = 1.0 0.90 for $\dot{e}$ = 10	56 for $\dot{e}$ = 0.1 60 for $\dot{e}$ = 1.0 66 for $\dot{e}$ = 10

[a] $\dot{e}$ = elongation rate; all materials tested at $\dot{e}$ = 0.1, 1.0, and 10 in./min. Materials other than plastics showed no elongation rate effects.

Source: Ref. 35.

POTENTIAL FOR THE RECLAMATION OF USEFUL MATERIALS AND ENERGY FROM SOLID WASTE

The total mixed municipal solid waste generated annually in the United States is about 126 million tonnes. Theoretically, it might be possible to reclaim about

    9,000,000 tonnes of steel.
        9,000 tonnes of aluminum.
   12,000,000 tonnes of glass.

In addition, we could save 1200 quadrillion J (1.2 quadrillion Btu) per year of energy, equal to 564,000 barrels of oil per day (arithmetic equivalent).

Obviously, such theoretical figures are unrealistic. Waste is not generated conveniently at one or several central locations where the processing can take place, nor is this reclamation possible without an expenditure in materials and energy.

If energy recovery from MSW had been practiced in all SMSAs in 1973, 950,000 trillion J (900 trillion Btu) would have been available (less processing costs). This figure is equivalent to about 4.6% of all the fuel consumed by all utilities, 10% of all the coal consumed by all utilities, and about 20% of the electrical energy demand of the private sector of a municipality.

Although these figures must obviously be adjusted to reflect the losses incurred in producing electricity, the use of refuse as a source of energy clearly has tremendous potential.

The materials recovery potential is similarly a "slumbering giant," as shown in Table 1-16. Interestingly, except for aluminum, the fractions of recovered matrials seem to have been decreasing up to 1975. There is little doubt that this picture will change.

Considering the potentials in economic terms adds credence to the wisdom of resource recovery. Table 1-17 is a summary of gross receipts for recovered materials in the United States and indicates the potential economic gains from resource recovery: over $1 billion per year. Although the prices are subject to change, there is little doubt that a large potential exists, provided that the projected costs of separating these materials and transporting them to the markets can be accomplished.

Taking into account transportation and processing charges, it still appears that the economics for large processing plants in metropolitan areas (close proximity to refuse and markets) are quite favorable [35]. The proof of this, of course, is the impressive number of processing facilities in operation or under construction. Estimates of the number of such facilities vary greatly, but all estimates show substantial increases [36].

TABLE 1-16

Materials Recovery in 1975

Material category	Gross discards	Material recycled	
		Quantity	Percent
Paper	44.1	6.8	15.4
Glass	13.7	0.4	2.9
Metals	12.7	0.6	4.7
Ferrous	(11.3)	(0.5)	(4.4)
Aluminum	(1.0)	(0.1)	(10.0)
Other nonferrous	(0.4)	(0.0)	(0.0)
Plastics	4.4	0.0	0.0
Rubber	2.8	0.2	7.1
Leather	0.7	0.0	0.0
Textiles	2.1	0.0	0.0
Wood	4.8	0.0	0.0
Other	0.1	0.0	0.0
Total nonfood product waste	85.4	8.0	9.3
Food waste	22.8	0.0	0.0
Yard waste	26.0	0.0	0.0
Miscellaneous inorganic wastes	1.9	0.0	0.0
Total	136.1	8.0	5.9

Source: Ref. 8.

TABLE 1-17

Economic Potentials for Resource Recovery

Waste Material	Total U.S. Generation (million tons/yr)	Practical Recovery (million tons/yr)	Price of secondary material ($/ton)	Gross annual Revenue (millions)
Paper	44.1	35.3	$ 20	$ 705
Ferrous	11.3	9.0	35	316
Aluminum	1.0	0.8	200	160
Glass	13.7	11.0	7	77
Nonferrous (excluding aluminum)	0.4	0.32	120	38
Total revenue				$1296

OBSTACLES TO RESOURCE RECOVERY

The preceding section is devoted to a demonstration of the vast potential of resource recovery. The fact that we are only beginning to tap this potential suggests that there must be obstacles blocking the implementation of resource recovery systems.

The obstacles are diverse in nature, and any list will be at once incomplete and redundant. It is instructive, however, to name some of the major problems associated with resource recovery.

Heterogeneity of Wastes

As discussed earlier, the purity of a material greatly influences its value, and waste, especially MSW, is notoriously heterogeneous.

Putrescibility of the Wastes

Resource recovery facilities must be designed with the knowledge that the raw material is unstable with respect to time. Storage and handling thus become an important and expensive consideration.

Location of the Wastes

The transportation costs of the waste often prohibit the implementation of larger recovery facilities, both before and after processing. The freight rates on scrap iron and steel are higher than on virgin iron ore, even on the basis of equivalent tons of iron produced.

Low Value of Product

The reason that an item is considered "waste" is often that the products (even when pure) have little value. Fly ash and gypsum from air pollution control in coal-fired power plants are prime examples.

Uncertainty of Supply

The production of solid waste depends on the willingness of collectors to transport it, the cooperation of consumers to throw things away according to a predictable pattern, and the economics of marketing and product substitution, which may significantly influence the availability of a material. Conversion from aluminum to glass beverage container, whether by legislation, marketing options, or consumer preferences, will significantly change the available aluminum in solid waste. Potential solid waste processors thus have little control over their raw material.

Unproven Technology

Many of the large systems presently contemplated and in operation involve capital expenditures of over $100 million per 1000-tons/day capacity. Such investments require either rapid returns or low risk, neither of which is possible in resource recovery.

Administrative and Institutional Constraints

Local ordinances and in some cases state laws must be amended to allow municipalities to implement resource recovery systems. For example, cities are seldom allowed to enter into long-term (20-year) contracts, although these are often necessary before private capital is willing to invest in resource recovery.

Legal Restrictions

Often, technically and economically recoverable materials have no market because of legal restrictions. Liquor bottles, for example, cannot legally be refilled and sold.

Uncertain Markets

Recovery facilities must depend on the willingness of customers to purchase the end products—materials or energy. Often, such markets are fickle, being either small, fragile operations or large, vertically integrated corporations that purchase the products on margin so as to satisfy unusually heavy short-duration demand. The wastepaper market is especially unstable, and large price fluctuations are normal, because of the use of waste fiber as a supplement in virgin pulp and paper operations.

PARTING SHOTS

Large-scale implementation of resource recovery systems would in effect reduce both the extraction of raw materials and the quantities of waste disposed of into our environment. Increased recovery would mean less landfilling (hence less water pollution and other detrimental effects of landfills), less incineration (hence less air pollution), and less blatant open dumping (hence reduction in visual affronts and public health problems). The rate of the increased use of natural resources for both energy and materials production will at least be slowed by a wider use of secondary materials. It is now commonly accepted that the supply of natural raw materials will be restricted worldwide, and that the United States will feel disproportionately greater pressure as a result of our large consumption and limited domestic supplies.

We thus recognize that resource recovery is reasonable, logical, prudent, economical, and feasible. But we must also ask, "Is it right?", which is possibly the most difficult question to answer.

Over the course of the development of our basic Western philosophical and ethical framework, environmental concerns have played a minor role. With the exception of Thoreau, St. Francis of Assisi, and a few others, we have not wrestled very long or hard with the problem of environmental ethics. The question posed above is thus one that does not yet have a simple answer, nor are there many philosophers even willing to tackle it. Much work remains to be done in this area, as is the case with economics, engineering, and the other sciences. The absence of an answer should not, however, deter us from thinking out the "good" in this endeavor, and to seek understanding.

REFERENCES

[1] MEADOWS, D. H., et al., *The Limits of Growth*, Potomac Associates, Washington, D.C., 1972.

[2] BOUGHEY, A. S., *Strategy for Survival*, W. A. Benjamin, Inc., Menlo Park, Calif., 1976.

[3] KESLER, S. E., *Our Finite Mineral Resources*, McGraw-Hill Book Company, New York, 1976.

[4] *Third Report to Congress: Resource Recovery and Waste Reduction*, U.S. EPA-OSWMP, Washington, D.C., 1975.

[5] Glass Containers Manufacturers Institute, *Glass Containers*, 1971 ed., May 1972.

[6] National Center for Resource Recovery, *Bulletin*, 3 (2) (1973).

[7] Midwest Research Institute, *Economic Study of Salvage Markets for Commodities Entering the Solid Waste Stream*, U.S. EPA-ASWMP, Washington, D.C., 1970.

[8] *Fourth Report to Congress: Resource Recovery and Waste Reduction*, U.S. EPA-OSWMP, Washington, D.C., 1977.

[9] BOYD, G., and M. HAWKINS, *Methods of Predicting Solid Waste Characteristics*, U.S. EPA-OSWMP, SW-23c, Washington, D.C., 1971.

[10] SHELL, R. L., and D. S. SHUPE, "A Study of the Problems of Predicting Future Volumes of Wastes," *Solid Waste Manage.*, Mar. 1972.

[11] GROSSMAN, D., J. H. HUDSON, and D. H. MARKS, "Waste Generation Models for Solid Waste Collection," *J. Environ. Eng. Div. ASCE*, **100** (EE6) (1974).

[12] QUAN, J. E., M. TANAKA, and A. CHARNES, "Refuse Quantities and Frequency of Service," *J. Sanit. Eng. Div. ASCE*, **94** (SA2), 403 (1968).

[13] SHUSTER, K. A., *A Five-Stage Improvement Process for Solid Waste Collection Systems*, U.S. EPA, Cincinnati, Ohio, 1974.

[14] HUDSON, J. A., and D. H. MARKS, "Solid Waste Generation and Service Quality," *J. Environ. Eng. Div. ASCE*, **103** (EE2), 245 (1977).

[15] BOND, R. G., and C. P. STRAUB, *Handbook of Environmental Control*, Vol. II: *Solid Waste*, CRC Press, West Palm Beach, Fla., 1973.

[16] *Incinerator Guidelines*, U.S. Dept. of Health, Education, and Welfare, 1969.

[17] VESILIND, P. A., J. A. NISSEN, and F. MCALISTER, "Data Generation for Solid Waste Allocation Models," *Solid Waste Manage.*, Nov. 1977.

[18] WESTERHOFF, G. P., and R. M. GRONINGER, "Population Density vs. per Capita Solid Waste Production," *Public Works*, **101** (2), 86 (1970).

[19] HAGERTY, D. J., J. L. PAVONI, and H. E. HEER, Discussion of "Sample Weights in Solid Waste Composition Studies," by A. J. Klee and D. Carruth, *J. Sanit. Eng. Div. ASCE*, **97** (SA3), 384 (1971).

[20] DAVIDSON, G. R., *Residential Solid Waste Generation in Low Income Areas*, U.S. EPA-OSWMP, Washington, D.C., 1972.

[21] NIESSEN, W. R., and A. F. ALSOBROOK, "Municipal and Industrial Refuse: Composition and Rates," *Proc. ASME Natl. Incin. Conf., 1972*.

[22] *Office Paper Recovery: An Implementation Manual*, U.S. EPA-OSW Sw-571c, Washington, D.C., 1977.

[23] WINKLER, P. F., and D. G. WILSON, "Size Characteristics of Municipal Solid Waste," *Compost Sci.*, **14**(5) (1973).

[24] *Municipal Solid Waste: Its Volume, Composition and Value*, National Center for Resource Recovery, Washington, D.C., 1973.

[25] "World Survey Finds Less Organic Matter," *Solid Waste Manage.*, **10**(26) (1976).

[26] ALTER, H., "European Materials Recovery Systems," *Environ. Sci. Technol.*, **11**(5), 444 (1977).

[27] WHITE, M., Auckland Regional Authority, Auckland, New Zealand, personal communication.

[28] KIROV, N. Y., "Principles of Waste Management: Unit Operations and Processes," Univ. of New South Wales, Sydney, Australia.

[29] KLEE, A. J., and D. CARRUTH, "Sample Weights in Solid Waste Composition Studies," *J. Sanit. Eng. Div. ASCE*, **96**(SA4), 945 (1970).

[30] O'BRIEN, J., Unpublished data, Duke Environmental Center, Duke University, Durham, N.C., 1977.

[31] RIGO, G., Systech, Inc., Dayton, Ohio, personal communication.

[32] National Center for Resource Recovery, Washington, D.C.

[33] WILSON, D. L., *Determination of Heat of Combustion of Solid Waste from Ultimate Analysis*, U.S. EPA RS-03-68-17, Washington, D.C., 1971.

[34] ALTER, H., "Energy Equivalents," *Science*, **189**(4198) (1975).

[35] TREZEK, G., D. HOWARD, and G. SAVAGE, "Mechanical Properties of Some Refuse Components," *Compost Sci.*, **13**(6) (1972).

[36] ABERT, J. G., H. ALTER, and J. F. BERNHEISEL, "The Economics of Resource Recovery from Municipal Solid Waste," *Science*, **183** (Mar. 15, 1974).

[37] WOODYARD, J. P., and A. J. KLEE, "Solid Waste Characterization for Resource Recovery Design" *Proc.* IIT-Bureau of Mines Conference, Chicago, 1978.

[38] MARTELLO, W. "Heat Value of Refuse Derived Fuel" M.S. Thesis, Duke University, 1979.

PROBLEMS

1-1. A large furniture manufacturer, with 100 new employees, has decided to locate in a small community of 1500 residents. If the wastes from this plant go to the town landfill, which has an estimated life of 4 years, and if the cost of operating the landfill to the town is $3 per ton, how much more will this cost the town, and how much sooner will the town need a new landfill?

1-2. Estimate the quantity of paper of different categories that might be obtained from the building where you work.

1-3. Discuss the accuracy of the data in Table 1-8. Why is there such a wide discrepancy in the percentages?

1-4. Take a photograph of refuse in a living hall or office refuse container. Calculate the composition as described in Example 1-1.

1-5. A landfill operation uses a tractor to compact the waste. The tractor weighs 8 tons and has two tracks, each 2 ft × 10 ft. Calculate the maximum compaction attainable in the landfill.

2 COLLECTION OF SOLID WASTE

SOLID WASTE COLLECTION SYSTEMS

In the United States, as in most other parts of the developed world, solid waste collection systems are invariably person/truck systems. With only a few minor exceptions, the collection of MSW is done by men or women who traverse a town in trucks and then ride with the truck to a site at which the truck is emptied. This may be a disposal site (landfill) or an intermediate stopover where the refuse is transferred from the small truck into larger vans, barges, or railway cars for long-distance transport. Such facilities are known as *transfer stations*.

The process of collection must not, however, be thought of as a single-phase process, since it is possible to define at least five separate phases, as shown in Fig. 2-1. First, the individual homeowner must transfer whatever is considered waste (defined as material having no further value to the occupant) to the refuse can, which may be inside or outside the home. The second phase is the movement of the refuse can to the truck, which is usually done by the collection crew, at times assisted by the occupant. If the can is moved to the street by the occupant, the system is called *curb side collection*.

The truck must collect the refuse from many homes in the most efficient way possible (phase 3) and when it is full (or at the end of the day), it must travel to the disposal or transfer site (phase 4). The fifth phase of the

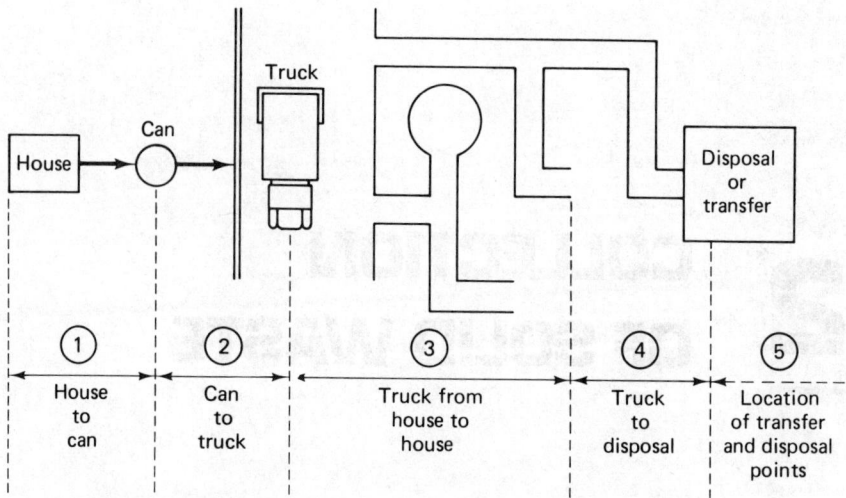

Figure 2-1. Various phases of the municipal solid waste collection system.

collection system involves the location of the disposal site or transfer station. This is a planning problem, often involving more than one community. Each of these phases is discussed more fully below.

Phase 1

The house-to-can phase has received almost no attention or concern by researchers or government because the efficiencies and conveniences gained here are personal and not communal. As discussed later under source separation, one major drawback of collecting separated material is the inconvenience suffered by the individual.

One exception to the general lack of innovative technology in this first phase is the home compactor. These devices, usually under the kitchen counter, compress about 9 kg (20 lb) of refuse into a convenient block within a special bag which then can be placed at the curb.

Phase 2

The second phase of collection, moving the refuse to the truck, can be done by collection crews or by residents. Curbside collection (or alleyway collection in some towns) can be facilitated by placing the cans on rollers, or by using specially constructed plastic cans on wheels which are hydraulically lifted into the truck. These cans hold up to 300 liters (80 gal) and

Solid Waste Collection Systems

Figure 2-2. The can-on-wheels or "green can" system for getting the refuse into the truck. (Courtesy of Rubbermaid Applied Products, Inc.)

can be used for garden waste as well, thus eliminating a special trash pickup. The system is illustrated in Fig. 2-2.

Other methods of transferring refuse to the truck have been attempted with various degrees of success [1]. The motivation for mechanical means of emptying cans into trucks has been both efficiency and safety. It should not be a surprise that solid waste collection is the most hazardous job in America. As shown in Table 2-1, which lists accident figures compiled by the National Safety Council, collection crews suffered over 3000 days lost per million hours worked [2].

The trucks used for residential and commercial refuse collection are covered compacters called *packers*, and vary in size and design, with 12- and 15-m^3 (16- and 20-yd^3) loads being common. Other types of trucks, such as simple open-bed vehicles, are often used, when the refuse is prepackaged in paper bags. In most towns, however, some type of packer is used. These trucks have hydraulically operated compaction mechanisms

TABLE 2-1
Injuries of Solid Waste Collection Crews Compared to Other Professions

	Days lost from work per million worker-hours
All industries	650
Municipal employees	900
Solid waste disposal	1400
Solid waste collection	3000

Source: National Safety Council.

Figure 2-3. When hopper is full, the loading cycle starts. The packer blade moves up as the hopper rises, and then bulldozes the refuse into the body. When the body is full, the tailgate is raised and the packed refuse is ejected.

which can compress the refuse from a loose density of about 35 kg/m³ (60 lb/yd³) to about 200 to 240 kg/m³ (350 to 400 lb/yd³). The rear-loading vehicle has a low loading platform which makes can dumping convenient. These trucks can compact to high densities, equal to allowable axle loads, and are the most popular vehicle used. The compaction (packing) mechanism of one manufacturer is shown in Fig. 2-3.

Phase 3

The most studied phase of solid waste collection is the movement of the truck from house to house. It is obvious that the selection of a route can greatly affect the efficiency of collection, and changes in routes can result in substantial cost savings to a community. The routing of a vehicle within its assigned collection zone is often called *microrouting* to distinguish it from the larger-scale problems (phase 4) of routing to the disposal site and the establishment of the individual route boundaries. The latter problem is commonly known as *macrorouting* or "districting", [3 and is discussed later.

The present question is how to route a truck through a series of one- or two-way streets so that the total distance traveled is minimized. Put another way, the objective is to minimize *deadheading*, traveling without picking up refuse. The assumption is that if a route can be devised that has the least amount of deadheading possible, it is the most efficient collection route.

The problem of designing a route so as to eliminate all deadheading was actually addressed as early as 1736. The brilliant mathematician Leonard Euler, was asked to design a route for a parade across the seven bridges of Königsberg, a town in eastern Prussia, such that the parade would not cross the same bridge twice but would end at the starting point. The problem is illustrated in Fig. 2-4, together with a schematic diagram. The routes are shown by lines called *links* and the locations are known as *nodes*. The system shown has four nodes and seven links.

Euler not only proved that the assignment was impossible, but he generalized the two conditions that must be fulfilled for any network to

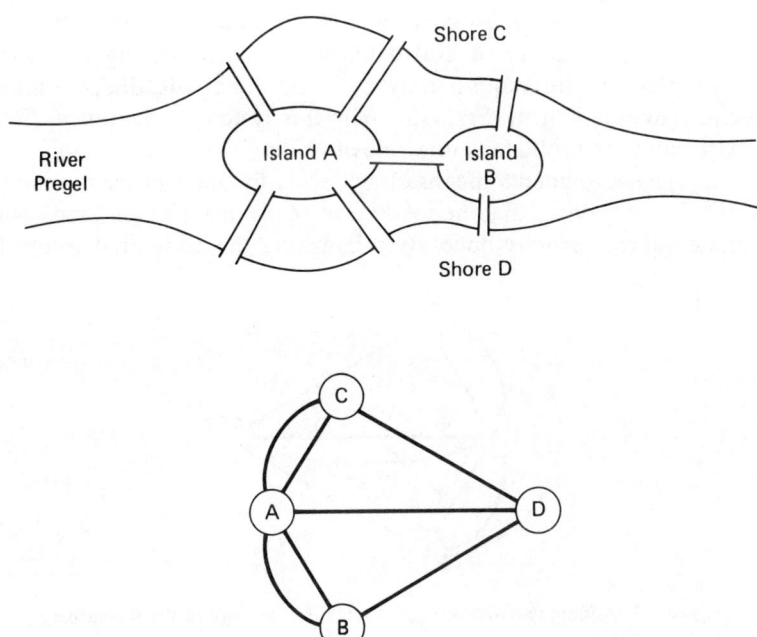

Figure 2-4. Euler's parade routing problem—the seven bridges of Königsberg.

make it possible to traverse a route without traveling twice over any road. These two conditions are:

1. All points must be connected (one must be able to get from one place to another).
2. The number of links to any node must be of an even number.

The first condition is logical. The second similarly makes sense, in that if one travels to a location such as island A or B in Fig. 2-4, one must be able to get off again—hence two roads. Euler's parade problem had all the nodes with an odd number of links, a clearly impossible situation.

The number of links connecting a node designates its degree, and the existence of any odd-degree nodes in a system indicates that a route without deadheading is impossible. A system that has all nodes of even degree is know as a *unicursal network*, and an "Euler's tour" is theoretically possible. [4].

In the real world, one-way streets, dead ends, and other restrictions can often make a practical application of the theoretical analysis difficult. One-way streets can be considered in Euler's theory by recognizing again that one must be able to get to a node exactly as many times as one leaves it. Thus a node with three one-way streets leading to it and a single one-way street leading away from it immediately makes a network nonunicursal, even though the number of links at that node is even.

The development of a least-cost route involves making a system unicursal with the least number of added links. For example, the Königsberg bridge problem would require only two additional (deadhead) links to make the system unicursal (Fig. 2-5). With this system, a theoretical Euler's tour exists, and the problem is now one of finding the proper route.

Kwan [5] has provided a means of achieving the most efficient unicursal network* by observing that networks are really a series of loops where each node appears exactly once. By minimizing the additional connecting

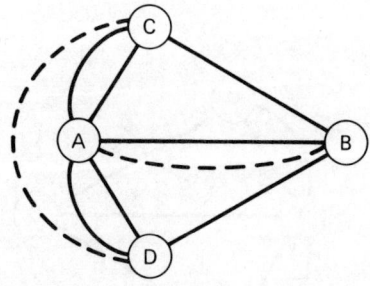

Figure 2-5. Adding two links creates a unicursal network of the Königsberg bridge problem.

*And also provided the name for this procedure—the Chinese postman problem.

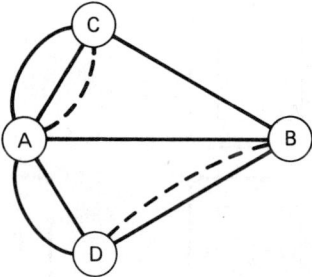

Figure 2-6. Second unicursal Königsberg bridge network.

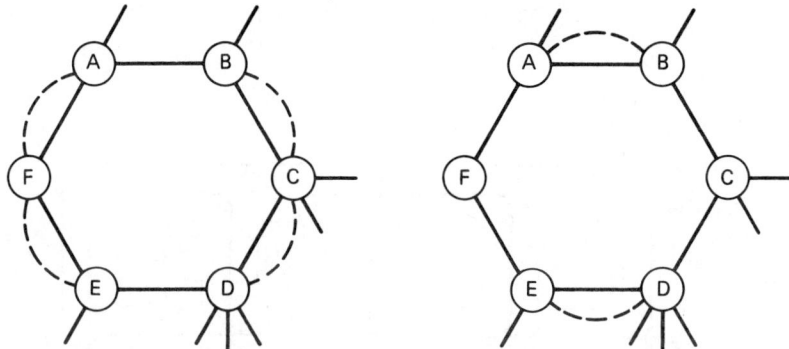

Figure 2-7. Two solutions to creating a unicursal loop. The dashed lines are deadheads.

links (deadheads) necessary to achieve a unicursal system, one can in fact achieve an overall optimum system. For example, the unicursal network of the Königsberg bridge problem shown in Fig. 2-5 is clearly a poor choice. (It, in fact, requires a new bridge.) It would make much more sense to trade the two deadheads shown in Fig. 2-5 for the two in Fig. 2-6. The latter is an obviously more efficient solution. Similarly, the large loop shown in Fig. 2-7 can be made unicursal in one of two ways. The skill of the route planner must come into play in such trade-offs, since a shorter street with many traffic problems may in fact be a more expensive alternative to a longer but clear street.

Once a unicursal network has been designed, it remains to route the truck through this network. The method of heuristic (commonsense) routing developed by Shuster and Schur has found wide application [3]. They promulgated the following set of rules for microrouting, some of which are pure commonsense judgment, and some are useful guidelines for determining overall strategy when attacking a network.

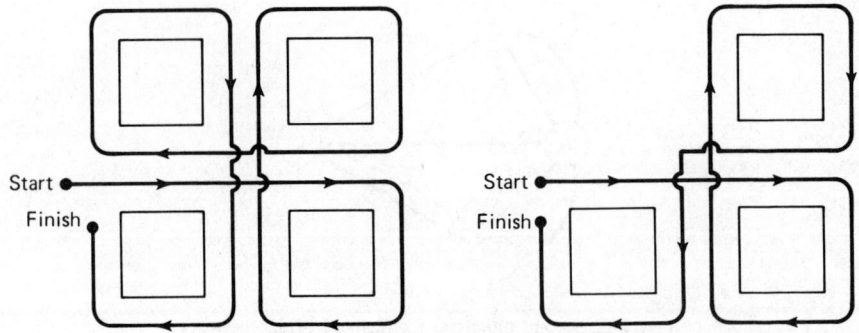

Figure 2-8. Specific routing patterns for four- and three-block configurations.

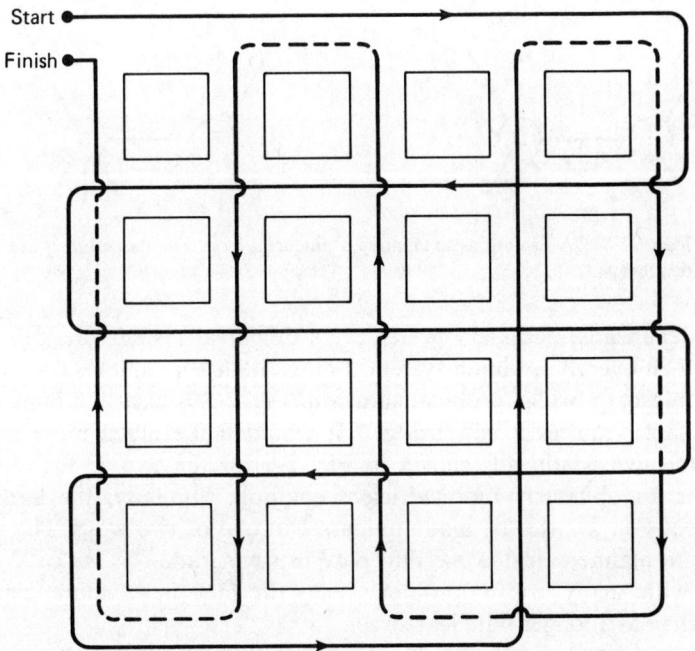

Figure 2-9. Example of heuristic routing of large networks; collection on two sides of the street.

1. Routes should not overlap, but should be compact and not fragmented.
2. The starting point should be as close to the truck garage as possible.
3. Heavily traveled streets should be avoived during rush hours.
4. One-way streets that cannot be traversed in one line should be looped from the upper end of the street.
5. Dead-end street should be collected when on the right side of the street.
6. On hills, collection should proceed downhill so that the truck can coast.
7. Clockwise turns around blocks should be used whenever possible.
8. Long, straight paths should be routed before looping clockwise.
9. For certain block patterns, standard paths, as shown in Fig. 2-8, should be used.

Liebman et al. [4] have added another rule:

10. U-turns can be avoided by never leaving one two-way street as the only access and exit to the node.

These rules can be used to develop effective routes, with minor deadheading. Figure 2-9 is an example of some of these routing rules applied to a large area.

Elegant computer programs have been developed by a number of researchers [6, 7], but in practice it has been found that "the tours constructed manually are almost always better than those done by mechanical tour-building codes" [4].

Phases 4 and 5

For smaller isolated communities, the macrorouting problem reduces to one of finding the most direct road from the end of the route to the disposal site. For regional systems or large metropolitan areas, however, macrorouting in terms of developing the optimum disposal and transport scheme can be used to great advantage. The available techniques, called *allocation models*, are all based on the concept of minimizing an objective function subject to constraints, linear programming being the most common technique.

The simplest allocation problem is the optimization of solid waste disposal to more than one disposal site. Often the solution is obvious—the closest sources are allocated first, followed by the next closest, and so on. With more complex system, however, it becomes necessary to use optimization techniques. The most appropriate one is the *transportation algorithm*, which is a type of linear programming.

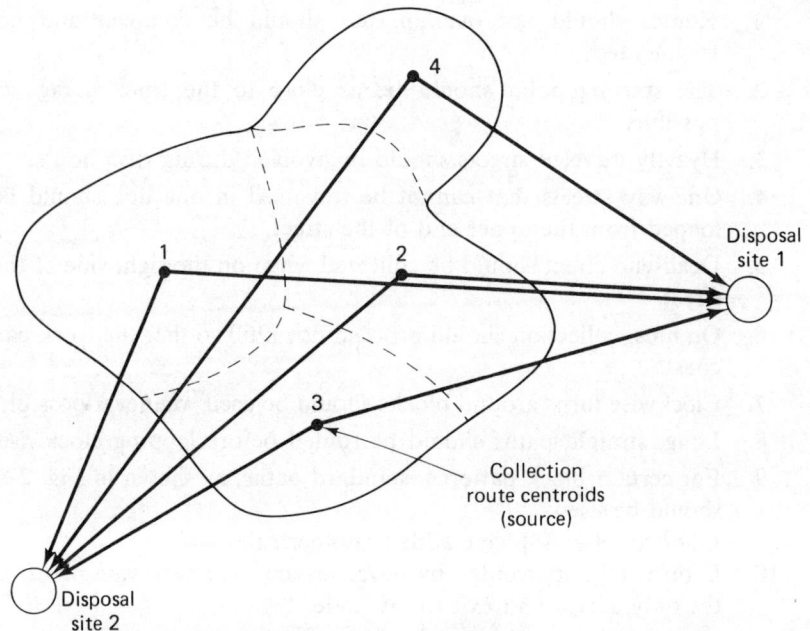

Figure 2-10. Waste allocation scheme. In this case, $N = 4$, $K = 2$ (see the text for a key to the symbols).

Consider the simple system pictured in Fig. 2-10. The waste generated at four sources (denoted by centroids of the collection area, a poor assumption, especially if the disposal sities are close to the collection routes) are to be allocated to two disposal sities. The objective is to achieve this in a minimum-cost manner.

At the same time, several requirements must be met (constraints in an optimization model).

1. The capacity of each disposal site (e.g., a landfill) is limited.
2. The amount of refuse disposed of must equal the amount generated.
3. The collection route centroids cannot act as disposal sites; or the total amount of refuse hauled from each collection area must be greater than or equal to zero.

The following notation is adopted:

x_{ik} = quantity of waste hauled from source i to disposal site k, per unit time

c_{ik} = cost per quantity of hauling the waste from source i to disposal site k

F_k = disposal cost per waste quantity at disposal site k (capital plus operating)

Solid Waste Collection Systems

B_k = capacity of disposal site k, in waste quantity per unit time
W_i = total quantity of waste generated at source i, per unit time
N = number of sources, i
K = number of disposal sites, k

The problem then boils down to minimizing the following objective function:

$$\sum_{i=1}^{N} \sum_{k=1}^{K} x_{ik} c_{ik} + \sum_{k=1}^{K} \left(F_k \sum_{i=1}^{N} x_{ik} \right)$$

subject to the following constraints:
Constraint 1 above requires that

$$\sum_{i=1}^{N} x_{ik} \leq B_k \quad \text{for all } k$$

Constraint 2 above requires that

$$\sum_{k=1}^{K} x_{ik} = W_i \quad \text{for all } i$$

Constraint 3 above requires that

$$x_{ik} \geq 0 \quad \text{for all } i, k$$

The first term in the objective function is transportation costs and the second term is disposal costs.

For the case shown in Fig. 2-10, the objective function is:

Minimize $[x_{11}c_{11} + x_{21}c_{21} + x_{31}c_{31} + x_{41}c_{41} + x_{12}c_{12} + x_{22}c_{22} + x_{32}c_{32}$
$+ x_{42}c_{42} + F_1(x_{11} + x_{21} + x_{31} + x_{41}) + F_2(x_{12} + x_{22} + x_{32} + x_{42})]$

subject to the following constraints:

$$x_{11} + x_{21} + x_{31} + x_{41} \leq B_1$$
$$x_{12} + x_{22} + x_{32} + x_{42} \leq B_2$$
$$x_{11} + x_{21} + x_{31} + x_{41} = W_1$$
$$x_{12} + x_{22} + x_{32} + x_{42} = W_2$$
$$x_{11} \geq 0, \; x_{12} \geq 0, \; \ldots, \; x_{23} \geq 0, \; x_{24} \geq 0$$

This problem can be solved using any linear programming algorithm— but the transportation algorithm in particular.

Example 2-1

Assume the solid waste generation and disposal figure for the system pictured in Fig. 2-10 is as follows:

		Cost of transport, c_{ik}	
Source i	Generation, W_i (tonnes/week)	To site 1 ($/tonne)	To site 2 ($/tonne)
1	100	5	12
2	130	7	5
3	125	4	8
4	85	13	6

Disposal site, k	Capacity, B_k	Cost, F_k ($/tonne)
1	250	4
2	200	6

Using a transportation algorithm program, we find the following:

From Source, i	To Disposal Site, k	Waste Hauled (tonnes/week)	Transport Cost ($/week)	Processing Cost ($/week)
1	1	100	500	400
2	1	25	175	100
2	2	105	525	630
3	1	125	500	500
4	2	85	510	510
Total cost			2210	2140

Therefore, the total system minimum cost is $4350/week. Note that 10 tonnes/week capacity remains unutilized in landfill 2.

The complexity of the problem can be increased by including an intermediate facility such as an incinerator or a transfer station in the scheme, as shown in Fig. 2-11. If one or more such intermediate facilities are placed in the system, the waste generated can be hauled to these stations, or it can go directly to the disposal sites. From the intermediate facilities, it must go to one of the disposal sites. In more complicated models, waste flow between intermediate facilities can be included.

In this situation, the trucks have K disposal points and J intermediate facilities. As before, these facilities have processing costs F_j and F_k (capital and operating). The problem now is to minimize the following objective

Solid Waste Collection Systems

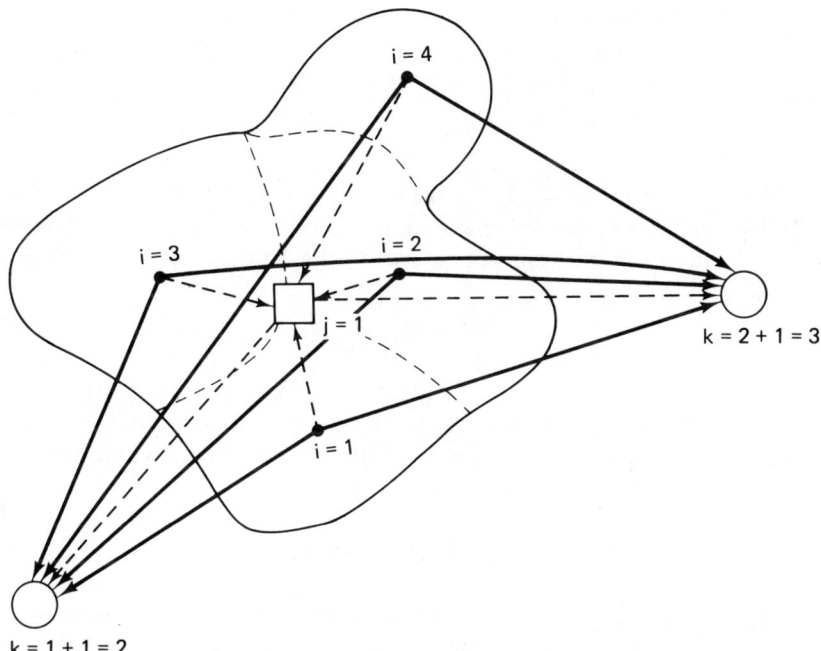

Figure 2-11. Waste allocation scheme, with an intermediate facility. In this case, $N = 4$, $J = 1$, $K = J + 2 = 3$.

function:

$$\sum_{i=1}^{N} \sum_{j=1}^{J} c_{ij} x_{ij} \sum_{i=1}^{N} \sum_{k=1}^{K} c_{ik} s_{ik} + \sum_{j=1}^{J} \sum_{k=1}^{J} c_{jk} x_{jk}$$

$$+ \sum_{j=1}^{J} F_j \sum_{i=1}^{N} x_{ij} + \sum_{k=1}^{K} F_k \sum_{i=1}^{N} X_{ik}$$

where c_{ij} = cost per quantity of hauling the waste from source i to intermediate facility j
 c_{jk} = cost per quantity of hauling the waste from intermediate facility j to final disposal facility k
 x_{ij} = quantity of waste hauled from source i to intermediate facility j, per unit time
 x_{jk} = quantity of waste hauled from intermediate facility j to final disposal facility k, per unit time
 B_j = capacity of intermediate facility j, in waste quantity, per unit time

P_j = proportion of waste at intermediate facility j which, after processing, remains for disposal; $P_j = 1.0$ if the facility is a transfer station, but $P_j \simeq 0.2$ if it is an incinerator

J = number of intermediate facilities, j

The rest of the variables are defined as before.

This object function is subject to the following constraints:

1. The quantity of waste generated at source i, W_i, must equal the sum of all the waste hauled from that source to the J intermediate sites and K disposal points.

$$\sum_{j=1}^{J} x_{ij} + \sum_{k=1}^{K} x_{ik} = W_i \quad \text{for all } i$$

2. The capacity of the jth intermediate site, B_j, must be more than or equal to the total waste brought to it. If this constraint is omitted, the model can be used to determine the required capacity,

$$\sum_{i=1}^{N} x_{ij} \leqslant B_j \quad \text{for all } j$$

3. B_k, the capacity of the final disposal site k, which might be influenced by the number of trucks a tipping floor or landfill site can handle, the compaction capacity at a landfill, and so on, should not be exceeded by the waste brought in directly from the collection sites or from the intermediate facilities,

$$\sum_{i=1}^{N} x_{ik} + \sum_{j=1}^{J} x_{jk} \leqslant B_k \quad \text{for all } k$$

4. Whatever waste is shipped in to an intermediate site must be shipped out to a disposal site. The proportion of waste that remains for disposal after any processing is denoted by P_j:

$$P_j \sum_{i=1}^{N} x_{ij} - \sum_{k=1}^{K} x_{jk} = 0 \quad \text{for all } j$$

5. The nonnegativity constraints are

$$\left. \begin{array}{l} x_{ij} \geqslant 0 \\ x_{ik} \geqslant 0 \\ x_{jk} \geqslant 0 \end{array} \right\} \quad \text{for all } i, j, k$$

Similar models have also been developed [8–11], which hold great promise in their application to large-scale collection systems.

In situations where the sophistication of linear programming models is not warranted, a brute-force technique using a simple grid system can be of value [12]. In this case, the region is divided into equal blocks on an

$X-Y$ grid, and the solid waste generation is then estimated based on population. The sites for transfer stations and final disposal facilities are initially screened to eliminate obviously inadequate areas (e.g., urban areas for landfills). Trial-and-error siting of facilities is then used to obtain the most reasonable combination of solid waste disposal facilities.

Solid Waste Transportation and Materials Recovery Models

The next logical step in the development of solid waste transportation models is to include a recovery component. One such widely used model is WRAP (Waste Resource Allocation Program). This model can be applied to a regional system and can be used to provide a presentation of technical and economic data that will clearly outline the implications of alternative regional approaches. From an economic point of view, regionalization of at least the transfer and disposal options reduces the cost through economies of scale achieved by large central processing facilities, but it also increases costs through additional vehicle and haul costs. The WRAP model is used to establish an optimal balance (a minimum cost solution) between alternatives.

The model is a fixed-charge linear programming model, using as the optimizer an algorithm developed by Walker [13]. The fixed-charge capability of the model permits the representation of economies of scale in process costs. Figure 2-12 illustrates a concave total-cost function, typical of solid waste processing, which is represented by several linear segments. Because the model seeks a minimum-cost solution, it attempts to determine the lowest-cost segment at any level of tonnage. The model can thus treat costs in two parameters (fixed and variable, or intercept and slope) and thus represent economies of scale at any level of accuracy desired. A major shortcoming of fixed-change linear programming models, the solution being a local optimum, is overcome by Walker's algorithm. In addition to a local optimum, global optimum is also computed.

Simply stated, WRAP allows the user to compare the costs of various feasible levels of system centralization in order to determine a minimum-cost regional plan. While the model does not consider collection functions within a community in analyzing system alternatives, the following aspects are considered by the model: transfer functions, types of resource recovery technology, source separation, locations and sizes of facilities, feasible alternatives, overall system costs, and the cost sensitivity of design changes.

WRAP consists of a series of equations that consider the sources of solid waste generation in a given planning region, a set of sites (and the processing options to be considered at those sites), and site and process

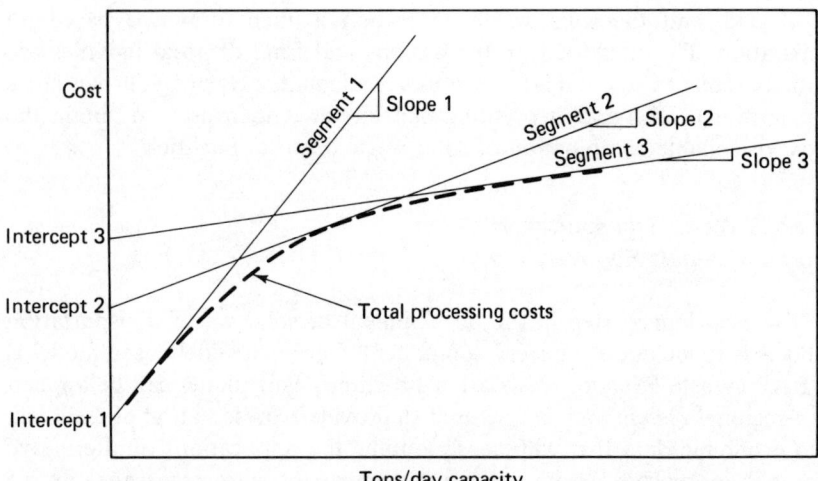

Figure 2-12. Piecewise linear approximation of concave function.

capacity constraints. The model accommodates such processes as transfer stations, primary processes such as incinerators, resource recovery processes (including source separation), secondary processes which receive the residue of the primary process as input, and various disposal processes. WRAP then integrates these components with transportation considerations. The model considers the many transportation route alternatives from sources of waste generation to sites, and from sites, given traffic constraints at any site. If transportation links are not specified explicitly, then a "crow-fly" option introduces possible links by computing distances given the latitude and longitude of sources and sites.

The economics of a system configuration are also considered. Processing costs are input to WRAP so as to reflect the economies of scale for the processes considered, haul costs, and the revenues that are generated from marketing recovered materials.

WRAP has three important components: structure, cost, and procedure. The model's *structure* assures that each alternative considered is feasible in the sense that all wastes generated are entered into transportation, that all residues generated are processed at the site or go to transportation, that no process exceeds the indicated tonnage maximums, and that traffic constraints are considered. The model's *cost* functions assure that each alternative is properly costed, including the economies of scale where appropriate. The model also utilizes *procedures* to permit an organized mathematical search which allows those options that improve the solution to be separated from less desirable options. The model finally indicates when the least-cost solution has been identified.

WRAP operates in the static or dynamic mode. In the dynamic mode, the total planning period is divided into two to four model periods. The model periods must be any combination of integer years that add up to the planning period. This enables the program to handle variations in data with time.

The model thus achieves what hand solutions of simple linear optimization models cannot—a sorting of *complex* regional-type solutions to the solid waste problem. But the model must be used correctly [14–16] to be of any value, and the user is cautioned to first consider if such a tool is really needed to analyze a problem. If its use is indicated, careful consideration must be given to the processes chosen, the cost functions used to describe the systems, and the transportation links provided. WRAP and other such models can then be used to effectively model a region's solid waste management problem and be expected to produce a reasonable set of recommended solutions. The decision maker can then determine which solution set is most practical and feasible.

Types of Information Required for Solid Waste Collection Models

Almost all of the contemporary solid waste allocation models use a grid system as a means of specifying the location of the various components, such as landfills and transfer stations. Such a grid system should, if possible, be compatible with other planning efforts being used by state agencies or regional councils.

Another basic requirement is population data, both present numbers and future projections. In some models, other demographic information, such as income level or education, is required, but this is unusual. The usual source for these data is the U.S. Census, although some states have developed grid systems containing population levels and projections.

In one study [17] of solid waste disposal in a rural North Carolina region, the population changes, based on historial evidence, were estimated as

$$\Delta X = (7.9 \times 10^{-5})(X)^{1.5} + 0.05$$

where ΔX = annual change in population per square mile
X = population per square mile at the start of a year

As the empirical equation indicates, the percentage increase in urban population can be expected to be considerably higher than for rural areas. In fact, the population of some rural areas will in all probability decline, but since the total overall population density (and hence solid waste generation) in these areas is low, the potential inaccuracies due to these small reductions in population will have minor effects on the overall accuracy of results from these models.

A second problem with this equation is that for densely populated urban areas, such as the inner-city cores, population increases may be

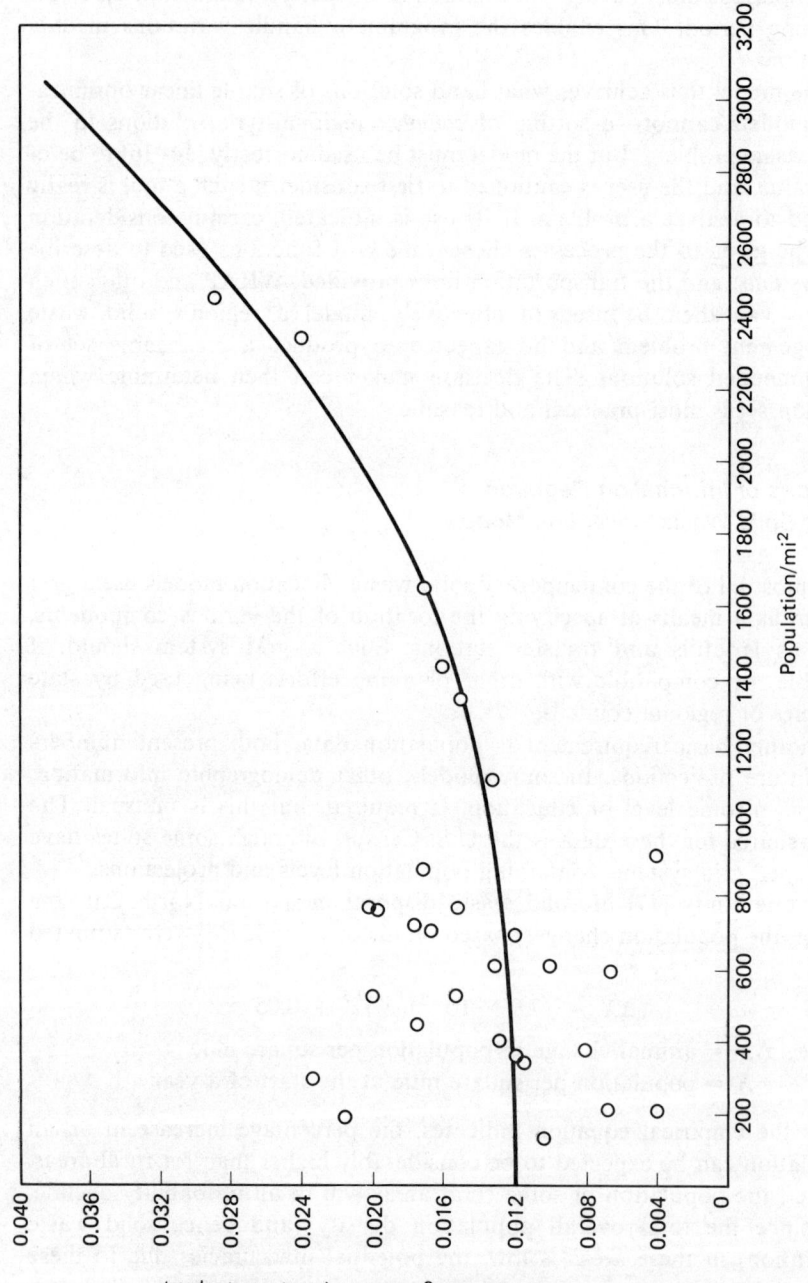

Figure 2-13. Solid waste generation as influenced by population density for a rural region. All solid wastes entering a landfill are included. (From Ref. 17).

considerably lower than increases that would be calculated by the model. This function was developed for a rural area which did not include dense urban populations and should not be used for highly urbanized areas which would require a modified population projection function.

Whatever method is used, population projections can be calculated with a reasonable degree of precision, unlike the next information necessary—solid waste generation. The amount of solid waste generated per capita has been a subject of speculation and argument for some time. It has been demonstrated that such factors as income, life-style, age, and so on all influence the generation rate, and some authors have even attempted to quantify these differences, as discussed in Chapter 1. Figure 2-13 shows that for a rural region, the per capita production of solid waste increases as population density increases.

The function for this case is

$$Y = (1.08 \times 10^{-11})X^{2.7} + 0.012$$

where Y = solid waste generation, tons/week/capita
 X = population density, population/square mile

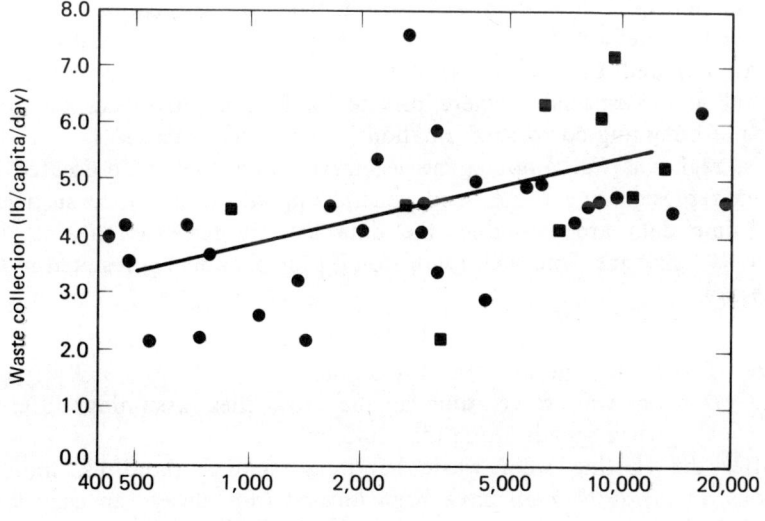

Figure 2-14. Solid waste generation as influenced by population density for a metropolitan region. Only collected MSW is included. (From Ref. 18.)

Although this function was derived for a mostly rural area, the increase in solid waste production with population density seems to hold for urban areas as well, as shown in Fig. 2-14

The solid waste generation data are of course obtained by actual measurements. In some cases it is possible to inspect each truckload entering a landfill to determine its source and quantity. In fortunate instances, scales are available and tonnage records can be quite accurate. It is not unusual, however, to find that a number of different haulers cover a single area, and that there is no record of trucked tonnage. If a landfill is not equipped with scales, the available records are of little value, and other procedures must be used for estimating solid waste generation.

After present solid waste generation is measured or estimated, the next requirement is to develop projections of per capita generation. Projections of national averages are of little use to local planners, since local conditions can play a major role in determining the amount of waste generated. Rural residents, for example, are well known for devising ingenious schemes for reusing material, while most urban dwellers, as a community, generate more than their national average per capita share. Some authors [17] have estimated that the per capita generation of solid waste will not increase, except as an indirect result of population increase. Others [9] have used incremental annual increases, such as 0.083 lb/day/capita.

Collection method, frequency, and routing are also necessary if baseline data are required. In urban areas, this information is usually available, but in rural areas, especially where private haulers are involved, the only means of obtaining correct information is to ride the trucks.

If a region is fairly homogeneous, travel times can be estimated by driving representative routes and generalizing the data. Once sufficient travel-time data are available, the data can be regressed against the "crow-fly" distance. One such regression [9] for New Jersey resulted in the expression

$$T = 1.5D - 0.65$$

where T = actual one-way travel time, min
D = one-way travel time as the crow flies, assuming different truck speeds along the route, min

After the existing solid waste handling facilities (landfills, transfer stations, incinerators, etc.) have been located (not always an easy task, since private operators often favor a nearby surreptitious dump to a longer drive to the landfill), the amount of material entering these facilities must be determined. With scales, this is an easy matter; without scales, the recollections of the operator are often the only available data, and such information estimates are always suspect. For privately operated landfills, the operators can, and have, refused to cooperate, thus making undercover surveillance necessary.

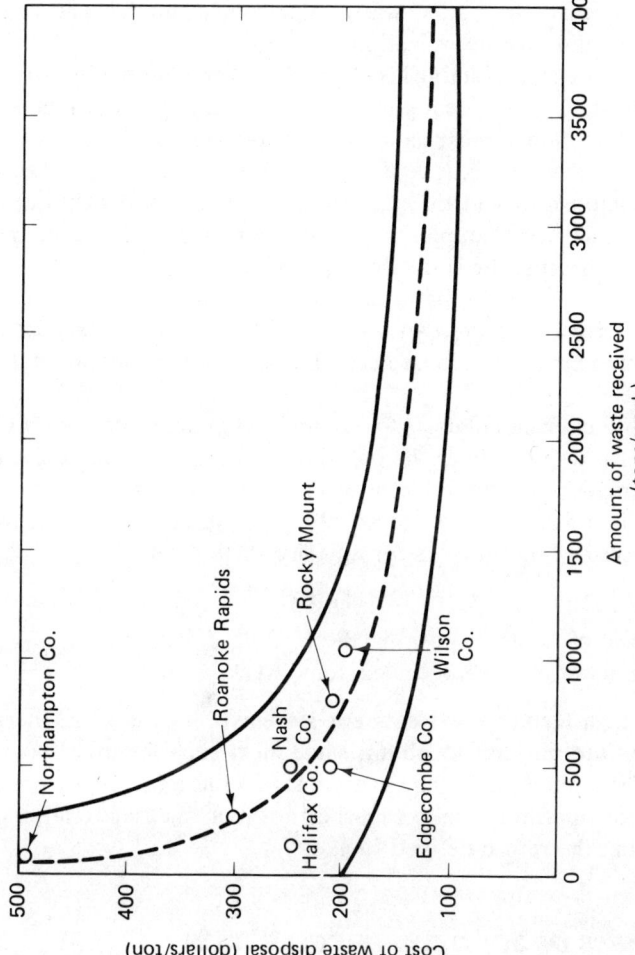

Figure 2-15. Landfill disposal costs. (The area bounded by the solid lines is from **Ref. 19**; the data points are from **Ref. 17**.)

With service areas defined, and the collection and disposal quantities at hand, it is necessary to compute a mass balance for the region. This often turns out to be an unnerving experience when the tons collected and tons disposed of do not agree—at times by a wide margin. Such a mass balance shows where data are deficient and/or questionable, and where field work (perhaps with wheel scales) may be necessary. The remaining data are all dollar items necessary to a solid waste management model, and essential for proper and valid cost optimization.

Collection cost can be established from fees if private haulers are used, and may be available from local government accounting departments if the service is public. Since most communities do not use the cost-center approach to cost accounting, it is often necessary to pore over budgets and accounts to establish a realistic estimate of collection and disposal costs. Truck maintenance, for example, may be in the motor pool budget and not directly charged against the solid waste account.

Landfill costs can usually be established fairly accurately. The cost for land as well as salvage values can often be obtained from local real estate agents, and municipalities are usually able to estimate operational costs fairly accurately.

Some landfill operational costs for a rural region are plotted on a widely reproduced EPA curve such as the one shown in Fig. 2-15. The EPA curve is several years old, but the low cost of land and labor in the rural region compensates for this. Based on these data, a continuous function can be derived to describe landfill costs for any size facility, as

$$C = (20.11)M^{-0.345}$$

where C = cost of waste disposal in the landfill, dollars/ton
M = amount of waste received, tons/week

Costs for high-technology treatment systems (shredding, incineration, pyrolysis, etc.) are difficult to obtain since most such facilities have been operational for a short time. The economics of these facilities depend on local conditions (quantities, labor, markets for reclaimed material, markets for energy from the refuse-derived fuel, etc.)

EFFECTIVENESS OF SOLID WASTE COLLECTION

Municipal officials as well as planners view solid waste collection in terms of "efficiency" (dollars expended per ton of refuse collected, or other such quantifiable measurement). In fact, all the models of solid waste collection and management discussed above seek a least-cost solution (i.e., the least dollars spent).

Although this is certainly an admirable goal, it is not sufficient. It is also necessary to be able to measure the *effectiveness* of the system—that is, how well the task is being done. It would do little good to optimize a system and have widespread public dissatisfaction with the service.

Unfortunately, limited research has been conducted in the area of the effectiveness of any public service. One problem with such attempts to measure effectiveness is that they must in part depend on public opinion, which is notoriously fickle.

One potential method of obtaining a measurable indication of solid waste collection effectiveness [20] is to evaluate a system on the basis of three types of variables:
1. The satisfaction of the system users (the people whose refuse is picked up).
2. The effects of the service on the community.
3. Societal values (what the community is willing to buy).

The first two variables can be estimated by using appropriate indexes, and the third can be related to the first two by a graphical correlation.

User satisfaction is measured by using a public questionnaire. The telephone survey has been found to be the best compromise between the door-to-door survey (the most accurate) and the mail survey (the cheapest). A stratified sample, where individuals are selected to represent larger groups, has been found to be superior to the random-dial method. The questions asked should be similar to the following:
1. Which term would you say generally describes your neighborhood?
   (a) Clean
   (b) Moderately clean
   (c) Moderately dirty
   (d) Dirty
2. How would you rate your refuse collection service?
   (a) Very good
   (b) Good
   (c) Fair
   (d) Poor
3. How often have the collectors missed picking up your garbage?*
   (a) Seldom
   (b) Often
4. During the past year, have the collectors created excessive noise?
   (a) Yes
   (b) No

*It is usually expeditious to use the inaccurate term "garbage" in place of "refuse" when speaking to the general public.

5. During the past year, have the collectors scattered your garbage?
    (a) Yes
    (b) No

These questions can then be given point values, with the last four responses being worth 10 points for each "no" answer, and the first two yielding values of 30, 20, 10, or 0.*

The responses from these questionnaires can be conveniently analyzed by calculating the *user satisfaction index,* defined as

$$U.S.I. = \frac{\sum_{i=1}^{N} R_i}{N}$$

where R_i is the sum of the values from the ith questionnaire and N the total number of people interviewed. The U.S.I. has a maximum value of 100 if the numerical values stated above are used.

The adequacy of a solid waste collection service can physically be measured by evaluating the cleanliness of the streets. Street cleanliness has been measured in several American cities, with the ratings of the Urban Institute being the most widely used [21]. In this system, cleanliness is rated on a scale of 1 to 4, and the measurements can be quite precise.

In measuring street cleanliness as a physical effect attributable to efficiency of solid waste collection; the following more convenient scale can also be used:

100 = very clean street, with no visible litter

75 = moderately clean street with only one accumlation of trash not set out for collection

50 = litter scattered along most of street with several significant accumulations

25 = heavily littered with enough trash to fill at least one standard garbage can

0 = extreme conditions, more trash than would fill one standard garbage can.

Examples of streets with 100, 75, 50, and 25 ratings are shown in Fig. 2-16.These values are then corrected for the presence of unusual items or accumulations of uncollected trash. For example, 10 points can be subtracted from street cleanliness rating for any one of the following:
1. Littered vacant lots.
2. Health hazards.
3. Fire hazards.

---

*These values have no other justification other than that they seem reasonable.

Figure 2-16. Street cleanliness ratings.

4. Uncollected garbage cans set out a day too early, or empty cans left at the curb too long.
5. Abandoned motor vehicles.
6. Overflowing public receptacles.
7. Brush piles.

The list might also include a number of other conditions dictated by local conditions.

Combining these two parameters, it is possible to define a *community effects index*:

$$\text{C.E.I.} = \frac{\sum_{i=1}^{N}(S-P)_i}{N}$$

where S is the street cleanliness rating, 100 to 0, and P is the presence of any of several special conditions, each one being worth 10 points. The total number of streets rated is N. The C.E.I. has a maximum value of 100

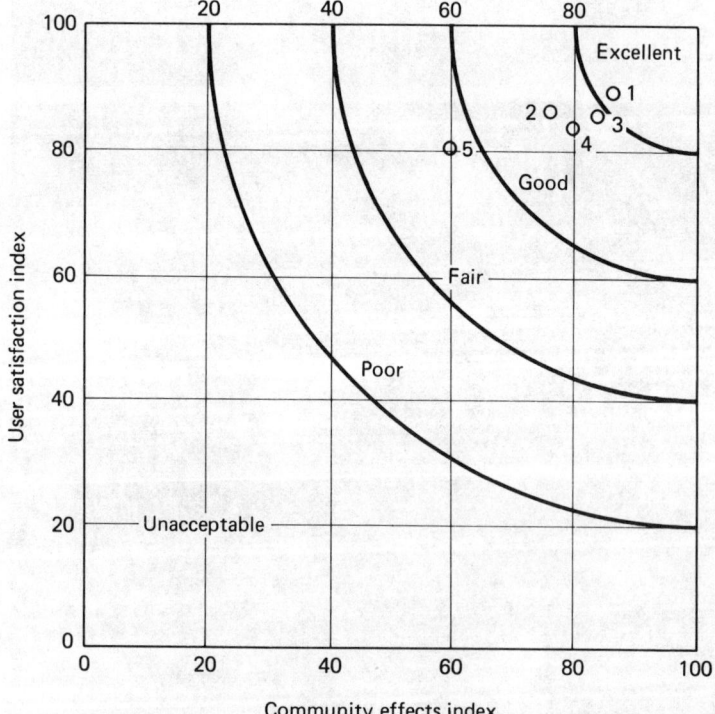

Figure 2-17. Correlation of user satisfaction and community effects, defining the societal values of solid waste collection. The data points are for five neighborhoods in a city of 100,000 people.

(extremely clean streets and no unusual trash conditions) and it is theoretically possible to have a negative value.

The most difficult value to establish is society's perception of what it might consider to be "good" or "bad" solid waste collection. It is suggested that the value society places on the effectiveness of waste collected be defined as shown in Fig. 2-17. In this figure, a high value of the user's satisfaction index, concurrent with a high community effect index, will result in a high level of service, and low values in either or both indices is an unacceptable level of service. The level of service denoted as "excellent" refers to an area that requires only continual surveillance to assure that the level of service would not decline. Areas with a "good" level of service would be those that require minimum improvement; those with a "fair" level of service require somewhat more attention. If left unattended, they would probably deteriorate. Areas with a "poor" level of service require major attention, which might include changing collection crews, increasing frequency of collection, additional street sweeping, or citizen education efforts. "Unacceptable" levels of service should not be allowed to continue, and a major overhaul of the quality of service is needed.

Also plotted in Fig. 2-17 are the results of a study of solid waste collection effectiveness in five different neighborhoods in one city [20]. Such numerical results can be used to compare neighborhoods within a city, as done here, or to estimate the effect of new collection programs. It is also possible to compare solid waste collection effectiveness among different cities.

COLLECTION OF SOURCE-SEPARATED MATERIALS

It seems eminently reasonable that if pure materials have value, one should not mix the various materials that comprise MSW in the first place. Would it not be better, the argument goes, to collect material that has been already separated into various components by the individual householder? This *source-separation* potential holds great promise, and has received support by governmental agencies. Indeed, it would be remiss not to consider the source separation of materials as a legitimate unit operation in the resource recovery process.

Although source separation makes theoretical sense, it is not without some serious problems when practically applied. One of these is that many of our waste products are made of different materials, and in this "dirty" state none of the materials are of much value. For example, a small kitchen appliance contains several different metals, each of which in its pure state would be valuable. Bolted, glued, welded, and tied together, however, none of the metals has market value, since each is contaminated

TABLE 2-2
Recoverable Materials in Solid Waste

Component	Percent of Refuse	Sale Price of Separated Material ($/ton)	At 4 lb/capita, Total Value ($/day)
Garbage	10	0	0
Paper			
Corrugated	10	20	0.004
Newsprint	10	10	0.002
Other	20	0	0
Glass			
Clear	4	40	0.003
Colored	4	2	0
Metal			
Iron and steel	4	30	0.002
Copper	0.5	200	0.002
Aluminum	4	340	0.027
Other		0	0
Total			0.040

Source: National Center for Resource Recovery.

by the others. It is unreasonable to ask the householder to separate the various metals in an appliance before its disposal.

For any waste material to be separated practically (efficiently), it must be

1. Plentiful.
2. Valuable.
3. Separable.

The last requirement is discussed above. The first two can be estimated by using common production figures and market prices, as shown in Table 2-2.

It is immediately obvious that 4 cents/day is not a great deal of money to a single individual. However, multiplied by the population in a medium-size city, of say 100,000 people, it can be a significant figure. An annual income of $1 million from the sale of recovered material is not unimpressive.

The design and operation of a source-separation program must be a three-pronged effort, with the final effort using elements of three solutions; technical, social and cognitive.*

---

*This discussion is limited to the solutions for achieving separation by the individual household. It is recognized that products could also be changed, eliminated, or substituted to promote source-separation, but this is "waste reduction" and not discussed in this chapter.

Technical Solutions

Technology can be of service in designing the hardware for the short term storage of separated material, transport to the collection vehicle, and the transport to a collection depot. Unfortunately, only meager research and development efforts have been directed toward this aspect of solid waste management.

Racks on regular packer trucks for the collection of newspaper have been tried in several places, with varying degrees of success. The common problem seems to be that the racks fill up faster than the rest of the packer trucks [22]. Some imaginative approaches, such as a three-compartment rolling container, or a rolling container with a rack for newspapers, have been suggested but have not been implemented [23].

The design and construction of a separate truck for collecting separated materials has been slow, although the vehicles developed for EPA-sponsored programs in Somerville and Marblehead, Massachusetts, have proven that such a design is practical [24].

Much remains to be done in terms of technical solutions for promoting source separation.

Social Solutions

Classed under social solutions are all those approaches which require laws, ordinances, or other social constraints for implementation. For example, some New England towns have strict ordinances against including garbage with the remainder of the refuse, thus achieving source separation. In some towns, the collectors will empty only one can twice a week, thus making it necessary that high-volume items such as newpapers be separately packaged and disposed of.

It is also theoretically possible to pass an ordinance requiring each household to separate various materials before pickup. Such mandatory separation laws have been tried at various places, with modest success [22]. Some people find ingenious methods to circumvent these restrictions, and some do not even care if their refuse is picked up, much less whether it is source-separated.

Cognitive Solutions

A cognitive solution is one that will attain a change in behavior as the result of a change in attitude. Attitude measurement is not easy, and changes in public attitudes are difficult to achieve. There are many ways of measuring attitude, ranging from measurements where the data are self-reports ("I believe that . . . "), to the opinions of others, to measurements based on actual behavior when confronted with a choice [25]. The method of attitude measurement used will yield vastly different results, and thus it must be true that one's attitudes on any issue are complex, and the resulting behavior can vary greatly depending on the various conditions and choices presented.

In a study of attitudes toward source separation, metropolitan housewives were asked to indicate their beliefs about recycling [26]. Some results were:

94% agreed that "we need to do something about pollution now."

81% believed that "recycling of solid waste would help greatly in cutting down pollution."

90% said that they would be willing to separate some of the refuse. The results were clearly encouraging. These percentages have been confirmed by similar surveys—always indicating enthusiasm for source separation.

These numbers must be contrasted with the results of participation (attitude, as measured by self-evaluation vs. behavior). In one widely publicized program [22], the participation was dismal. In all, less than 5% of the materials in the refuse were recovered by source separation. The participants (defined as the number of householders who placed separated material out at least once during the program) were predominantly high-income, professional, well-educated people.

Research on how to convince people to source-separate is continuing with governmental support. There is no doubt that source separation is the least-expensive, least-energy-consuming method of material recovery. The problem is that it is also the most difficult to implement.

An additional problem with source separation is the effect that such programs might have on the economics of resource facilities. If a substantial fraction of paper, for example, is removed from the waste stream, what effect will this have on the energy value of shredded and classified refuse?

Studies [27] by the EPA have shown that the effect of the recovery of say 20% of the newsprint (a practical maximum in most towns) will result in a reduction of about 10% of the fuel value. Similarly, a ban on nonreturnable beverage containers (source separation and reuse) would result in a revenue loss of $0.50 per ton of refuse for ferrous materials, and a loss of about $0.40 per ton for aluminum and glass. These can be compared to the estimated total gross revenue received for recovered ferrous aluminum and glass of $4.40 per ton of refuse. Hence the separation of materials at the source will have some effect on the economics of a resource recovery facility.

LITTER

Litter is a special type of MSW. It is distinct from other types of MSW in that it is a solid waste that is not deposited in proper receptacles. We usually think of litter as existing in public places, but litter could be on private premises as well.

Although litter is usually considered to be a visual affront only, it may also be a health hazard. Broken glass and food for rats are but two examples. It is also a drain on our economic resources, because the public must pay to have it collected and removed when it is on public property.

The collection of litter is of secondary importance to a community because it does not represent a critical public service, as does police and fire protection, water treatment, or collection of refuse from residence and commercial establishments. Yet litter removal is expensive, costing U.S. municipalities over $500 million annually [28].

The composition of roadside litter can vary considerably from place to place, as can the method of data collection. One major problem with any litter data analysis is that the data fail to specify the guidelines used in the collection and indentification of litter, and seldom specify the way in which the percentages of the various components were calculated. For example, a broken bottle can be counted as either one item or many items, depending on the guidelines. Similarly, the results can be calculated as a percent of the total of each of the following items:

Items by actual count.

Weight.

Volume.

Visible items by actual count.

Because of the problem of not having a standard counting technique, the following guidelines for conducting litter studies are suggested [29]:

1. Count as one item all pieces larger than 2.5 cm (1 in). This includes removable tabs from beverage cans.
2. Do not count rocks, dirt, or animal droppings.
3. Count as one item all pieces of any item clearly belonging together, such as a broken bottle. Otherwise count each piece of glass, newspaper, and so on, singly.
4. Do not count small, readily decomposable material, such as apple cores.
5. For roadside litter surveys, measure all items within the officially designated right-of-way.
6. Empty liquids out of all bottles and cans before collecting them.

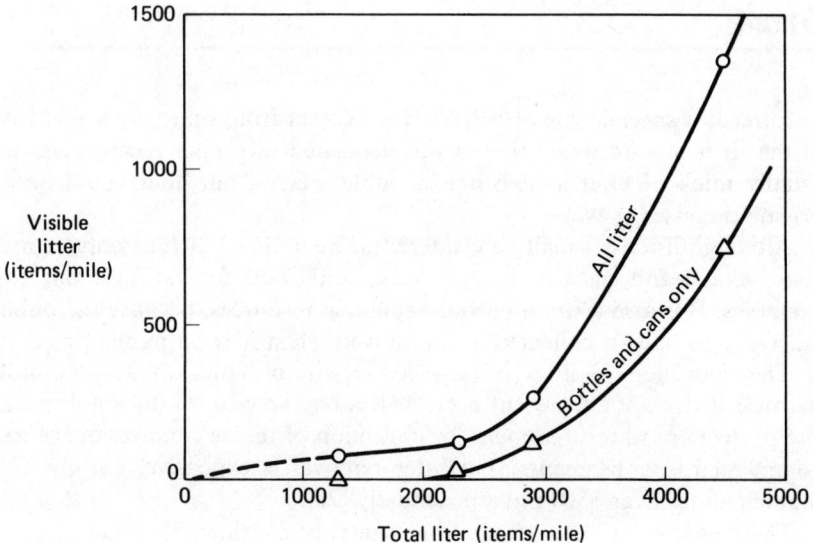

Figure 2-18. Visible roadside litter is only a small fraction of the total litter. (From Ref. 25.)

The litter survey, if conducted along a road, should be started by driving along the road at a slow speed and having the passenger record the visible items into a tape recorder for future transcription. Next, the litter is identified, recorded, and manually collected. The items should be separated during collection into as many components as feasible. The collected items are then weighed and the volume measured. The relationship between visible items and total items along a roadside is shown in Fig. 2-18. It is interesting that along fairly clean roads, the visible fraction is only about 6% of the total litter count! These data also confirm that a large fraction of our visible litter is bottles and cans.

For community litter surveys, the photometric technique developed for Keep America Beautiful, Inc., has found wide acceptance. The block faces of a community are first numbered, and a preliminary sample size established. About 5% of the blockfaces are usually adequate. Using the random number technique, the block faces and the locations on those block faces to be measured are selected.

As shown in Fig. 2-19, a marker is located in the front center of the survey area, and a chalkboard is used to identify the location and date. To facilitate the counting of litter from the developed photographs, a picture is taken of a clean pavement laid out with white marking tape in a 1-ft grid, 6 ft wide, 16 ft long, parallel to a street curb. A transparency of the grid is

Litter

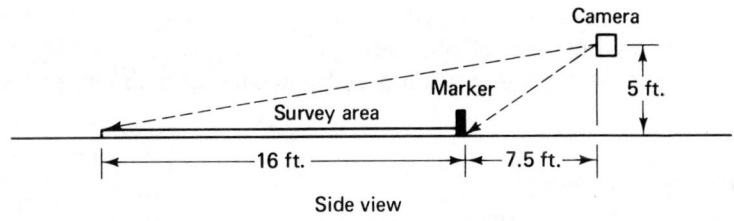

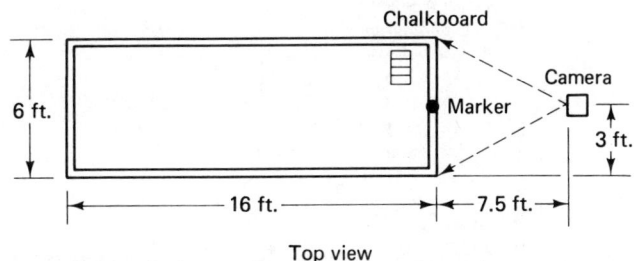

Figure 2-19. Keep America Beautiful's technique for photogrammetric measurement for litter.

prepared and the resulting 96-square grid is placed on top of each litter photograph. The litter is then counted and classified using a magnifying glass.

The first photographs are used for establishing the baseline litter conditions. The *litter index* (L) is calculated for each picture (location) as

$$L = \frac{\text{number of squares containing some litter}}{\text{total number of squares (96)}}$$

After initial baseline photographs are analyzed and the L is calculated, the number of sampling sites necessary can be calculated as

$$N = 22.8(S^2)$$

where N = sample size needed to make a 0.5-point difference between two average litter ratings (L's) in an area, significant at the 90% confidence level
 S^2 = variance of the litter ratings (L's) of the initial photographs

The variance is calculated as

$$S^2 = \frac{\sum_{i=1}^{n} f(L_i)^2}{n-1} - \frac{\left(\sum_{i=1}^{n} fL_i\right)^2}{n(n-1)}$$

where L_i = letter rating of the ith photograph
n = total number of photographs
f = frequency, or the number of photographs with any one L

Example 2-1

Suppose that a town has 600 blockfaces, and that a 5% sample, or 30 blockfaces, is photographed as explained above. The L's are as shown:

L	Number of photographs, f	fL	fL^2
1	6	6	6
2	3	6	12
3	1	3	9
4	4	16	56
5	1	5	25
		36	108

The fL and fL^2 are calculated and summed. The S^2 is calculated as

$$S^2 = \frac{108}{30-1} - \frac{(36)^2}{30(29)} = 2.0$$

If $S^2 = 2.0$, the number of sampling sites necessary is

$$N = 22.8(2.0) = 45.6$$

In other words, 15 more sites are needed in order to have a statistically satisfactory baseline.

Litter can theoretically be controlled by cognitive, social, and technological means: a cognitive solution would be convincing people not to litter; a social solution would be depriving the public of items that might become litter, or fining them heavily if they are caught; and a technical solution would be simply cleaning up after littering has occurred.

The first option demands an explanation of why people litter, a question requiring studies on the psychology of litterers. In one study [30], the actions of 272 persons were observed when they bought a hot dog wrapped in paper. Of interest was the final deposition of the wrapper. Ninety-one people chose to dispose of the wrapper improperly (they littered). The probability of any one person littering, based on this sample, could be calculated* as

$$E = 0.190 + 0.4141(A) + 0.1654(C) + 0.1532(D)$$

*This is slightly modified from the original expression, which included a term relating to the race of the person observed.

where E is the probability that a person would litter; and A, C, and $D = 0$, except that $A = 1$ if the person is 18 years old or younger, $C = 1$ if there are no trash cans conveniently located, and $D = 1$ if the area is already dirty with litter.

From the study, it is clear that age is quite important, with younger people being much more likely to litter than are older persons. There was no statistical difference between 19- to 26-year-olds and persons older than 26 years. Economic status was buried in the race term and was found not to be independently significant. Similarly, sex was found to be statistically insignificant.

Example 2-2

Calculate the probability of a 40-year-old person littering a dirty street that has no convenient trash cans.

$$E = 0.190 + 0 + 0.1654 + 0.1532 = 0.5080$$

That is, of 100 people answering that description, 50 would probably litter the street.

Such studies yield clues as to how persons might be induced not to litter (e.g., put out more trash cans and clean up the street) and who the target population is (e.g., young people).

Other litter psychology studies have been directed at finding out what motivates people not to litter. In one study [31], movie theater patrons during Saturday matinees were asked by several means not to litter the theater. The total quantity of litter was then measured and used as an indicator of the success of that control approach. The results showed that measures such as personal exhortation for cooperation and antilitter cartoons had no effect on litter, but that payment of money for pieces of litter at the end of the showing resulted in about a 95% reduction in litter. The clear indication is that self-interest, such as placing a substantial deposit on beverage containers, is an effective force in convincing people not to litter.

The second method of litter control is to prevent items that might become litter from ever reaching the consumer. For example, in the earlier example of the hot dog wrapper, it would seem reasonable to suggest that 100% litter-free results could be obtained by not giving customers a paper wrapper around their hot dog. The banning of tear away metal tabs on beer and soft drink cans is a practical means of controlling this type of litter.

The third method of litter control is to clean up the mess once it has occurred. This system is commonly used in sports stadiums and other public areas where no effort is made to ask people to properly dispose of their waste. For roadside litter, it seems that the most economical litter control alternative is actually frequent cleanup, and attempts have been made at designing mechanical litter collection machines. One towed device has proven both inexpensive and effective. It works by having a series of rotating plastic teeth which fling the litter into a collection basket, much like a leaf collector connected to a lawn mower [32]. A more sophisticated and ambitious unit, developed by a major manufacturer of beverage containers, uses a vacuum arm on a truck which sucks up the roadside litter while cruising at highway speed [33].

Whatever the method of attack, the task is prodigious. In the United States, alone, the litter resulting from one holiday weekend would fill a row of garbage trucks 43 miles long, or the litter could cover a four-lane highway from Boston to Detroit [28].

PARTING SHOTS

The collection of municipal solid waste accounts for about 80% of the total cost of refuse management. It is thus little wonder that considerable effort has been directed toward cost-saving schemes in collection practices. Even a 5% savings in an annual national cost of $4 billion would be a significant savings.

Because of the close interrelationship of person and machine, any increase in collection productivity and hence decrease in cost must come from either the design of systems in which the machines and methods enhance the speed of collection, or from the increased work output of the collection crews without a change in their collection technique or method.

Given the unsavory nature of the job, an increasing standard of living, and alternative job opportunities, it is unlikely that collection crews can be spurred on to increased productivity. Hence it is in the area of design of new hardware and collection techniques that future savings will probably occur.

In addition to changes in the area of equipment design, future efforts will be directed at reducing collection costs. Such future devices as cable-television newspapers, preprocessed foods, and edible containers will all reduce the waste stream.

The day when the per capita increase in solid waste generation begins to decrease is far away, however. The present problems of collection will be with us for many years, and improvements in collection systems technology will still be of great benefit to all.

REFERENCES

[1] National Center for Resource Recovery, *Municipal Solid Waste Collection*, Lexington Books, Lexington, Mass., 1973.

[2] "Refuse Pickup Personnel Injuries Are Nine Times National Industrial Average," *Solid Waste Manage.*, Jan. 1975, pp. 9–48.

[3] SHUSTER, K. A., and D. A. SCHUR, *Heuristic Routing for Solid Waste Collection Vehicles*, U.S. EPA OSWMP SW-113, Washington, D.C., 1974.

[4] LIEBMAN, J. C., J. W. MALE, and M. WATHNE, "Minimum Cost in Residential Refuse Vehicle Routes," *J. Environ. Eng. Div. ASCE*, 101, (EE 3), 339–412 (1975).

[5] KWAN; M-K., "Graphic Programming Using Odd or Even Points," *Chinese Math.*, 1, 207–218 (1962).

[6] EDMONDS, J. "Maximum Matching and a Polyhedron with 0.1 Vertices," *J. Res. Nat. Bur. Stand.*, **69B** (Jan.–June 1965).

[7] WHITE, L.J. "A Parametric Study of Matchings and Coverings in Weighed Graphs," Ph.D. thesis, University of Michigan, Ann Arbor, 1967.

[8] WALKER, W., M. AQUILINA, and D. SCHUR, "Development and Use of a Fixed Charge Programming Model for Regional Solid Waste Planning," *Modeling of Environmental Systems*, U.S. EPA, Washington, D.C., 1976.

[9] GREENBERG, M. R., et al., *Solid Waste Planning in Metropolitan Regions*, the Center for Urban Policy Research, Rutgers University, New Brunswick, N.J., 1976.

[10] BERMAN, E. B., "WRAP: A Model for Regional Solid Waste Management Planning," *Modeling of Environmental Systems*, U.S. EPA, Washington, D.C., 1976.

[11] KRABBE, D. M., "St. Louis: An Application of WRAP," *Modeling of Environmental Systems*, U.S. EPA, Washington, D.C., 1976.

[12] *Solid Waste Disposal for Region L*, Wiggins-Rimer & Associates, Durham, N.C., 1975.

[13] WALKER, W., *Adjacent Extreme Point algorithms for the Fixed Charge Problem*, Tech. Rep. 40, Cornell University, College of Engineering, Department of Operations Research, Ithaca, N. Y., Jan. 1968.

[14] *Wrapping up the Solid Waste Problem: A Model for Regional Solid Waste Management Planning*, U.S. EPA, Apr. 1976.

[15] *WRAP—A Model for Regional Solid Waste Management Planning—Users Guide*, EPA/530/SW 574, Feb. 1977.

[16] "*WRAP—A Model for Regional Solid Waste Management Planning — Programmer's Guide*," U.S. EPA/530/SW 575, Feb. 1977.

[17] VESILIND, P. A., J. A. NISSEN, and J. F. MCALISTER, "Data Generation for Regional Solid Waste Allocation Models," *Solid Waste Manage.*, Nov. 1977.

[18] *New York Solid Waste Plan*, U.S. EPA OSWMP, Washington, D.C., 1970.

[19] BRUNNER, D. R., and D. J. KELLER, *Sanitary Landfill Design and Operation*, EPA OSWMP SW-65ts, Washington, D.C., 1972.

[20] DAJANI, S. J., P. A. VESILIND, and G. HARTMAN, "Measuring the Effectiveness of Solid Waste Collection," Duke Environmental Center, Duke University, Durham, N.C., 1976.

[21] BLAIR, L. H., and A. I. SCHWARTZ, *How Clean Is Our City?* The Urban Institute, Washington, D.C., 1972.

[22] *Analysis of Source Separation Collection of Recyclable Solid Waste— Separate Collection Studies*, U.S. EPA OSWMP SW-95, c. 1, Washington, D.C., 1974.

[23] *Report on Solid Waste Collection for the Town of Chapel Hill, N.C.*, Wiggins-Rimer & Associates, Durham, N.C., 1976.

[24] HANSEN, P., and J., RAMSEY, "Demonstration of Multi-Material Source Separation in Somerville and Marblehead, Mass.," *Waste Age*, 7(2) (1976).

[25] KLEE, A. J., "Attitude Concepts and Applications to Solid Waste Management," Unpublished.

[26] *Metropolitan Housewives Attitudes toward Solid Waste Disposal*, U.S. EPA OSWMP (NTIS EPA-R5-72-003), Washington D.C., 1974.

[27] SKINNER, J. H. "The Impact of Source Separation and Waste Reduction on the Economics of Resource Recovery Facilities," *Resour. Recovery Energy Rev.*, Mar.–Apr. 1977.

[28] *Community Litter Prevention Guide*, U.S. Brewers Association, Inc., Washington, D.C.

[29] VESILIND, P. A., "Measurement of Roadside Litter," Duke Environmental Center, Duke University, Durham, N.C., 1976.

[30] FINNIE, W. C., "Field Experiments in Litter Control," *Environ. Behav.*, 5(2) (1973).

[31] BURGESS, R. L., et al., "An Experimental Analysis of Anti-Litter Procedures," *J. Appl. Behav. Anal.*, 4(2) (1971).

[32] HART, F. D., et al., "Design and Development of a Machine to Remove Litter from the Roadside," School of Engineering, North Carolina State University, Raleigh, N.C., 1973.

[33] National Center for Resource Recovery, *Municipal Solid Waste Collection*, Lexington Books, Lexington, Mass., 1973.

[34] TCHOBANOGLOUS, G., H. THEISEN, and R. ELIASSEN, *Solid Wastes*, McGraw-Hill Book Company, New York, 1977.

PROBLEMS

2-1. Using the data in Example 2-1, optimize each waste allocation scheme.
 (a) The intermediate facility is an incinerator with a capacity of 150 tonnes/day, $P_j = 0.2$, at a cost of \$10 per tonne.
 (b) The intermediate facility is a transfer station with a capacity of 500 tonnes/day at a cost of \$2 per tonne.

2-2. Design an innovative system for transporting refuse from the kitchen to the curb. Cost should not be a major consideration. Imaginative ideas that just might work are required.

2-3. On a map of your campus (or any other convenient map), develop an efficient route for refuse collection, assuming that each blockface must be collected.

2-4. Visit City Hall and Obtain the accident records for city employees. Report on the relative accident rate of solid waste workers.

2-5. The haul distance is 15 miles as the crow flies and the anticipated average truck speed is 35 mph. Estimate the one-way haul time.

2-6. Using the street cleanliness criteria on page 66 and the photographs in Fig. 2-16, rate a predetermined route by driving it. This should be done by a number of people and the results compared.

2-7. Using a study hall or social lounge as a laboratory, study the prevalence of litter by counting the items in the receptacles vs. the items improperly disposed of. Each day vary the conditions as follows:
Day 1: Normal (baseline).
Day 2: Remove all receptacles except one.
Day 3: Add additional receptacles; more than normal.
Plot the percent properly disposed of vs. number of receptacles. Discuss the implications of your results.

2-8. Using the principles of heuristic routing, develop a collection route for the streets shown in Fig. 2-20. Each blockface must be collected (i.e., one side of street collection). Eliminate all blind blockfaces (it is possible!) and minimize left-hand turns.

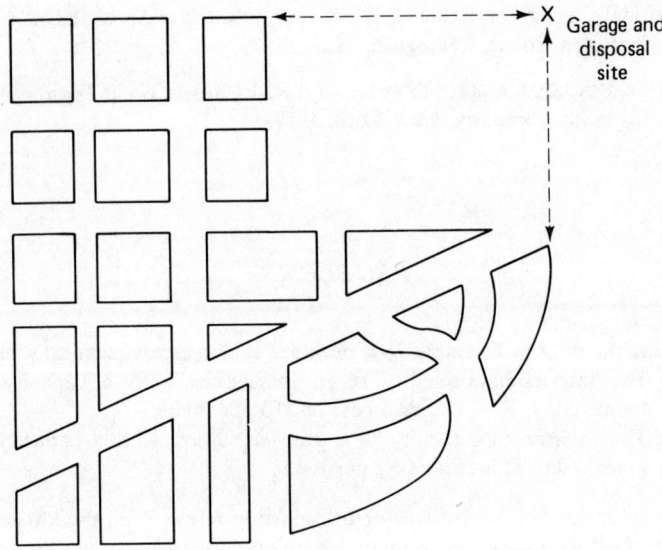

Figure 2-20. Blocks for routing trucks.

3 STORAGE, HANDLING, AND SHREDDING OF SOLID WASTE

In almost all engineering disciplines, the designs of the processes used to produce a desired end product are based on the reasonable assumption that the nature of the raw material is known and can be defined accurately and precisely. This condition lends a feeling of confidence to the analysis of unit operations and breeds sophistication in the design of the process.

Unfortunately, the resource recovery engineering profession is not so blessed. Solid waste processing and recovery facilities must be able to accept almost all manner of solid waste. Some types of solid waste are easy to process, but occasionally materials that are difficult and/or dangerous to handle are also found in refuse. As a result, solid waste processing operations have large factors of safety, and must be designed for extraordinary contingencies. This requirement often results in overdesign and underutilization in order to be able to process all of the feed material.

In this chapter solid waste handling prior to further materials separation or processing is discussed. None of the unit operations discussed here actually accomplishes materials separation, but rather prepares the refuse for the separation operations that follow.

STORAGE AND CONVEYING

Municipal refuse, either in its original or shredded state, has some materials properties which make its processing and conveying hazardous and difficult. Although the art and science of solids conveying and storage is well studied in the mining and chemical engineering field [1], the application of most of that knowledge is inappropriate for a heterogeneous, unpredictable, and time-variable material such as refuse.

Hickman [2] has listed a number of materials characteristics that must be considered in the design of any storage and conveying equipment. Among these are:

Material size—Because of the nature of refuse, particle size is difficult to define. Sieves define size by two dimensions, so that a piece of wire could pass a sieve and still prove troublesome in conveying. The problems of defining particle size are discussed further in Chapter 4.

Material density—Shredded refuse, when stored in a silo, can achieve densities as high as 400 kg/m^3 (250 lb/ft^3). The variation in density has been found to be significant, as shown in Fig. 3-1. This variation in density is an important factor in designing the side walls of storage silos and the retrieval systems for removing material from those silos.

Angle of repose—Because of variable density, moisture, and particle size, the angle of repose of shredded refuse can vary between 45° to greater than 90°.

Material abrasiveness—Prepared refuse consists of many types of abrasive particles, including sand, glass, metals, and rocks. Removal of this abrasive material is often necessary before some operations such as pneumatic conveying can become practical.

Moisture content—All of the foregoing properties are influenced by moisture content. The extent of this effect depends on the material. When the moisture level exceeds 50%, the high organic fraction can undergo spontaneous combustion if the material is allowed to stand undisturbed.

Storage

The storage of refuse has long been a serious problem, especially at large incineration facilities. Most incinerators must be continuously fired and require sufficient storage for at least 2 days, to allow for the unavailability of refuse over weekends. Similarly, energy recovery systems must store shredded material in sufficient quantities to even out the fluctuations in supply (see Chapter 8).

Storage and Conveying 85

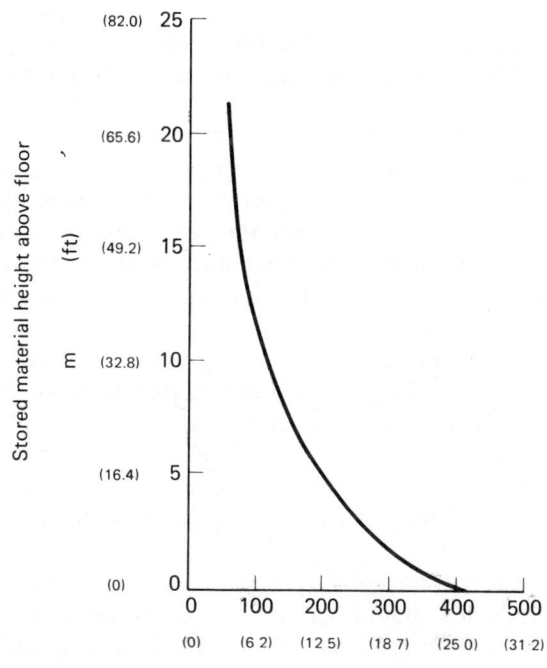

Figure 3-1. Density of shredded refuse as influenced by storage. (From Ref. 2.)

Two major considerations in the design of refuse storage facilities are public health and fire. Rats and other rodents can inhabit storage areas unless special precautions are taken. The odor of slowly decomposing garbage can be overwhelming and can cause public relations problems downwind.

Spontaneous combustion is possible with the storage of shredded refuse. The rule of thumb is that 2 days of storage is considered the safe maximum, with a week being dangerous. A fire in a storage silo is not only difficult to extinguish, but the resulting wet shredded refuse presents new disposal problems.

All storage silos should be constructed as "first-in/first-out" systems. Unfortunately, this is not a simple task, and many of the existing storage systems tend to result in the long-term storage of some fraction of the refuse.

The common live bottom bin used in chemical engineering solids handling operations has been found to be inadequate, since the refuse easily bridges the space above the bottom conveyors. One successful alternative is to use a flat bottom instead of a hopper configuration and spread either fixed or moveable conveyors across the bottom of the bin [3]. The sides of the bin are slanted outward toward the bottom and bridging is unlikely.

Some manufacturers of storage silos install screw conveyors with a variable pitch and blade diameter. The superiority of these conveyors over constant pitch and diameter models has not been proven. The total feed discharged by the conveyor per rotation, and thus its capacity, is the volume within the last pitch. The remainder of the conveyor is needed simply to move the material to that point, and the conveyor shape should not influence its capacity.

Some screw conveyors on the bottom of live bins are fitted with digging bars, welded to the flights. These assist in breaking up bridging and moving the material between the screw blades. As with all screw conveyors, however, the conveyor works only if the material does not rotate within the blades, but has only longitudinal motion, slipping on the blade surfaces. Digging bars may cause rotation of the material and may thus reduce capacity.

The design of better storage silos requires not only a knowledge of the theory of materials flow, but also a means of experimentally evaluating the flow of solid material in a storage chamber. The use of velocity probes, especially for solid waste, is clearly unacceptable. A number of potentially effective techniques, as reviewed by Resnick [4], are stereophotogrammetry, radio pills (transmitters that move with the solids in the bin), radiological tagging (e.g., with cesium 137), and X-ray methods. With nonhomogeneous materials such as refuse, the radio pill or photogrammetry seem to be most applicable.

Conveying

Four basic types of conveyors are used for refuse: (1) rubber-belted, (2) vibratory, (3) pneumatic, and (4) screw conveyors.

Rubber-belted conveyors have been found to be inadequate for moving unshredded raw material, but are acceptable for less abrasive and less rugged loads such as air-classified organics. Rigid interlocking belts have been successfully applied to raw refuse conveying. The skirts (sides) of the conveyor must be vertical, or even wider at the bottom than at the top, to eliminate bridging and jamming. A typical feed conveyor is shown in Fig. 3-2.

A conveyor with inclines greater than 20° will usually experience "tumble back" by some material. This can be considered advantageous if the conveyor is feeding a shredder, since the movement of the refuse can

Storage and Conveying

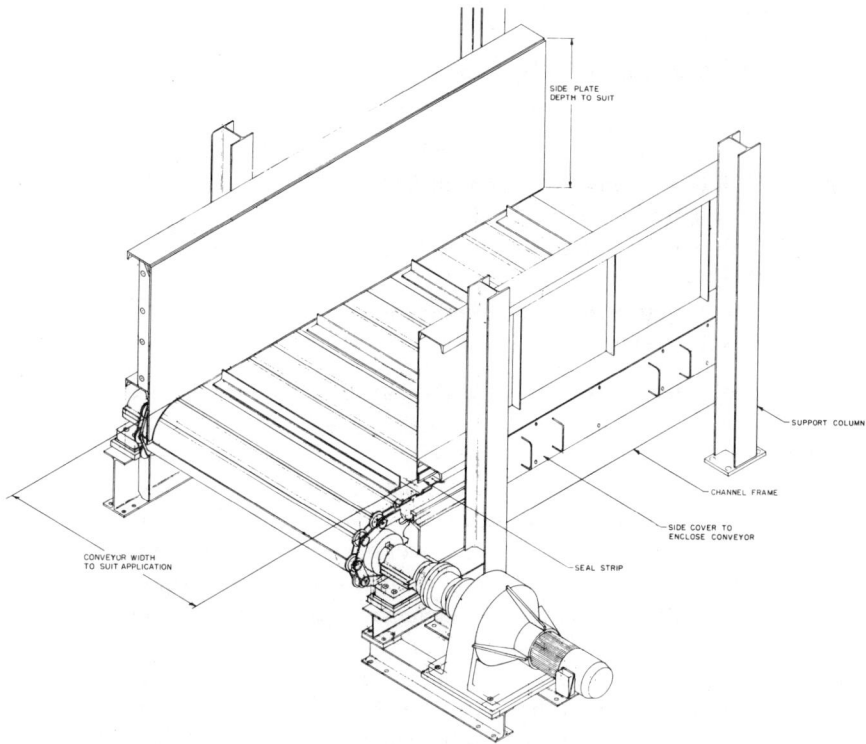

Figure 3-2. Typical feed conveyor. (Courtesy of Allis-Chalmers.)

even out the feed to the shredder. One manufacturer, in fact, recommends angles from 38° to 40° to achieve a more even feed rate [5].

The power requirements of belt conveyors can be estimated by a number of empirical equations, such as the following [6]:

$$\text{horsepower} = \frac{LSF}{1000} + \frac{LTC}{990} + \frac{TH}{990} + P$$

where L = length of the conveyor belt ft
S = speed of belt, ft/min
T = capacity, tons/h
H = lift, ft
F = speed factor
C = idle resistance factor
P = pulley friction

The first two terms on the right-hand side of the equation represents the power necessary to move the load horizontally, while the third term represents vertical movement. The last term represents power loss due to friction.

Another equation [7] reads

$$\text{horsepower} = \frac{P(1.07L + 50)(1.25T + 0.03QS)}{1000} \pm \frac{1.25(TH)}{1000}$$

where P = coefficient of friction on bearings
L = length of conveyor, ft
T = capacity of conveyor, tons/h
Q = factor
H = height of material lifted or lowered, ft
S = conveyor speed, ft/min

The coefficient of friction (F) and the factor Q must be established from experience.

Vibrating feeders are advantageous because they also even out materials flow. These devices have been used at right angles to belt conveyors which feed shredders, so that a fairly constant flow of refuse is accepted by the belt conveyor. At other installations, vibrators are used to feed shredders. Raw refuse can absorb considerable energy, and thus vibrating conveyors must supply sufficient energy to achieve material movement. A 2-cm stroke at a frequency of 900 strokes/min is common.

Pneumatic conveyors have been used mainly for collecting raw bagged refuse in hospitals and other large buildings and in feeding shredded organic fractions to boilers as supplemental fuel.

It is possible to estimate the required air velocities in pneumatic tubes by means of empirical relationships. For vertical tubes, the material velocity is

$$v_m = v_a - v_f$$

where v_m = material velocity
v_a = velocity of the air stream
v_f = the "floating velocity," or terminal velocity when falling in still air

The floating velocity can be calculated by an empirical equation such as

$$v_f = 3250\sqrt{(SG)(d)}$$

where d is the aerodynamic diameter of a representative particle, in inches, and SG the specific gravity of the material (relative to water) [8].

The material velocity must be sufficient to be able to dislodge stuck particles and to even out the flow, and can be estimated as $v_m = 585\sqrt{W}$, where W is the bulk weight of the material, in lb/ft^3. Because of the problems of measuring specific gravities in heterogeneous mixtures, it has been suggested that the specific gravity can be estimated as SG =

$0.1(W)^{2/3}$. The total recommended air velocity is thus

$$v_a = 1030\sqrt[3]{W}\sqrt{d} + 585\sqrt{W}$$

The problem of accurately estimating the representative diameter of a particle makes the practical use of this equation difficult. Experience has shown, however, that maintaining air velocities of about 1400 m/min (4500 ft/min) is sufficient for maintaining the materials flow in vertical tubes. Table 3-1 lists some recommended air velocities for common materials.

Operational experience has shown that a materials-to-air concentration of 0.1 (e.g., 0.1 kg of paper/1.0 kg of air) is reasonable [9]. The friction losses within a duct due to materials flow is less than 10%, well within a factor of safety commonly used in fan design.

Screw conveyors, the last of the four methods commonly used for moving refuse, are widely used in storage bins, as discussed above.

The volume of material moved by screw conveyors can be estimated by recognizing that the capacity of the conveyor in the "flooded" condition (i.e., all the space between the blades is full, as might occur when a screw conveyor is used in the bottom of a hopper) is

$$Q = CNRV$$

where Q = volume of refuse delivered by the screw conveyor per minute
C = the efficiency factor
R = rpm
V = volume of refuse between each pitch
N = number of conveyor leads

These terms are defined further in Fig. 3-3.

TABLE 3-1

Recommended Air Velocities for the Pneumatic Conveying of Some Representative Materials

Material	Minimum Air Velocity (ft/min)[a]
Coal, powdered	4000
Cotton	4500
Iron oxide	6500
Shavings	3500
Vegetable pulp	4500
Paper	5000
Rags	4500

[a]To obtain m/sec, multiply ft/min by 0.00508
Source: Ref. 10.

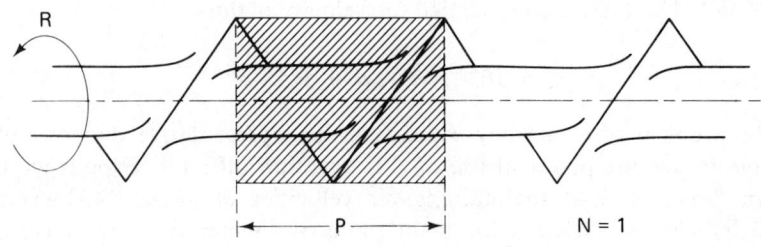

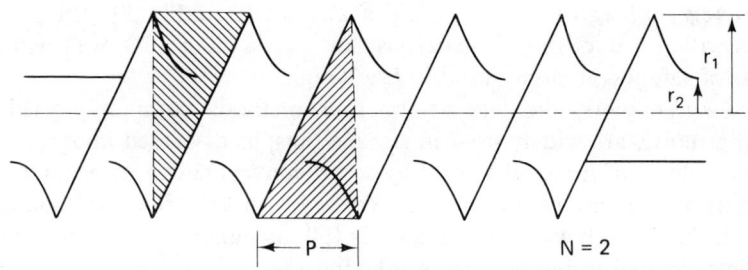

Figure 3-3. Definition of conveyor capacity parameters. The top screw has one lead ($N = 1$) and the bottom screw has two leads ($N = 2$).

The volume within each pitch can then be calculated approximately as

$$V = P\pi(r_1^2 - r_2^2)$$

where P = pitch (distance between adjacent conveyors' blades if $N = 1$)
r_1 = radius to the conveyor tip
r_2 = radius to conveyor hub

The dimensionless efficiency factor, C, is a function of the amount of slippage that occurs. Ideally, a screw conveyor operates by allowing the material to slide freely on the blade and to prevent radial rotation. Any rotation by the material (sticking to the blade) lowers efficiency. In live bottom bins, where the conveyor screws are not in individual troughs, considerable slippage is expected.

If the screw conveyor is not flooded, its capacity cannot be determined theoretically because the rate is influenced by a large number of variables [11].

Example 3-1

A live bottom bin has eight screw conveyors, each with $r_1 = 15$ cm, $r_2 = 6$ cm, $N = 1$, $P = 50$ cm, and $R = 10$ rpm. Assuming that $C = 0.5$, calculate the total material flow.

$$V = P\pi(r_1^2 - r_2^2) = 50(3.14)(225 - 36) = 29{,}106 \text{ cm}^3$$

For each conveyor,

$$Q = CNRV = (0.5)(1)(10)(29{,}106)(10^{-6}) = 0.14 \text{ m}^3/\text{min}$$

or total flow = $8 \times 0.14 = 1.2 \text{ m}^3/\text{min}$.

As a general rule, refuse should be conveyed and transferred as little as possible. In many refuse-processing facilities, the weak link in the operation has been conveying and rubber belt conveyors have been especially susceptible to breakdown. It is thus good engineering design to eliminate, or at least minimize, points of transfer and conveying within a facility.

COMPACTION

One problem in the disposal of MSW is the low density of the material, which requires large volumes for its collection, handling, and final disposal. Compacting MSW can thus lead to significant economies.

The structure of refuse can be pictured as an assemblage of particles interspaced with open air spaces called *voids*. Because these voids are large, and since many of the particles are absorbent, any moisture is absorbed in the material and is not in the voids. The total volume of material is thus made up of the solids plus the voids, or

$$V_m = V_s + V_v$$

where V_m = volume of material
  V_s = volume of solids (including the moisture)
  V_v = volume of voids

The *void ratio* is defined as

$$e = \frac{V_v}{V_s}$$

and the *porosity* is

$$n = \frac{V_v}{V_m}$$

By weight, the total material is made up of the solids plus moisture, or

$$W_m = W_s + W_w$$

where W_m = weight of material, including moisture
  W_s = weight of solids
  W_w = weight of moisture

The *wet density* is defined as

$$\rho_w = \frac{W_m}{V_m}$$

and the *dry density* is

$$\rho_d = \frac{W_s}{V_m}$$

The dry density is also known as the *bulk density*. Most densities in compaction literature are expressed in terms of bulk density, mainly because it is easy to measure and can be readily used in comparative

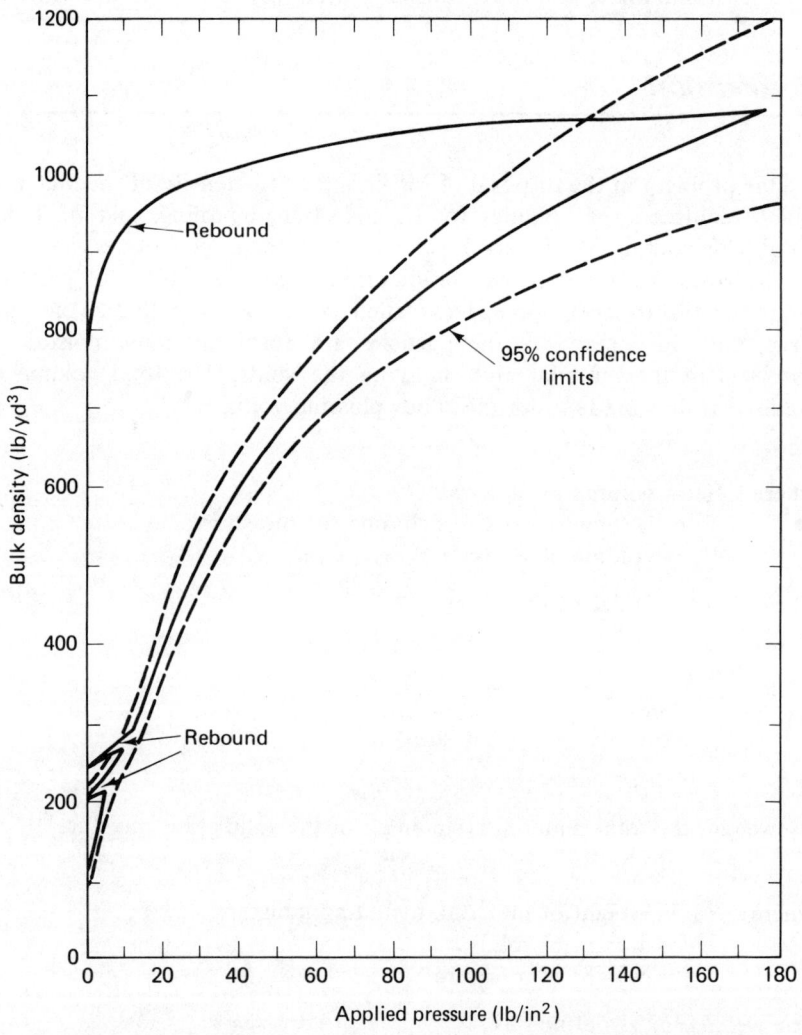

Figure 3-4. Compressibility of refuse in a laboratory press. (From Ref. 12.)

Shredding 93

studies. The compaction of refuse is expressed as the increase in bulk densities with compression energy expended.

When refuse is compacted, the density is increased as a result of the crushing, deforming, and relocating of individual items in the refuse. Hollow containers such as bottles and cans begin to collapse at different pressures, depending on their orientation and strength. For example, cans collapse at pressures of 0.1 to 0.3 N/m^2 (10 to 30 psi) and bottles at 0.05 to 0.35 N/m^2 (5 to 35 psi) [12].

The compaction of some materials is irreversible, in that when the pressure is released, the material does not spring back to its original volume. Refuse, however, contains many items that contribute to reversible compaction. One study has shown that at normal compaction pressures, 20% expansion can occur within a few seconds after the release of pressure, and this expansion can be as much as 50% after a few minutes [13]. The greater the pressure, the greater will be the bale integrity (its resistance to falling apart). A typical compression curve, using a small sample of refuse in a laboratory press, is shown in Fig. 3-4. Other compression curves are shown on pages 32 and 33 and the baling of refuse for the purpose of conserving landfill volume is discussed in Chapter 10.

SHREDDING

Many types of shredders are presently on the market, and almost all of them were developed originally for an application and feed material other than refuse. Most of our present refuse shredding technology comes from the mining industry, which has for many years used shredders for ore processing. The application of this technology to refuse, however, is not an easy matter, because these devices were developed for homogeneous feeds having well-established breakage characteristics. Considerable research is presently under way to develop a better understanding of MSW size reduction, establish greater control over the particle size of the product, reduce power consumption, and eliminate other problems associated with shredders (e.g., explosions, dust, hammer wear, etc.)

Shredding for Refuse Disposal

The first applications of shredders to MSW were to facilitate disposal, with little consideration for materials recovery. The pioneering work on shredding for disposal was done by Robert Ham and his colleagues at the University of Wisconsin. They found that shredded MSW had a more uniform particle size, was fairly homogeneous, and compacted more readily than unshredded waste, mainly because the larger voids had been eliminated [14].

After shredding, MSW looks not unlike confetti, and has a light, bulky nature. In fact, the overall density of the material is decreased by over 50%, from 200–240 kg/m^3 (350–400 lb/yd^3) to 75–90 kg/m^3 (125–150 lb/yd^3). Shredding reduces required landfill volume since shredded refuse compacts better within the landfill, and the need for a daily earth cover is eliminated. The effective density* by landfilling shredded refuse can be increased by 25 to 60% over unshredded MSW [15]. Shredding becomes an especially attractive alternative where cover material must be trucked in.

Experience with landfilling of shredded MSW indicates that shredded refuse does not require an earth cover. Extensive testing has shown that the conditions that make an earth cover necessary in a conventional landfill no longer exist with shredded refuse. Earth cover in a landfill is necessary because of:

Odor—The shredded refuse is so well mixed, and retains its aerobic character when spread in reasonably thin layers, that odor is not a problem.

Rats—There are no large food particles in shredded refuse that could support a rat population.

Insects—The drier refuse, regularly covered with new layers, suppresses insect breeding, and all of the fly maggots are killed during shredding.

Fires—When set on fire, shredded refuse cannot support combustion when on the ground and will eventually be extinguished.

Blowing paper—Small pieces are not caught by the wind and do not blow away.

Because of the lack of cover, leachate from shredded refuse is produced sooner, and this leachate is at a higher pollutional concentration than leachate from normal landfills. Accordingly, provisions must be made when landfilling shredded MSW to either reduce leachate production by controlling drainage or to capture and treat the resulting leachate.

Because of the superior compaction characteristics of shredded MSW, shredders have also been used before high-compression bailing [16].

Shredding for Materials Recovery

The breaking apart of the various constituents within MSW achieves two main goals: (1) the resulting material is easier to handle and thus extraction of the various components is facilitated, and (2) shredded waste

---

*Effective density = weight of refuse/(volume of refuse + volume of cover materials).

Shredding

burns more readily, thus increasing its value as a fuel. Because of these two factors, shredding is employed in almost all large-scale resource recovery facilities.

Types of Shredders Used for Solid Waste Processing

The list of size-reduction equipment used in both the mining and chemical industries is surprisingly long. One well-known mining and ore dressing handbook, for example, lists over 50 different devices that could conceivably be applied to solid waste shredding [6]. A review of size-reduction equipment widely used in chemical engineering applications lists 21 different devices [17].

The application of a specific device for MSW has not been a simple matter, however, since the material is significantly different from ore and other homogeneous feeds encountered in these industries. For example, coal can be counted on to shatter upon impact, and thus a shredder* that would process coal would also shatter glass bottles rather well. On the other hand, the same shredder probably would not shred metal cans, which must literally be cut or torn apart within a shredder.

Experiences with full-scale shredding installations have shown that the two types of size-reduction devices most applicable to MSW are the vertical and horizontal shaft swing hammermills. These two shredders are illustrated in Fig. 3-5.

The hammermill consists of a central rotor on which are pinned radial hammers which are free to swing on the pins. The rotor is enclosed in a heavy-duty housing, an integral part of the shredding operation.

In the horizontal shaft mill, the rotor is supported by bearings on either end and the feed is by gravity (free drop) or conveyor (force fed). A discharged grate placed below the rotor determines the size of the product, since a particle cannot pass through this grate until it is smaller than the grate opening in two dimensions. Some hammermills are symmetrical so that the direction of the rotor can be changed to alternate wear surfaces without necessitating hammer maintenance after each run.

The vertical hammermill, as the name implies, has a vertical shaft, and the material moves by gravity down the sides of the housing. These mills usually have a larger clearance between the housing at the top of the mill, and progressively smaller clearances toward the bottom, thus reducing the size of the material in several steps as it moves through the machine. Since there is no discharge grate, the particle size of the product must be controlled by establishing a proper clearance between the lower hammers and the housing.

---

*As before, the term "shredder" is considered synonymous with "mechanical size reduction" in resource recovery—an equation not universally accepted, especially in the mining and ore dressing industry.

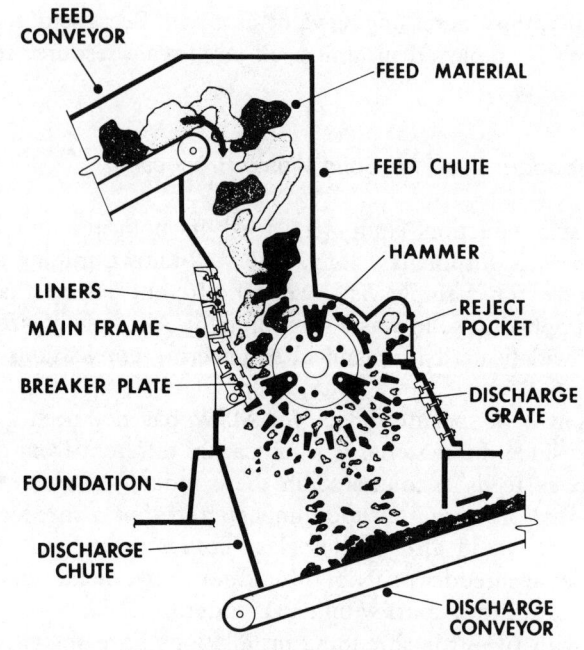
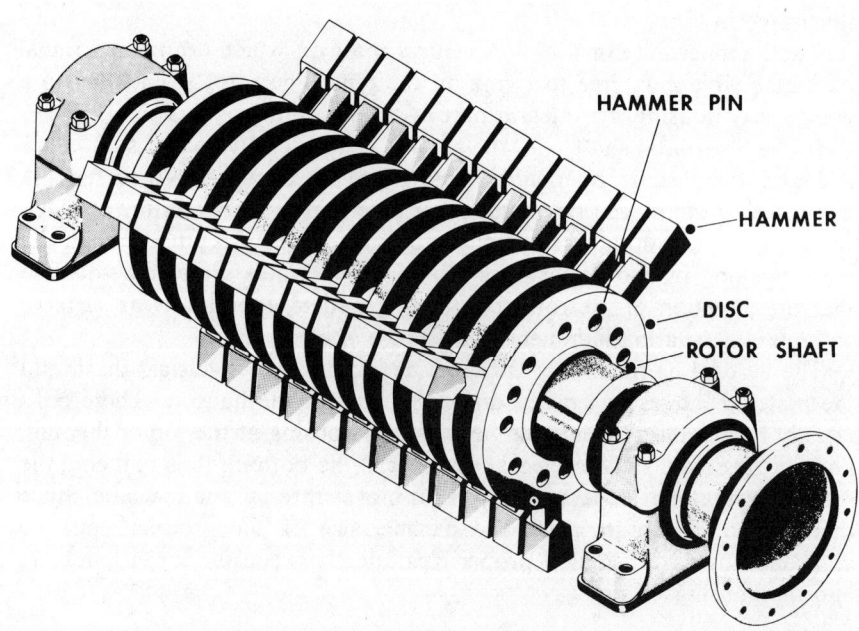

Figure 3-5. Horizontal and vertical shredders (From Ref. 43.)

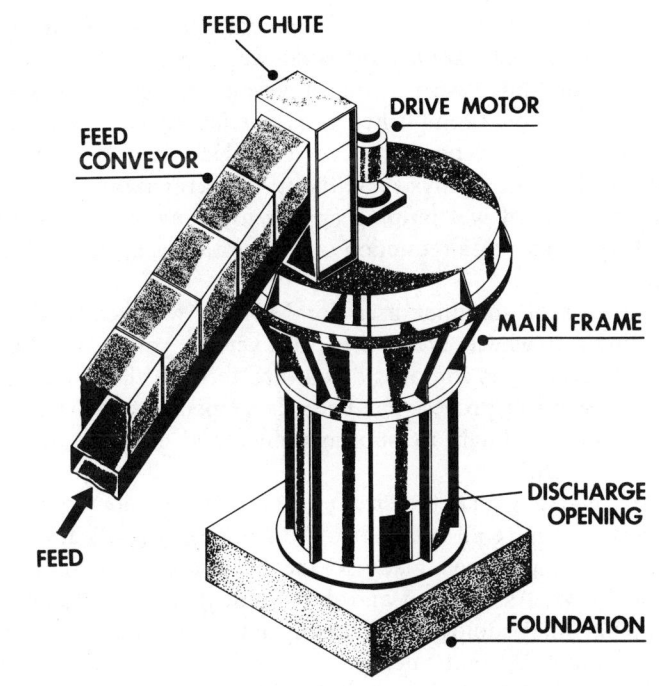
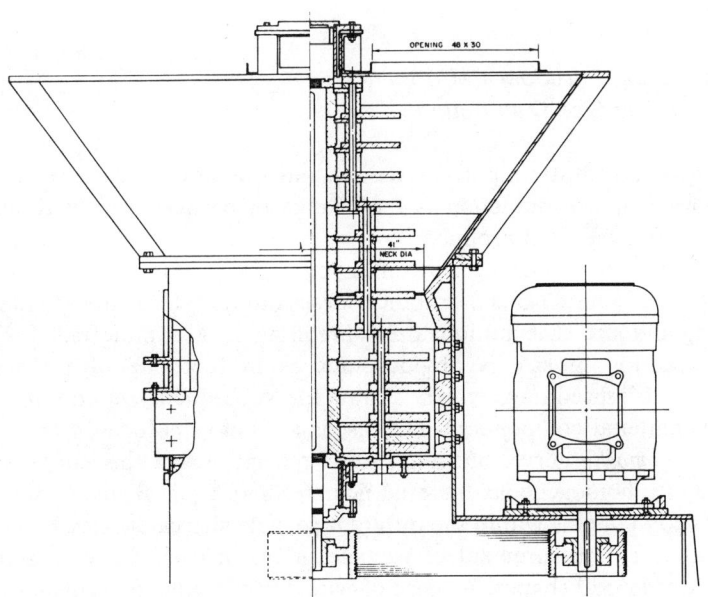

Figure 3-5. (*Continued*).

Other types of size-reduction equipment are also available on the market and have been applied to solid waste processing. Although not classified as shredding, wet pulping of sewage sludge and MSW together has been successfully accomplished in a large facility [18], and the economics of this system can be attractive if sludge disposal alternatives are included in the cost analysis. Without the availability of the sludge, however, the costs of wet pulping appear to be more than those for dry systems. Pulping as a size-reduction unit operation is discussed later in this chapter.

Another type of shredder used for MSW is technically a grinder, in which rolling star wheels are pinned to a vertical rotor, and the refuse is progressively ground as it moves downward through the machine.

Finally, a semiwet process has been developed where the MSW is wet with water only enough to suppress dust and improve its shredding characteristics [19].

It bears repeating that the size-reduction devices on the market today for the most part have all been borrowed from other applications, and only minor changes have been made to the hardware when they have been applied to MSW. As research and development work continues, however, size-reduction methodology will rapidly improve. There is no reason to doubt, for example, that such new technologies as lasers, ultrasonics, cryogenics, and controlled explosions will find useful application in the future.

Describing Shredder Performance by Particle-Size Distribution

One of the most important design and operational parameters to be considered in size reduction is the change in particle-size distribution of the feed and the final product.

The effect of shredding can differ for various material components in solid waste. Figure 3-6 is a graphical description of 13 different categories showing the size distribution after shredding in a hammermill [12]. The wide variation in size is obvious, and is in fact one of the primary attributes of shredding, which allows for subsequent separation of the various material components. The discussion below is focused on describing the *composite* curve, but one should recognize that this curve is not a picture of a homogeneous material nor are there equal distributions of the various components within the different particle-size categories.

Laboratory measurement of particle size is in itself difficult, since the material is in odd shapes. A piece of wire, for example, presents a difficult problem in classification because it is clearly quite small in two dimensions (and thus can escape further reductions in size in the shredding operation)

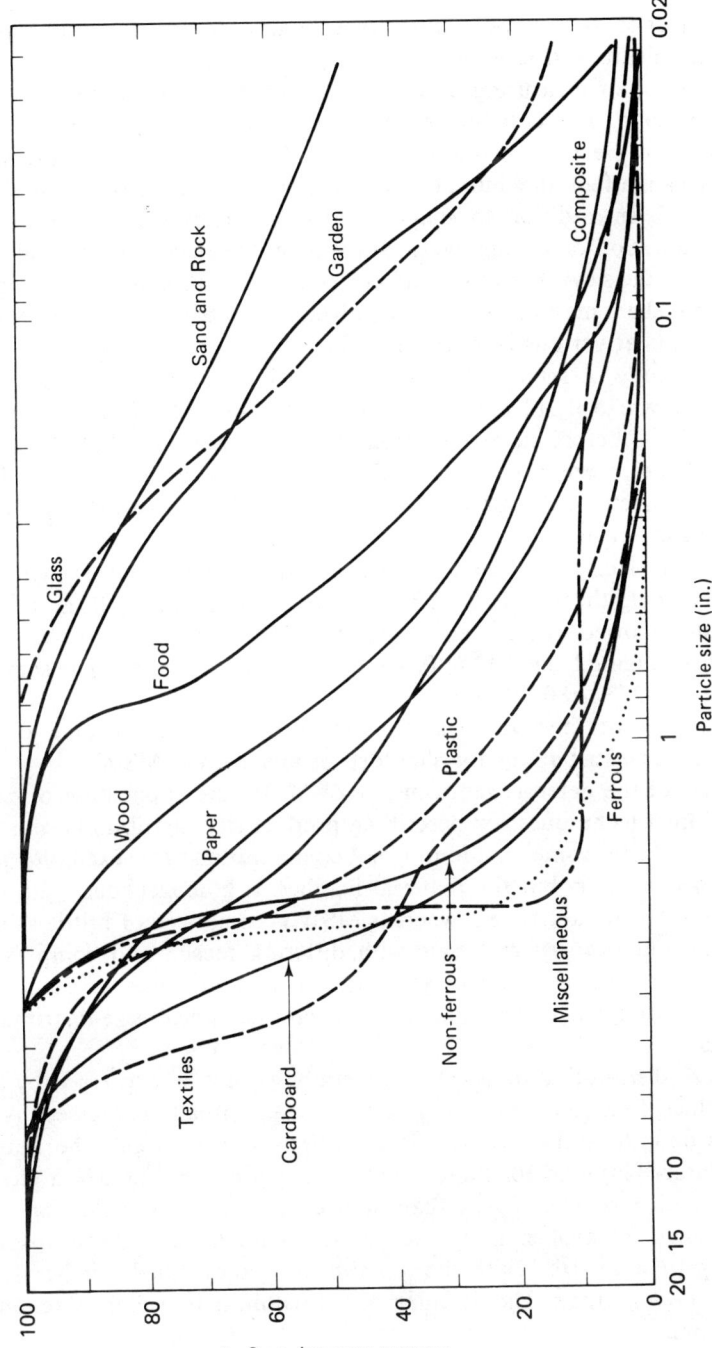

Figure 3-6. Cumulative distributions of various materials in shredded MSW. (From Ref. 12.)

but the effect of its length on subsequent separation operations such as air classification can be troublesome.

The method of measuring particle size can also influence the results of any given study. The common procedure for measuring particle-size distribution is by sieving, yet the shape of both the particles and the sieve openings can affect the number of particles that can pass through an opening [12]. In addition to shape factors, problems with providing an adequate duration of sieving, wear and tear on the sieve and the material, variations in the sieve apertures, and errors in observation and sampling all suggest that there may be problems involved in comparing size distribution data obtained at various laboratories [20].

In one study, a set of 13 sieves ranging in size from 5.1 cm (2 in.) openings down to 0.012 cm (0.0049 in.) were used [21]. The sample was dried at 104°C before sieving. Comparative tests indicated that about 100 to 125 shakes were necessary, and these could be either vertical or horizontal direction. Samples of about 100 g (0.22 lb) were found to be adequate for sieving.

In another study [22] the screening was done in two stages, with the coarse screening through 10.2-, 5.1-, 2.5-, and 1.5-cm (4-, 2-, 1-, and $\frac{3}{8}$-in.) wire mesh, followed by a 20-min sifting through a standard geometric-ratio sieve series ranging from 9.42 to 0.147 mm (0.24 to 0.0058 in.). The samples were dried at 27°C (80°F).*

At the present time, there is no universally accepted method of determining size distributions for shredded or unshredded MSW.

Because of the heterogeneous nature of MSW, the application of existing analytical techniques for describing product-size distribution must be approached with some caution. Coal comminution, for example, is a well-studied process, but the material handled is homogeneous, with uniform breakage characteristics, whereas MSW is comprised of brittle as well as ductile and rubbery materials with differing tensile and compressive strengths (see Chapter 1). Nevertheless, the first attempts at describing MSW shredding have been based on such theoretical size-distribution functions.

The size distribution of particles generally cannot be expressed by any single-valued function. Therefore, particle size is usually expressed by an equation describing the distribution of various size fractions. The general nomenclature [23] used for these equations is defined in Fig. 3-7, a plot of the cumulative weight Y less than size x plotted vs. particle size. The particle sizes are broken into an arbitrary number of intervals (usually according to sieve sizes) with the top size (or grade) called number 1, on down to the nth grade. The ith interval has within it W_i weight fraction of the material.

*The low drying temperatures will prevent some items from melting and sticking together.

Shredding

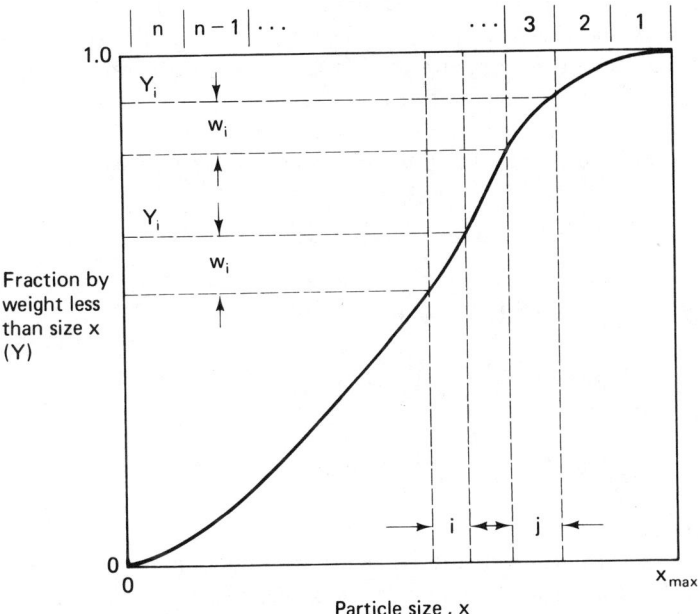

Figure 3-7. Nomenclature for describing particle-size distribution.

In the breakage process, as the curve is shifted to the left (greater number of smaller particles) some fraction of the material in the ith fraction remains there (are not broken) and some fraction originates from some larger sizes, or some general interval j. Y_j is the cumulative fraction by weight less than size j.

A number of equations have been proposed for describing the particle-distribution curve.

Gaudin [24] suggested the following equation to describe the product-size distribution for brittle materials:

$$Y = \left(\frac{x}{q}\right)^p$$

where Y is the cumulative fraction of material by weight less than size x, and q and p are constants specific to the material processed and the conditions under which the breakage occurs. In this case, q is the theoretical maximum size and p defines the slope of the line on log-log coordinates.

The most widely used particle size descriptor is the Rosin–Rammler model [25], first proposed in 1933, and stated as

$$Y = 1 - \exp\left(-x/x_0\right)^n$$

where n is a constant and x_0 is the "characteristic particle size," defined as the size at which 63.2% ($1 - 1/e = 0.632$) of the particles (by weight) are smaller. The Rosin–Rammler equation is, in fact, a generalized expression for sigmoidal curves, such as those in Fig. 3-6.

Note that the constant n is the slope of the line $\ln [1/(1 - Y)]$ vs. x on log-log coordinates, since the linear form can be derived as

$$Y = 1 - \exp\left(\frac{-x}{x_0}\right)^n$$

$$\ln\left(\frac{1}{1-Y}\right) = \left(\frac{x}{x_0}\right)^n$$

$$\log\left[\ln\left(\frac{1}{1-Y}\right)\right] = n \log\left(\frac{x}{x_0}\right)$$

$$\log\left[\ln\left(\frac{1}{1-Y}\right)\right] = n[\log(x) - \log(x_0)]$$

The value of x_0 is also defined as that size where $\ln [1/(1 - Y)] = 1.0$, or equivalently where $1/e$ of the particles are larger than x_0. The equation suggests that for a specific value of x_0, as the constant n increases due to changes in machine or feed characteristics, the value Y decreases, meaning that a coarser product is obtained. Conversely, for a given n, a larger x_0 also defines a coarser particle size of the product.

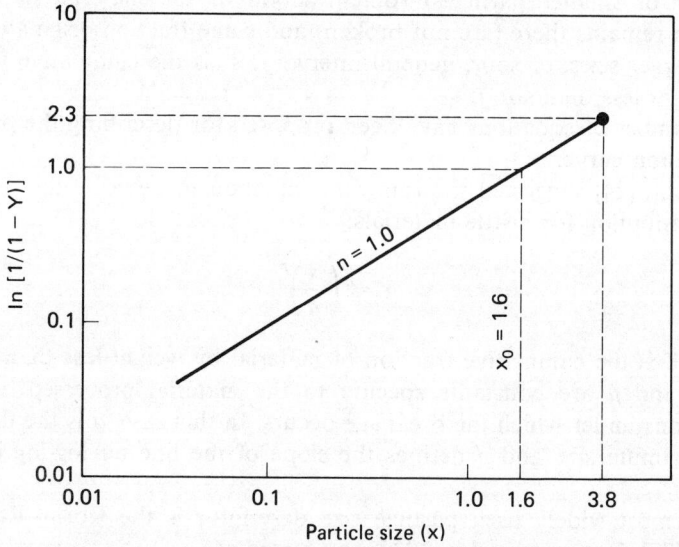

Figure 3-8. Rosin–Rammler particle-size distribution equation.

Shredding

TABLE 3-2
Rosin–Rammler Exponents for Shredded Refuse

	Rosin–Rammler Exponents	
Shredder Location	n (dimensionless)	x_0 (cm)
Washington, D.C.	0.689	2.77
Wilmington, Del.	0.629	4.56
Charleston, S.C.	0.823	4.03
San Antonio, Tex.	0.768	1.04
St. Louis, Mo.	0.995	1.61
Houston, Tex.	9.639	2.48
Vancouver, Wash.	0.881	2.20
Pompano, Fla.	0.587	0.67
Milford, Conn.	0.923	1.88
St. Louis, Mo.	0.939	3.81

Source: Ref. 27.

The Rosin–Rammler equation is plotted on log-log coordinates in Fig. 3-8 to illustrate the definition of both n and x_0. Table 3-2 presents a compilation of some Rosin–Rammler coefficients for various refuse shredders.

The characteristic size can be calculated from a specification such as "90% passing a given size," common in ore comminution practice. The following example, after Trezek and Savage [26], illustrates the procedure.

Example 3-2

Suppose that a sample of refuse must be shredded so as to produce a product with 90% passing 3.8 cm. Assume that $n = 1$. Calculate the characteristic size. Since $Y = 0.90$, $\ln [1/(1 - Y)] = \ln [1/(1 - 0.9)] = \ln 10 = 2.3$

Plot $x = 3.8$ cm vs. $\ln [1/(1 - Y)] = 2.3$ on log-log paper, as shown in Fig. 3-8. For $n = 1$, the slope of the line is 45°, which can be drawn. Lines for any other slope (n) can similarly be constructed by measuring the slope with a ruler. The characteristic size x_0 is then found at $\ln [1/(1 - Y)] = 1.0$, as $x_0 = 1.6$ cm.

The conversion from x_0 to 90% passing is possible by recognizing that

$$x_0 = \frac{x}{\left[\ln\left(\frac{1}{1-Y}\right)\right]^{1/n}}$$

If $Y = 90\%$, this expression reduces to

$$x_0 = \frac{x_{90}}{2.3^{1/n}}$$

where x_{90} = screen size where 90% of particles pass. If the value of n is 1.0,

$$x_{90} = 2.3 x_0$$

These expressions are convenient for design purposes, as illustrated later in this chapter.

Assuming only a single fracture of materials, the Gaudin–Meloy model has found some use in describing the breakage process. This model is written as

$$Y = 1 - \left(1 - \frac{x}{x'_0}\right)^r \tag{3-1}$$

where x'_0 is the "characteristic size" *of the feed material* (note the distinction between x'_0 and the Rosin–Rammler x_0) and r is the size ratio defined as the size of the broken piece to the size of the original piece.

A number of other particle-size-distribution functions have been suggested [29, 30, 31], but none of these appears to improve the required precision over the models discussed above and merely add complexity to a problem where the lack of precision in measurement of particle size makes further sophistication questionable [22].

One important application of these equations, and the major reason for their development, is the prediction of the particle-size distribution after breakage has occurred. Breakage is described by the breakage function, $B(x, y)$, which is defined as the fraction by weight of products that have a size less than x when particles of original size y are broken once [23]. For example, $B(20,500) = 0.4$ means that 40% by weight of an original particle of size 500 fall below size 20 after a single breakage. The breakage function can also be normalized, so that $B(x/y)$ describes the particle-size distribution regardless of the size of the original particle. For example, $B(0.04) = 0.26$ means that 26% by weight of the product falls in sizes below 0.04 of the original size.

Much of the original development in breakage theory is credited to Epstein [32], with the matrix application discussed below by Broadbent and Callcott [33]. Trezek at the University of California at Berkeley was the first to apply this technique to MSW [22]. This technique has also been applied to coal breakage by Evans and Pomeroy [29].

The basic assumption in this development is that during the breakage process, a fraction of the particles entering as feed are not broken, and exist as part of the product. For any given particle size, therefore, the product consists of some particles which were originally that size and were not broken, and some which are the result of larger pieces breaking into smaller ones, the latter being referred to as *complement*.

A second assumption is that the size distribution of the product due to the breakage of any single particle can be described by a continuous function. In other words, a large particle will break into many smaller

Shredding

pieces which when analyzed by sieving, will yield a smooth particle-size distribution. In the matrix analysis technique developed by Broadbent and Callcott, the distribution of particles following such breakage is described by the function

$$B(x, y) = \frac{1 - \exp(-x/y)}{1 - \exp(-1)} \qquad (3\text{-}2)$$

where y is the particle size being broken and $B(x, y)$ is the cumulative fraction of the product equal to or smaller than x. Although this equation was chosen in the Broadbent–Callcott analysis, other equations may better describe the particle-size distribution as well or better, and may be found to be superior when this analysis is applied to MSW.

Consider now a series of sieves, where the sieve sizes are related by a constant factor a so that the range of sizes passing the largest sieve is 1 to a, the next sieve is a to a^2, the next sieve a^2 to a^3, and so on to a^{n-1} to a^n.*

The feed particle-size distribution can then be described by the fractions f falling into these size ranges or grades, so that fraction f_1 of the feed is comprised of particles between size 1 and size a, f_2 is between a and a^2, and so on. (Obviously, $\sum_{i=1}^{n} f_i = 1.0$.) Further, assume that the particle-size distribution within each grade can be described by the geometric mean, or $a^{1/2}, a^{3/2}, a^{5/2}, \ldots, a^{(2n-1)/2}$.

Assuming now, as discussed above, that the fraction of the feed within each grade that actually breaks is π, so that the amount of the various fractions that are broken are $\pi f_1, \pi f_2, \pi f_3, \ldots, \pi f_n$.

The fraction π of the topmost grade (f_1) breaks into smaller pieces according to Eq. (3-2) so that of the π fraction breaking, the cumulative fraction of those particles equal to or finer than those in the top grade after breakage is, after Eq. (3-2),

$$B_{11} = \left[1 - \exp\left(\frac{-a^{1/2}}{a^{1/2}}\right)\right] \Big/ \left[1 - \exp(-1)\right]$$

Note that $B_{11} = 1.0$, since $y = a^{1/2}$, the particle size being broken and $x = a^{1/2}$, the size of the product.

Some fraction of the particles in the top grade break into a smaller size $a^{3/2}$. The cumulative fraction of particles equal to or smaller than this size originating in the top grade of particle size $a^{1/2}$ is

$$B_{21} = \left[1 - \exp\left(-\frac{a^{3/2}}{a^{1/2}}\right)\right] \Big/ \left[1 - \exp(-1)\right]$$

---

*Such geometric gradation is not necessary for this analysis, but it happens to be convenient for illustrative purposes.

and so on. The breakage of the top grade is summarized as follows:

Grade	Cumulative fraction of original (πf_1) particles equal to or finer than the grade after breakage
1 to a	$B_{11} = \dfrac{1 - \exp(-a^{-1/2}/a^{1/2})}{1 - \exp(-1)}$
a to a^2	$B_{21} = \dfrac{1 - \exp(-a^{-3/2}/a^{1/2})}{1 - \exp(-1)}$
a^2 to a^3	$B_{31} = \dfrac{1 - \exp(-a^{-5/2}/a^{1/2})}{1 - \exp(-1)}$
$\vdots$	$\vdots$
a^{n-1} to a^n	$B_{n1} = \dfrac{1 - \exp(-a^{(2n-1)/2}a^{1/2})}{1 - \exp(-1)}$

Of the topmost feed grade, πf_1 was designated for breakage [$(1 - \pi f_1)$ did not enter the breakage process]. The cumulative mass of the original feed which is equal in size or smaller than $a^{1/2}$ after breakage is $B_{11}(\pi f_1)$. If $b_{11} = 1.0$, as we have assumed, all the particles of the product are equal to or finer than $a^{1/2}$.

Similarly, the products of the breakage of the topmost grade are now distributed throughout the smaller grades such that B_{21} represents the cumulative fraction of particles equal to or finer than $a^{3/2}$, B_{31} is the cumulative fraction equal to or finer than $a^{5/2}$, and so on. The fraction of the product particles in the top grade after breakage is thus $b_{11} = B_{11} - B_{21}$. This can also be interpreted as the fraction of particles which, although broken, remained in the top grade. Similarly, the fraction of particles in the next smallest grade is $b_{21} = B_{21} - B_{31}$, and so on.

The example given above is for the breakage of particles that were originally in the topmost grade only, with a geometric size $a^{1/2}$. But breakage occurs within all grades, and this relationship must apply equally well for any other grade of the feed. Summing up the particles in the product that originated from breakage of larger particles, we get the following equations:

Grade	Product resulting from breakage
1 to a	$P'_1 = \pi b_{11} f_1$
a to a^2	$P'_2 = \pi b_{21} f_1 + \pi b_{22} f_2$
a^2 to a^3	$P'_3 = \pi b_{31} f_1 + \pi b_{32} f_2 + \pi b_{33} f_3$
$\vdots$	
a^{n-1} to a^n	$P'_n = b_{n1} f_1 + b_{n2} f_2 + \ldots + b_{nn} f_n$

where the first b subscript refers to grade of product and the second subscript defines the origin of that fraction. For example, the product in the second grade (a to a^2; subscript 2) is now made up of feed particles that originated as the top grade (1 to a: subscript 1) and were broken (the quantity $\pi b_{21} f_1$) and the particles within the second grade that, although broken, remained there ($\pi b_{22} f_2$). The product of the third grade is comprised of particles that originated on the first grade ($\pi b_{31} f_1$), the second grade ($\pi b_{32} f_2$), and the third grade ($\pi b_{33} f_3$). To state it another way, the parameter b_{ij} is defined as the fraction of material in size interval j which falls into the size interval i after breakage (see Fig. 3-7). Thus the products of size interval 1 are distributed so that b_{21} falls into size interval 2, b_{31} falls into interval 3, and so on. The sum of the values of b_{ij} over all the values of i is 1.

These equations can be summarized in a single equation using matrix notation. If $\mathbf{P}'$ is the product vector, $\mathbf{f}$ is the feed vector, and $\mathbf{B}$ is defined as the breakage matrix, derived from Eq. (3-2), then

$$\mathbf{P}' = \mathbf{B}(\pi)\mathbf{f}$$

Recall that not all of the particles in any one size fraction were broken in the process. Hence the product contains a fraction that originally existed in the feed and were not broken, $(1 - \pi)f_n$. The final product within each grade consists of these original particles, plus those which resulted from the breakage process. Hence

Grade	Total Product
1 to a	$P_1 = \pi b_{11} f_1 + (1 - \pi) f_1$
a to a^2	$P_2 = \pi(b_{21} f_1 + b_{22} f_2) + (1 - \pi) f_2$
a^2 to a^3	$P_3 = \pi(b_{31} f_1 + b_{32} f_2 + b_{33} f_3) + (1 - \pi) f_3$
$\vdots$	
a^{n-1} to a^n	$P_n = \pi(b_{n1} f_1 + b_{n2} f_2 + \ldots + b_{nn} f_n) + (1 - \pi) f_n$

These equations can similarly be expressed in matrix form as

$$\mathbf{P} = \pi\mathbf{Bf} + (1 - \pi)\mathbf{f} \tag{3-3}$$

In equation (3-3), the feed vector $\mathbf{f}$ and the product vector $\mathbf{P}$ can be obtained from sieve analysis, and the breakage matrix $\mathbf{B}$ is defined by the quantities $b_1, b_2, b_3, \ldots$ and can be calculated. The only unknown quantity is thus the scalar value π, which could well be a valuable index for describing the shredding operation. It should be recognized, however, that Eq. (3-3) depends on the validity of the original assumptions. If π is indeed constant for a given shredding operation, any one size particle is as likely to be broken as any other. We immediately recognize that this is not altogether true, since particles tend to become more resistant to fracture as they become smaller.

The question of the breakage function is still open to argument. Trezek et al. investigated the applicability of several models and decided that

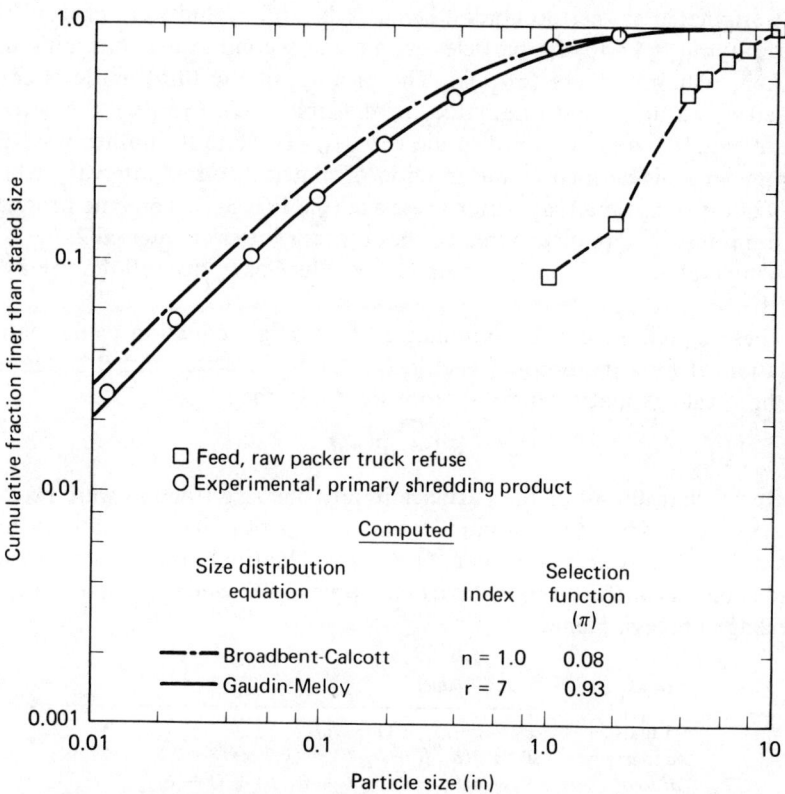

Figure 3-9. Experimental and calculated particle-size distribution: primary shredding. (From Ref. 26.)

either the Gaudin–Meloy function [Eq. (3-1)], or a modified Broadbent–Callcott function gave excellent results [22]. The latter was written

$$B_{(x)} = \frac{1 - \exp\left[-(x/x_0)^n\right]}{1 - \exp(-1)}$$

where n is a positive index requiring back calculations and varying from 0.845 to unity. The term x_0 is the "characteristic size" as previously defined. Figures 3-9 and 3-10 show the results of some experiments using these two functions. The first figure is an estimate of primary shredding, and the second shows the results from secondary shredding. In both cases, either breakage function seems to predict the particle-size distribution adequately.

Another refinement to the model is to eliminate the original assumption that the fraction of any particle size that enters the breaking process is

Shredding

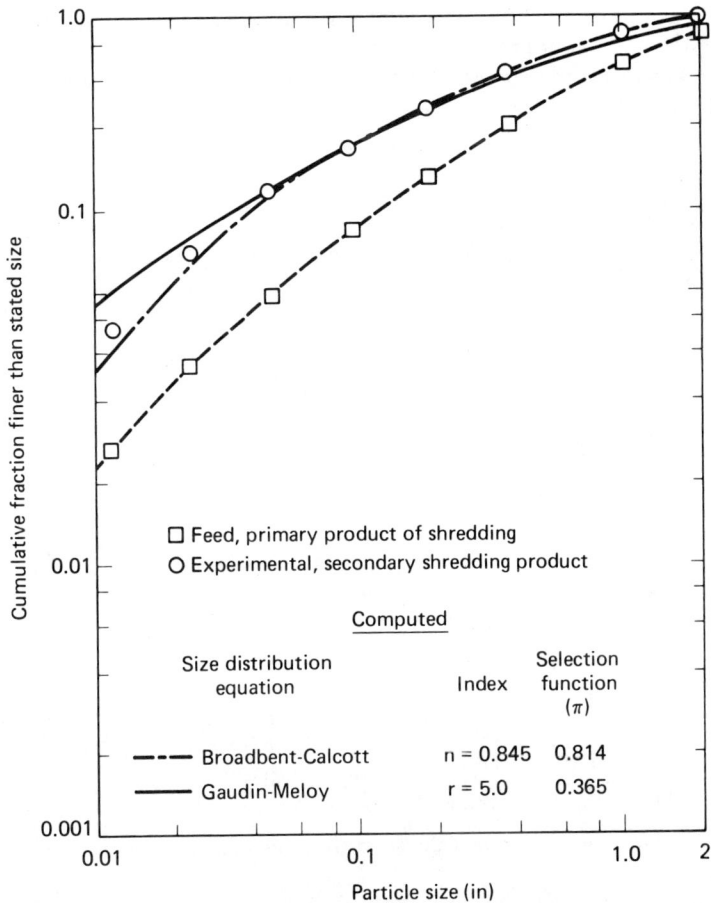

Figure 3-10. Experimental and calculated particle-size distribution: secondary shredding. (From Ref. 26.)

constant for all sizes—in other words, that π is constant. To do this, we can argue that π can be replaced by a matrix **S** such that it characterizes the size-reduction process, and is a diagonal matrix with entries $S_1, S_2, S_3, \ldots, S_n$ along the diagonal and zeros elsewhere. The final product equation can be expressed as

$$\mathbf{P} = \mathbf{BSf} + (\mathbf{I} - \mathbf{S})\mathbf{f}$$

where **I** is a unit matrix.

Gardner and Austin [34] used a radioactive tracer to measure the breakage process and to evaluate some of the assumptions made in the foregoing analysis. They found that the breakage function $B(x, y)$ did not vary with the time of grinding, given a certain feed size y. It was also found reasonable to normalize $B(x, y)$ to $B(x/y)$. Finally, they found by back-calculating the values for S and $B(x/y)$ that they were within reasonable agreement with experimental data.

Example 3-3

For the following feed, calculate the fraction of particles in each size range if it is assumed that $\pi = 0.5$ and the breakage function is $B(x, y) = [1 - \exp(-x/y)]/[1 - \exp(-1)]$.

Average size in a sieve (mm)	Fraction in feed
100	0.80
75	0.20
50	0
25	0

The basic question is

$$\mathbf{P} = \pi \mathbf{Bf} + (1 - \pi)\mathbf{f}$$

where **P** is the product vector, **B** the breakage matrix, and **f** the feed vector. Written in longhand:

$$\begin{vmatrix} P_1 \\ P_2 \\ P_3 \\ P_4 \end{vmatrix} = (\pi) \begin{vmatrix} b_{11} & 0 & 0 & 0 \\ b_{21} & b_{22} & 0 & 0 \\ b_{31} & b_{32} & b_{33} & 0 \\ b_{41} & b_{42} & b_{43} & b_{44} \end{vmatrix} \begin{vmatrix} f_1 \\ f_2 \\ f_3 \\ f_4 \end{vmatrix} + (1 - \pi) \begin{vmatrix} f_1 \\ f_2 \\ f_3 \\ f_4 \end{vmatrix}$$

The breakage matrix can be calculated from the breakage function:

$$B(x, y) = \frac{1 - \exp(-x/y)}{1 - \exp(-1)}$$

$$B(100, 100) = \frac{1 - \exp(-100/100)}{1 - \exp(-1)} = 1.0$$

$$B(75, 100) = \frac{1 - \exp(75/100)}{1 - \exp(-1)}$$

$$= \frac{1 - (0.472)}{1 - 0.367}$$

$$= \frac{0.528}{0.633} = 0.83$$

This means that of the particles in size interval 1 (100 mm), which enter the breakage process, 83% break down to size 75 mm or smaller. Hence the fraction of the particles that remains in the size interval 1 (even after breakage) is $b_{11} = 0.17$.

$$B(50/100) = \frac{1 - \exp(-50/100)}{1 - \exp(-1)} = 0.62$$

or 62% of the material that broke became 50 mm or smaller. Hence $0.83 - 0.62 = 0.21$, or 21% of the original size material is now in size interval 2 or $b_{21} = 0.21$. Similarly, $B(25, 100) - 0.35$ and $b_{31} = 0.62 - 0.35 = 0.27$, and the last size interval must have a contribution of 0.35. Note that these add up to 1.0.

The material which was at 75 mm also breaks, so that

$$B(50, 75) = \frac{1 - \exp(-50/75)}{1 - \exp(-1)} = 0.77$$

or

$$b_{22} = 0.23$$

and

$$B(25, 75) = \frac{1 - \exp(-25/75)}{1 - \exp(-1)} = 0.45$$

and

$$b_{32} = 0.77 - 0.45 = 0.32 \quad \text{and} \quad b_{43} = 0.45$$

There was no feed material at 50 and 25 mm, and therefore

$$b_{33} = b_{43} = b_{44} = 0$$

$$\begin{vmatrix} P_1 \\ P_2 \\ P_3 \\ P_4 \end{vmatrix} = (\pi) \begin{vmatrix} 0.17 & 0 & 0 & 0 \\ 0.21 & 0.23 & 0 & 0 \\ 0.27 & 0.32 & 0 & 0 \\ 0.35 & 0.45 & 0 & 0 \end{vmatrix} \begin{vmatrix} 0.8 \\ 0.2 \\ 0 \\ 0 \end{vmatrix} + (1 - \pi) \begin{vmatrix} 0.8 \\ 0.2 \\ 0 \\ 0 \end{vmatrix}$$

The solution is by matrix methods, facilitated by computers. For the simple case, the four equations can be written and solved directly as

$$P_1 = 0.47$$
$$P_2 = 0.20$$
$$P_3 = 0.15$$
$$P_4 = 0.18$$

Note that the sum of the product fractions equals 1.0, as it should.

Power Requirements of Shredders

Only limited information is available about the power requirements of shredders. Various estimates indicate that the hammermill is grossly inefficient, with only about 0.1 to 2.0% of the energy supplied to the machine appearing as increased surface energy of the product solids [35]. Part of the explanation for such low efficiencies lies in the plastic deformation and viscoelastic flow that accompany shredding; these deformation processes require many times more energy than the creation of new surface area. The efficiency of a shredding operation depends on how the energy is applied and on how the material reacts to it. For example, a roll mill, which crushes materials but does very little shearing, would waste energy trying to shred a newspaper. Also, since fracture in brittle materials occurs progressively, with flaws building up within the particle until the particle breaks up [35], the rate of application of the force is important. Because there is a time lag between the application of the force and eventual fracture, a machine that applies load rapidly is inherently inefficient, since more energy is required than if the force could be applied more slowly. It follows that for some high-speed machines, a slower speed could result in less energy use, at least to the point at which the rotor inertia is too low and the energy requirement once again increases with decreasing speed. This relationship, shown in Fig. 3-11, was confirmed by Trezek and Savage [26].

Higher speeds, nevertheless, produce the finest product particle size as shown in Fig. 3-12. This fine product, however, requires more energy

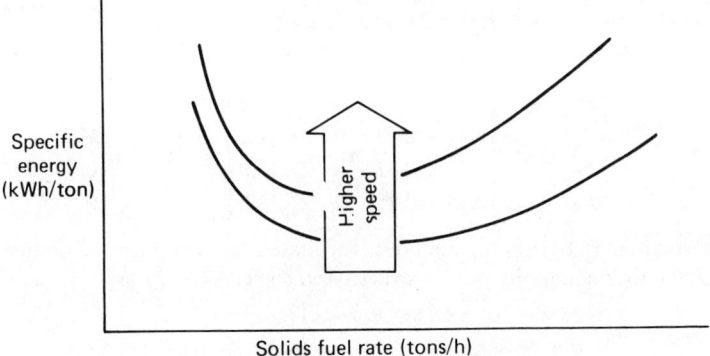

Figure 3-11. Specific energy used in shredding as influenced by rotor speed and solids feed rate.

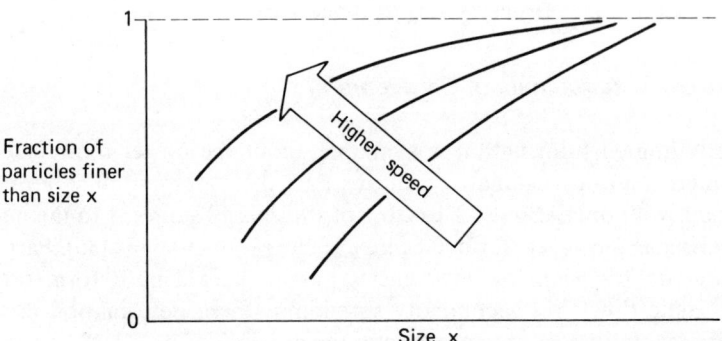

Figure 3-12. Particle-size distribution in shredding as influenced by shredder speed.

Shredding

consumption. As suspected, energy requirements increase with feed rate. Plotting the dry feed rate vs. the rate of energy use in kilowatts results in straight lines for both secondary and primary shredding [36].

Figure 3-13 shows the specific energy use (kWh/ton of refuse as dry weight) vs. feed rate. For low feed rates, there is simply not enough material going through and the machine is loafing. As the feed rate increases, the efficiency increases (lower specific energy use), and eventually increases until the machine becomes overloaded.

The amount of moisture in the feed refuse also influences energy use, as shown in Fig. 3-14, and the minimum energy requirement appears to be in the 35 to 40% moisture range [26]. Moisture also affects particle size, with drier feed resulting in smaller particles in the product [26] (Fig. 3-15).

To date, accurate estimates of the energy requirements for size reduction have not been possible, even for homogeneous materials. Several empirical relationships have been suggested. Those which consider only the

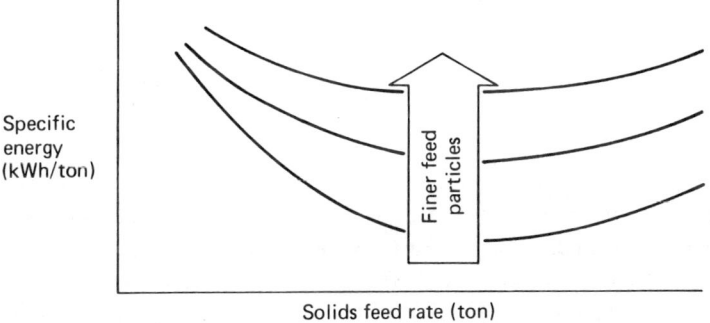

Figure 3-13. Specific energy used in shredding at various feed rates.

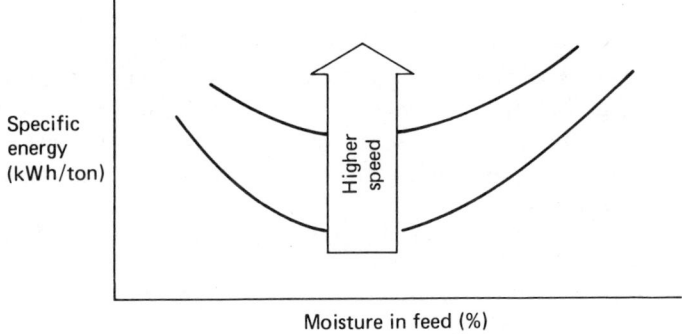

Figure 3-14. Specific energy used in shredding as influenced by moisture.

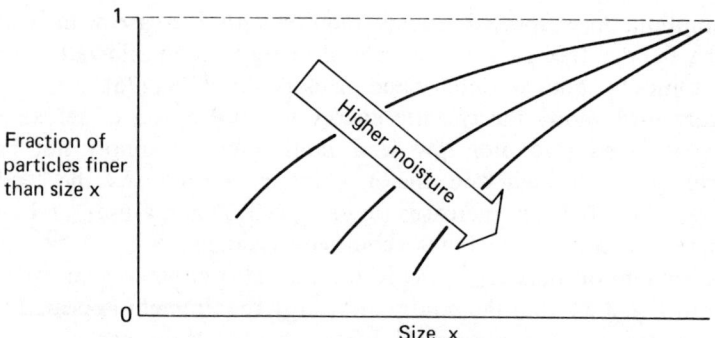

Figure 3-15. Effect of moisture on particle-size distribution in shredding.

fracture process are all based on the assumption that

$$\frac{dE}{dL} = -CL^{-n}$$

which states that the energy dE required to achieve a small size change dL in a unit mass of material is inversely proportional to the size of the particle, L. The symbols n and C are constants.

If $n = 1$, for example, the equation can be integrated to yield

$$E = C \log\left(\frac{L_1}{L_2}\right)$$

where L_1/L_2 is the size-reduction ratio and E is the work done to reduce the particles from size L_1 to size L_2. C, as before, is a constant. Physically, this expression states that the work done is proportional to the number of new smaller particles created from the larger ones. This is known widely as *Kick's law* [35].

If $n = 2$, the integrated form of the equation yields

$$E = C\left(\frac{1}{L_2} - \frac{1}{L_1}\right)$$

which is known as *Rittinger's law* and physically represents the assumption that the energy required is proportional to the amount of new surface formed by the size-reduction process [35]. Rittinger's law seems to agree better with the results of rough grinding operations, whereas Kick's law is a better approximation of fine grinding.

If $n > 1$ but is otherwise undefined, the integrated form of the equation is

$$E = \frac{C}{n-1}\left(\frac{1}{L_2^{n-1}} - \frac{1}{L_1^{n-1}}\right)$$

TABLE 3-3

Average Work Indices for Various Materials

Material	Work Index (kWh/ton)
Clay	7.1
Coal	11.4
Glass	3.1
Granite	14.4
Gravel	25.2
Sandstone	11.5
Silica sand	16.5
Slag	15.7

Source: Ref. 39.

The *Bond work index* [37] is based on the assumption that the work done in crushing and grinding is directly proportional to the total length of new cracks formed in the material being reduced in size [38]. In this case, it is suggested that $n = 1.5$, and that

$$E = E_i \frac{\sqrt{L_F} - \sqrt{L_p}}{\sqrt{L_F}} \sqrt{\frac{100}{L_p}}$$

where E = specific work (kWh/ton) required to reduce a unit weight of material with 80% finer than some diameter L_F, in micrometers, to a product with 80% finer than some diameter L_p, in micrometers.

E_i = work index, a factor that is a function of the material processed; this value is also the theoretical work required to reduce a unit weight from infinite size to 80% finer than 100 μm; units are (kWh/ton)

The expression above is more conveniently expressed as

$$E = 10 E_i \left(\frac{1}{\sqrt{L_p}} - \frac{1}{\sqrt{L_F}} \right)$$

The work indices for some common industrial materials are tabulated in Table 3-3.*

*The 10 in the equation is $\sqrt{100 \mu m}$, hence the units of E_i are KWh/ton.

The Bond work index can also be estimated by the dimensionally incorrect empirical relationship [39] as

$$E_i = 2.59 \frac{C_s}{SG}$$

where E_i = work index, (kWh/ton)
 C_s = impact crushing resistance, ft-lb/in. of thickness required to break
 SG = specific gravity

When ductile materials such as steel, aluminum, and copper are shredded, the energy requirements may be almost independent of the size reduction ratio L_1/L_2, but still depend on the feed rate. In this case, the work of shredding goes into deforming the material to the point of fracture, the fracture itself consuming an insignificant amount of energy by comparison. Consider the deformation of a strip of steel in tension until it fractures. The work of deformation is

$$W = \int F \, dl$$

where l is the distance through which the force F travels. Rewriting this equation in terms of stress ($S = F/A$) and strain ($d\varepsilon = dl/l$),

$$W = \int AlS \, d\varepsilon$$

Since volume is approximately constant, the work per deformed volume is simply the area under the stress-strain curve.

$$\frac{W}{V} = \int S \, d\varepsilon$$

And the work of shredding per ton is simply

$$\frac{W}{T} = C_1 \rho \int S \, d\varepsilon$$

where ρ is the density of the material and C_1 is a constant that depends on the units used. If this equation is valid, a plot of dW/dt (typical units: horsepower) vs. feed rate (typical units: tons/h) would exhibit a straight line with slope $C_1 \rho \int S \, d\varepsilon$ for any one material. Such a plot is presented in Fig. 3-16, with experimental points for a number of commercial shredders and with theoretical curves for auto-body steel sheet and wood [40]. The considerable discrepancy between the theoretical curves and the experimental data presumably lies in the concept of "redundant work." The theoretical concept presented above assumes that a material is deformed only as a prerequisite to fracture; in fact, any material being shredded will be bent, twisted, pulled, and compressed many times, both before and after fracture. This additional work may change the shape of the material fragments, but it does not contribute to size reduction in any substantial way.

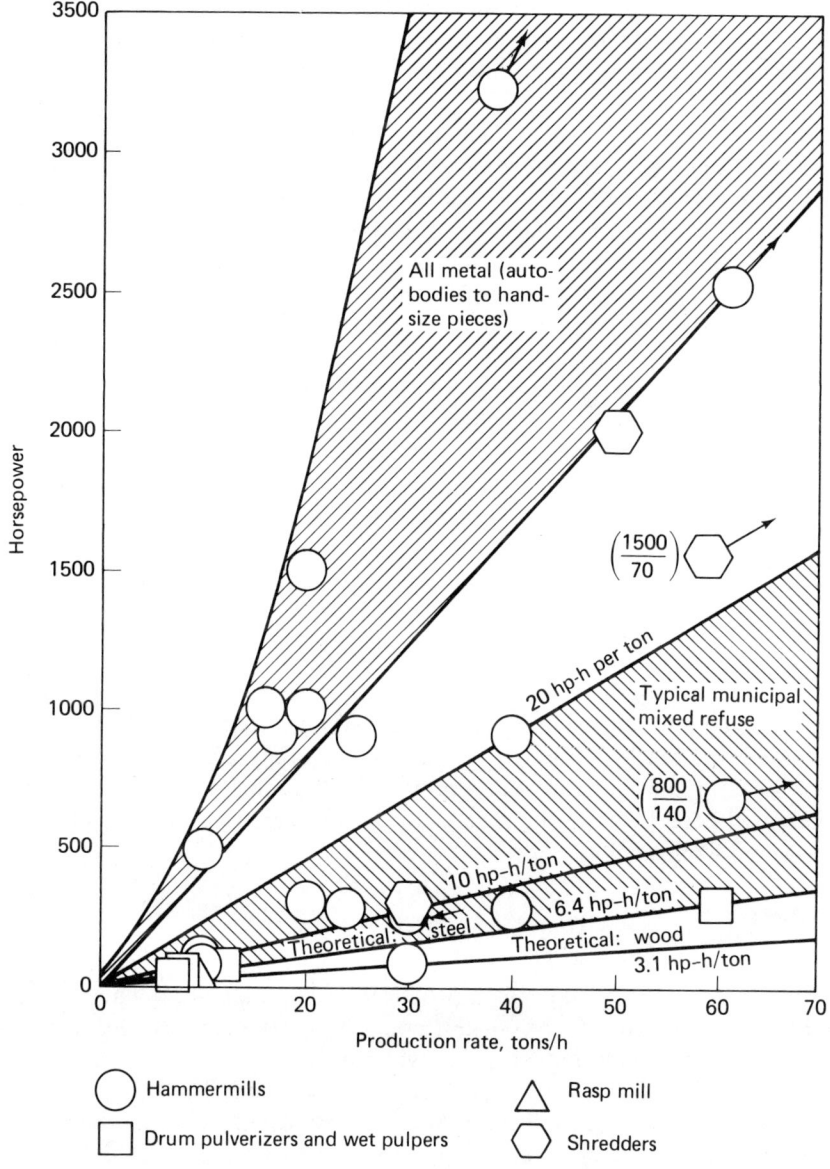

Figure 3-16. Typical energy requirements for shredders. (From Ref. 40.)

One way to minimize the work of deformation associated with shredding ductile materials is to maximize shear deformation and minimize tensile deformation. Shear deformation (as in cutting steel sheet with tinsnips or cutting copper wire with diagonal pliers) minimizes the deformed volume in the work equation by localizing it near the fracture surface. In addition, the possibility exists in shear of lowering the theoretical work requirements at high speed by adiabatic shear. *Adiabatic shear* involves deforming a volume of material fast enough that the heat of deformation (the area under the stress-strain curve, minus the elastic region) cannot diffuse away by thermal conduction; the heat generated thus raises the temperature and lowers the strength of the material being sheared.

Developing ways to minimize redundant work, decrease the deformed volume necessary to accomplish size reduction, and utilize adiabatic shear to increase feed rates may provide the keys to shredder design for MSW. For example, some of these techniques might be capable of both moving the minimum in Fig. 3-11 to a higher feed rate and reversing the effect of higher speeds.

Health and Safety

One need only reflect on the types of materials people thoughtlessly throw away into refuse cans to realize the serious health and safety aspects of shredding MSW. Although the health and safety aspects of size reduction deserve considerable study, only a brief summary of the problem is presented here. Specifically, shredders can be hazardous because of noise, dust, and explosions.

Noise

The noise levels around a 3-ton/h hammermill range from 95 to 100 dBA,* with much of the noise produced being a low-frequency rumble [22]. In addition to a high constant noise level, resource recovery facilities produce considerable impact noise which is difficult to measure, and the effect of which on human beings is poorly understood. The existing federal Occupational Safety and Health Act (OSHA) standard limits noise to 90 dBA over an 8-h working day. The corresponding limit set by the EPA is 85 dBA. It thus seems likely that shredder operators will need to wear ear protection, and noise reduction should be considered in the design of future resource recovery facilities.

---

*dBA is a standard method of noise measurement, and stands for "decibels on the A scale" of the sound-level meter. This scale is an attempt to duplicate the hearing efficiency of the human ear.

Dust

Dust can cause several problems—it can be a vector for the transmission of pathogenic microorganisms, it can itself have a detrimental effect on health by affecting the respiratory system, and it can explode. The latter problem is discussed in the next section.

OSHA standards presently limit dust inhalation to 15 mg/m^3 of total dust over a 8-h day. Limited studies of dust production in resource recovery facilities have shown that dust levels are from 7 to 13 times higher than the OSHA standard. This finding, which would not surprise shredder operators, dictates the use of face masks while working.

Plate counts at shredding operations have indicated that the total bacterial counts during the shredder operation are as much as 20 times greater than the ambient, which contains about 880 organisms/m^3 of air. Studies have shown that in resource recovery facilities where shredders are used, coliform counts can jump from 0 to 69 per cubic foot, and fecal streptococci from 0 to over 500 per cubic foot [41]. These indications clearly show the potential danger of disease transmission by the air route during shredding of MSW.

Explosion

The high temperature and metal-to-metal contact in shredders has caused numerous explosions at existing shredder installations. Actually, small explosions such as those due to the breakage of aerosol cans occur regularly, and shredders are designed to accept these without suffering damage. Larger explosions, however, can damage the shredder and even injure the operators.

Two types of explosions are generally recognized—dust explosions and those caused by explosive materials such as gunpowder and partially filled gasoline cans. Some operators will, however, argue that the dust will never become explosive by itself and that all explosions are caused by combustible materials. This contention remains to be investigated.

Dust can cause explosions when the concentration of a combustible dust is sufficiently high and there is adequate oxygen and a spark. Below a certain dust concentration, the heat of combustion is not sufficient to propagate combustion, and an explosion cannot occur.

If the concentration of oxygen is reduced to below about 10%, explosions should not occur. Other than continuously flooding the chamber with an inert gas, however, it is difficult to keep oxygen out of the shredder.

The only realistic means of preventing explosions is to maintain a high level of surveillance on what is being fed to the shredder. Accordingly, almost all shredder installations have people scanning the feed conveyor for such potentially dangerous items as cans of paint thinner, gasoline tanks from cars, lawn mower engines, and so on.

Even keen surveillance however is not foolproof. In one instance, a gas tank was removed from the conveyor belt, but some of the gasoline spilled on the refuse. This occurred just at the change of shifts and the new crew was not informed of the spilled gas. When the shredder was again fed, the gasoline vaporized and ignited from space heaters above the workers, sending a fireball back into the shredder.

Two methods presently used to reduce the damage when explosions occur in refuse shredders are venting and flame suppression. All shredders are constructed with blow-out doors, so that the pressure within the shredder housing can escape. Some are now being equipped with flame suppressors which are designed to be released as soon as the pressure

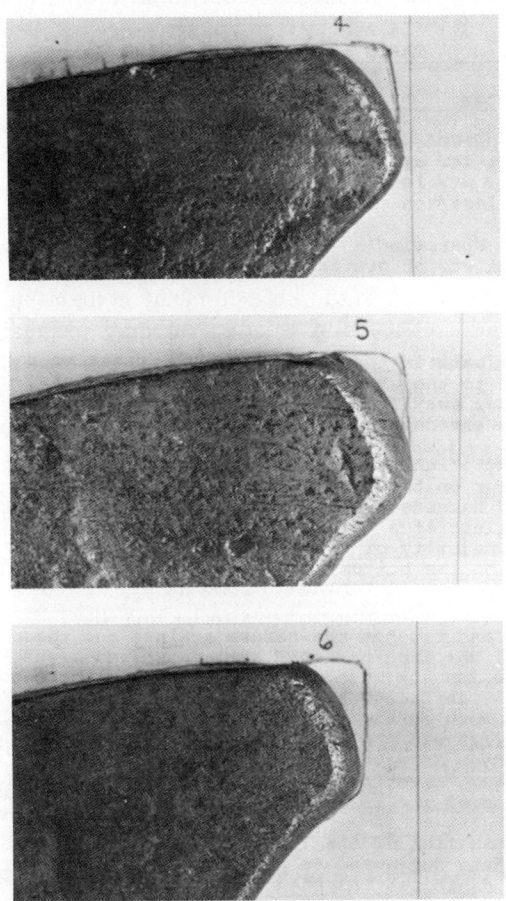

Figure 3-17. Non-hard-faced hammer after grinding 70 tons of refuse at 1200 rpm. The line shows the hammer shape when new. (Courtesy of G. J. Trezek.)

Shredding

builds up to a critical level. This system will work for most types of explosions, but it is not fast enough for dynamite, gunpowder, and other explosives.

Hammer Wear and Maintenance

Because of the relatively unsophisticated and brute-force nature of the shredding process, it can be expected that the wear and tear on shredders can be substantial. One of the major maintenance headaches (and expenses) at shredding facilities is the wear of the hammers.

The pattern of wear on hammers is illustrated in Fig. 3-17. The wear is almost exclusively on the bottom edge of the hammer's crushing face, since this is the area of impact as the material is crushed against the grate. The

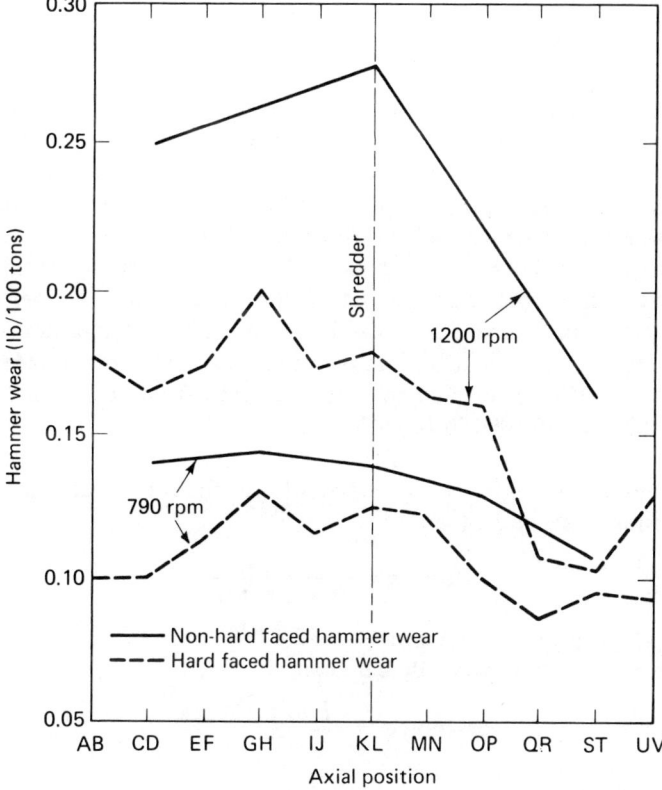

Figure 3-18. Effect of shredder speed and hard-facing of hammers on wear expressed as weight loss. (From Ref. 20.)

wear seems to be due mostly to abrasion, although severe impact with very hard objects can also contribute to wear.

Hammer wear can be reduced by both hard-facing the hammers with abrasion-resistant alloys and by slowing the speed of shredding, as shown in Fig. 3-18.

As the hammers wear down, the shredding performance decays. In one study [42], the percent of materials passing the sieve (Y) was related to the tons of refuse processed (T) by the equation

$$Y = b_0 + b_1 \exp(-b_2 T)$$

where b_0, b_1, and b_2 are all constants. For example, for 0.185-in. particle size, the relationship is

$$Y = 8 + 35 e^{-0.0081 T}$$

It has been suggested that hammer wear can be expected to be in the range 0.05 to 0.10 lb/ton when shredding MSW. Experience has shown, however, that much higher rates of hammer wear are also possible.

Shredder Design

Shredders are not, in the strict sense, designed by a consulting engineer or the purchaser. They are actually selected much as water pumps are selected for a specific application. Several guides for shredder selection have been published [3, 43]. Such specifications as the speed, motor horsepower requirements, and the rotor inertia must all be specified.

The motor inertia is usually expressed as WR^2, where W is the weight on the mass of the rotor assembly and R the radius to the hammermill tips. We recognize, of course, that this is not an accurate measure of rotor inertia, but is simply a convenient parameter for comparative purposes. A wide range of WR^2 is offered by manufacturers, from about 2000 to 6000 kg-m^2 (50,000 to 150,000 lb-ft^2) [43].

The motor horsepower is designed on the basis of starting horsepower. If the motor inertia WR^2 is expressed as lb-ft^2, the rotor speed in rpm, and starting time in seconds, then

$$\text{torque} = \frac{WR^2 \times \text{rpm}}{9.6 \times g \times t}$$

where g is the gravitational constant, 32.2 ft/s^2.

The horsepower is then calculated as

$$\text{horsepower} = \frac{T \times 2\pi \times \text{rpm}}{33{,}000}$$

where T is torque, in ft-lb.

Typically, a shredder used for MSW has a width/diameter ratio greater than 1.0, a hammer weight of 70 kg (150 lb), a rotor inertia of 1500 kg-m^2

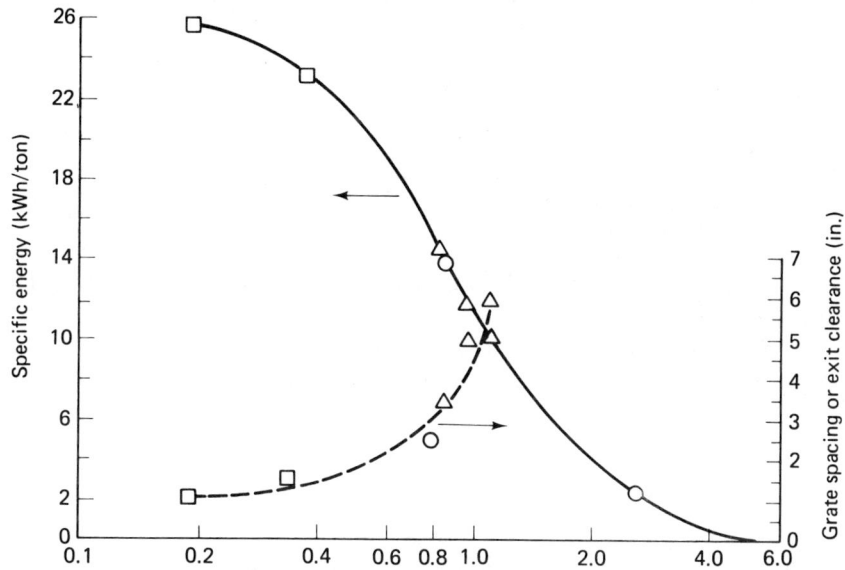

Figure 3-19. Specific energy used by various shredders. (From Ref. 26.)

(35,000 lb-ft^2), a hammer tip speed of 4260 m/min (14,000 ft/min), four rows of hammers, and a starting time of 30 s. As a rule of thumb [44], a MSW shredder must be designed for at least 15 kWh/ton.

For horizontal hammermills, the grate openings can be changed to achieve different size distribution of the product. Similarly, final clearance in vertical mills can be changed to produce various-sized products. Trezek and Savage [26] found that the "characteristic size," x_0 (see page 102) and the specific energy can be related to the grate spacing or exit clearance, as shown in Fig. 3-19. Note that the curves are drawn by connecting data points obtained from different types of shredders, with little overlap, which tends to cast some doubt on their general applicability. Even with this drawback, however, it offers the design engineer the opportunity to estimate the grate spacing or clearance, as well as the specific energy requirement (and hence the motor size, given the feed rate) as a function of desired particle size.

Example 3-4

For $x_0 = 0.64$ in. (1.62 cm) and $n = 1$ (as found in Example 3-2), find the grate spacing or clearance and the motor power requirements for a feed of 5 tons/h. Using Fig. 3-19, enter at $x_0 = 0.64$ and find the grate opening as 2.54 in. and a specific energy requirement as 18 kWh/ton, which translates to 90 kW (120 hp).

The Bond work index may be used for shredder design by noting first that the Rosin–Rammler equation can be written as

$$Z = \exp\left(\frac{-x}{x_0}\right)^n$$

where Z = cumulative fraction *greater than* some stated size x
  x_0 = characteristic size
  n = constant

That is, $Z = 1 - Y$, where Y was previously defined as the cumulative fraction finer than some size x.

At $Z = 0.2$ (20% of feed is larger), then $x = L_P$, the screen size through which 80% of the product passes. Solving for L_P,

$$0.2 = \exp\left(\frac{-L_P}{x_0}\right)^h$$

$$L_P = x_0(1.61)^{1/n}$$

Substituting this back into the Bond work index equation,

$$E = \frac{10E_i}{\left[x_0(1.61)^{1/n}\right]^{1/2}} - \frac{10E_i}{L_F^{1/2}}$$

TABLE 3-4

Bond Work Index for Shredding of Refuse and Refuse Components

Shredder Location	Material	Work Index (kWh/ton)
Washington, D.C.	Refuse	463
Wilmington, Del.	Refuse	451
Charleston, S.C.	Refuse	400
San Antonio Tex.	Refuse	431
St. Louis, Mo.	Refuse	434
Houston, Tex.	Refuse	481
Vancouver, Wash.	Refuse	427
Pompano, Fla.	Refuse	405
Milford, Conn.	Refuse	448
St. Louis, Mo.	Refuse	387
Washington, D.C.	Glass	8
Washington, D.C.	Paper	194
Washington, D.C.	Steel cans	262
Washington, D.C.	Aluminum cans	654

Source: Ref. 27.

Pulping

This equation now allows for an estimation of shredder power requirements, given a definition of the required product, and an estimation of the work index E_i.

Although no data on refuse shredder power consumption vs. particle size are yet available, using the method described in Example 3-4 and Trezek's curve (Fig. 3-19) makes it possible to back-calculate for the Bond work index. Table 3-4 is such a tabulation, and shows that on the average, the E_i for refuse is about 430 (kWh/ton).

Example 3-5

Assuming that $E_i = 430$, and $x_0 = 1.62$ cm, $n = 1.0$, and $L_F = 25$ cm (about 10 in.; a realistic estimate, as shown in Fig. 1-5), find the power requirement for a shredder processing 5 tons/h.

$$E = \frac{10 E_i}{\left[x_0(1.61)^{1/n}\right]^{1/2}} - \frac{10 E_i}{L_F^{1/2}}$$

(note: x_0 and L_F are in micrometers (μm) in this equation)

$$E = \frac{10(430)}{\left[16{,}200(1.61)^{1/1}\right]^{1/2}} - \frac{10(430)}{(250{,}000)^{1/2}}$$

$$= 18 \text{ kWh/ton}$$

or

$$18 \frac{\text{kWh}}{\text{ton}} \times 5 \frac{\text{tons}}{\text{h}} = 90 \text{ kW}$$

as before, in Example 3-4.

PULPING

Wet pulping, although a well-developed process in the pulp and paper industry, has been applied to solid waste processing by only one manufacturer. Their unit, pictured in Fig. 3-20 is a tub 3.6 m (12 ft) in diameter with a high-speed cutting blade on the bottom driven by a 300-hp motor. The raw refuse is mixed with sludge from a wastewater treatment plant and all pulpable and friable materials are reduced in size so as to fit through the holes immediately below the cutting blade. The resulting slurry has a solids content of about 4%. Pieces of metal and other nonbreakable materials are ejected from the pulper through an opening on the side of the tub, washed, and put through a ferrous recovery system. The slurry can be centrifuged to remove the organics [4].

Figure 3-20. Cutaway view of a wet-pulper. (Courtesy of Black-Clawson, Inc.)

ROLL CRUSHING

Roll crushers are used in resource recovery operations for the purpose of brittle materials such as glass while merely flattening ductile materials such as metal cans—hence allowing for subsequent separation by screening. Roll crushers were first employed in resource recovery for the reclamation of materials from incinerator residue, and have recently found use in processing partially source separated refuse comprised of glass containers and aluminum and steel cans. If this mixture is crushed, the glass can be readily removed with screens, the steel cans with magnets, and the remainder would be aluminum cans.

Roll crushers work by capturing and forcing the feed through two rollers operating in opposite directions, exactly like the wringers on older washing machines. The first objective of roll crushing is to capture the pieces that are to be crushed. This capture depends on the size and characteristics of the particles, and the size, gap, and characteristics of the

Roll Crushing

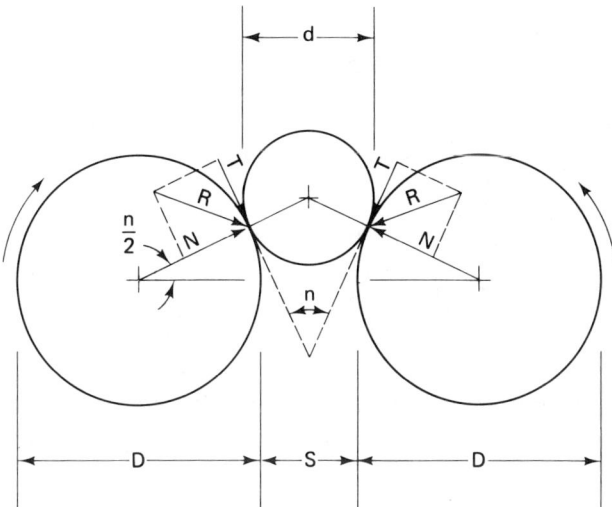

Figure 3-21. Roll crusher: definition of terms.

rollers. To illustrate the importance of this capture, imagine attempting to force a basketball through a clothes wringer. A small rubber ball, on the other hand, could be captured readily and flattened.

The variables involved in the analysis of roll crushing are shown on Fig. 3-21. The diameters of the two rolls are D while the diameter of the particle to be crushed is d. The normal force between the particle and the rollers is N, and the tangential force is T. If the resultant force R is pointed downward, the particle will be captured and crushed. If it points upward, the particle will ride on the rollers.

The vertical component of N is

$$N_v = N \sin \frac{n}{2}$$

where n is the angle between the two tangential forces and $n/2$ is the angle between the horizontal and the line connecting the centers of the feed particle and the roller. Similarly, the vertical component of the tangential force is

$$T_v = T \cos \frac{n}{2}$$

At the point where crushing is possible,

$$N_v = T_v$$

$$N \sin \frac{n}{2} = T \cos \frac{n}{2}$$

or

$$\tan\frac{n}{2} = \frac{T}{N}$$

In this instance, the angle n becomes known as the "angle of nip."

Since T/N = the coefficient of friction (ϕ), the necessary condition for crushing to take place is

$$\tan\frac{n}{2} \leq \phi$$

From Fig. 3-21 it can be shown that

$$\frac{S}{2} + \frac{D}{2} = \frac{D}{2} + \frac{d}{2}\cos\frac{n}{2}$$

where S is the separation between the rollers. It follows that

$$\cos\frac{n}{2} = \frac{D + S}{D + d}$$

Example 3-6

It is intended to crush pieces of glass of nominal diameter 5 cm (feed) to maximum diameter 0.5 cm (product). The coefficient of friction between steel and glass is 0.2. Find the diameter of the rollers that will just capture and crush this material.

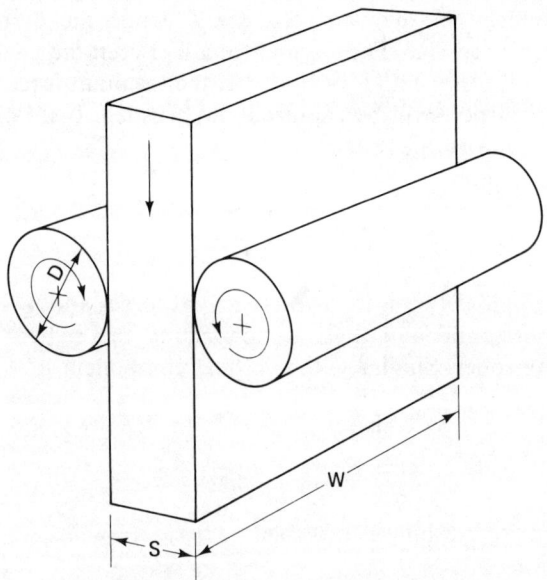

Figure 3-22. Capacity of a roll crusher.

$$\tan \frac{n}{2} = \phi = 0.2$$

$$\frac{n}{2} = 11.3$$

$$\cos \frac{n}{2} = 0.785 = \frac{D+S}{D+d} = \frac{D+0.5}{D+5}$$

$$D = 15.9 \text{ cm}$$

The capacity of a roll crusher can be estimated as the maximum volume squeezed through the rollers (Fig. 3-22):

$$C = kMDWS\rho$$

where C = capacity, tonnes/h
  k = dimensional constant = 60 if all dimensional units are as below
  M = speed of rollers, rpm
  D = diameter of rollers, m
  W = length of rollers, m
  S = separation, m
  ρ = density of material, g/cm^3

Example 3-7

For the rollers in Example 3-6, find the capacity if $M = 60$ rpm, $W = 0.5$ m, and $\rho = 2.5$ g/cm^3.

$$\begin{aligned} C &= kMDWS\rho \\ &= (60)(60)(0.159)(0.5)(0.005)(2.5) \\ &= 3.6 \text{ tonnes/h} \end{aligned}$$

PARTING SHOTS

People who have had extensive experience in the storage and handling of refuse are confirmed cynics. A new piece of equipment, no matter how highly touted by the manufacturer, is received with a wry, knowing smile. The question always is: "Yes, but will it work with refuse?"

Too many times they have seen the unsuccessful transfer of technology and hardware from another application. Always the problem is the same—solid waste is a heterogeneous and unpredictable material, and equipment designed for a simpler feed cannot handle refuse. With an increasing need for refuse processing, we will see more and more equipment specifically designed for refuse and proven in the field using the real stuff. Only then will these cynics be convinced.

REFERENCES

[1] Chemical Engineering, *Deskbook, Solids Separations*, Feb. 15, 1971.

[2] HICKMAN, W. B., "Storage and Retrieval of Prepared Refuse," *Proc. ASME Natl. Waste Process Conf.*, Boston, 1976.

[3] ROGERS, H. W., and S. J. HITTE, "Solid Waste Shredding and Shredder Selection," EPA OSWMP SW-140, Washington, D.C. 1974.

[4] RESNICK, W., "Flow Visualization Inside Storage Equipment," *Proc., Int. Conf. Bulk Solids—Storage, Handling and Flow*, Powder Advisory Centre, London, 1976.

[5] *Planning and Specifying a Refuse Shredding System*, Allis-Chalmers, Appleton, Wis.

[6] TAGGART, A. F., *Handbook of Mineral Dressing*, John Wiley & Sons, Inc., 1945.

[7] MITCHELL, J. R., "Designing for Batch and Continuous Weighers," *Chem. Eng.*, Feb. 28, 1977, p. 177.

[8] MARTIN, G., C. E. BLYTH, and H. TONGUE, *Trans. Brit. Cer. Soc. 23*, p. 61 1923 (as reported in Ref. 29).

[9] TANZER, E. K., "Pneumatic Conveying for Incineration of Paper Trim," *Proc. ASME Natl. Incin. Conf.*, New York, 1968.

[10] MADISON, R. D. (Ed.), *Fan Engineering*, Buffalo Forge Company, Buffalo, N.Y., 1948.

[11] BATES, L., "Performance Features of Helical Screw Equipment," *Proc. Int. Conf. Bulk Solids—Storage, Handling, and Flow*, Powder Advisory Centre, London, 1977.

[12] RUF, J. A., "Particle Size Spectrum and Compressibility of Raw and Shredded Municipal Solid Waste," Ph.D. thesis, University of Florida, 1974.

[13] American Public Works Association, *High Pressure Compaction and Baling of Solid Waste*, U.S. EPA-OSWMP SW-32d, Washington, D.C., 1972.

[14] HAM, R. K., "The Role of Shredded Refuse in Landfilling," *Waste Age*, **6** (12), 22 (1975).

[15] REINHARD, J. J., and R. K. HAM, *Solid Waste Milling and Disposal on Land without Cover*, EPA (NTIS PB-234 930 and PB-234 931), Washington, D.C., 1974.

[16] *Baling Solid Waste to Conserve Sanitary Landfill Space*, U.S. EPA-OSWMP, Washington, D.C., 1974.

[17] RATCLIFFE, A., "Crushing and Grinding," *Chem. Eng.* July 10, 1972, pp. 62–75.

[18] A. M. KINNEY, INC., *Franklin, Ohio's Solid Waste Disposal and Fiber Recovery Demonstration Plant*, U.S. EPA OSWMP SW-47d.1, NTIS PB-234 715, Washington, D.C., 1974.

[19] ITO, K., and Y. HIRAYAMA, "Semi-wet Selective Pulverizing System," *Resourc. Recovery Conserv*, **1**, 45–53 (1975).

[20] TREZEK, G. J., *Significance of Size Reduction in Solid Waste Management*, U.S. EPA-600/2-77-131, Cincinnati, Ohio, 1977.

[21] GAWALPANCHI, R. R., P. M. BERTHOUEX, and R. K. HAM, "Particle Size Distribution of Milled Refuse," *Waste Age*, 1973, pp. 34–45.

[22] TREZEK, G. J., D. M. OBENG, and G. SAVAGE, *Size Reduction in Solid Waste Processing*, U.S. EPA, Grant No. R801218, Second Year Progress Report, 1972–73.

[23] AUSTIN, L. G., "Introduction to the Mathematical Description of Grinding as a Rate Process," *Power Technol.*, **5**, 1 (1971–72).

[24] GAUDIN, A. M., "An Investigation of Crushing Phenomena," *Trans. AIME*, **73**, 253 (1926).

[25] ROSIN, P., and E. RAMMLER, "Laws Covering the Fineness of Powdered Coal," *J. Inst. Fuel*, **7**, 29–36 (1933).

[26] TREZEK, G. J., and G. SAVAGE, "Report on a Comprehensive Refuse Comminution Study," *Waste Age*, July 1975, pp. 49–55.

[27] STRATTON, F. E., and H. ALTER, "Application of Bond Theory to Solid Waste Shredding," *J. Environ. Eng. Div. ASCE*, **104** (EE1) (1978).

[28] GAUDIN, A. M., and T. P. MELOY, "Model and a Comminution Distribution Equation for Single Fracture," *Trans. AIME*, **223**, 40–43 (1962).

[29] EVANS, I., and C. D. POMEROY, *The Strength, Fracture, and Workability of Coal*, Pergamon Press, Inc., Elmsford, N.Y., 1966.

[30] HARRIS, C. C., "The Application of Size Distribution Equations to Multi-Event Process," *Trans. AIME*, **244**, 187–190 (1969).

[31] BENNET, J. G., "Broken Coal," *J. Inst. Fuel*, **10**, 22 (1936).

[32] EPSTEIN, B, "Logarithmico-Normal Distribution in Breakage of Solids," *Ind. Eng. Chem.*, **40**, 2289 (1948).

[33] BROADBENT, S. R., and T. C. CALLCOTT, "Coal Breakage Processes," *J. Inst. Fuel*, **29**, 524 (1956), and **30**, 13 (1957).

[34] GARDNER, R. P., and L. G. AUSTIN, "The Use of Radioactive Tracer Technique and a Computer in the Study of the Batch Grinding of Coal," *J. Inst. Fuel*, **35**, 174 (1962).

[35] COULSON, J. M., and J. F. RICHARDSON, *Chemical Engineering*, Pergamon Press, Inc., Elmsford, N.Y., 1955.

[36] DIAZ, L. F., "Three Key Factors in Refuse Size Reduction," *Resour. Recovery Conserv.*, **1**, 111–113 (1975).

[37] BOND, F. C., "The Third Theory of Comminution," *Trans. AIME*, **193**, 484 (1952).

[38] BOND, F. C., "Confirmation of the Third Theory," *Trans. AIME*, **217**, 139 (1960).

[39] PERRY, J. H. (Ed.), *Chemical Engineering Handbook*, McGraw-Hill Book Company, New York, 1963.

[40] DROBNY, N. L., H. E. HULL, and R. F. TESTIN, *Recovery and Utilization of Municipal Solid Waste*, U.S.EPA-OSWMP SW-10c, Washington, D.C., 1971.

[41] DIAZ, L. F., et al., "Health Aspect Considerations Associated with Resource Recovery," *Compost Sci.*, Summer 1976, pp. 18–24.

[42] FINAL REPORT ON A DEMONSTRATION PROJECT AT MADISON, WISCONSIN, *1966–1972*, U.S. EPS-OSWMP, Washington, D.C., 1973.

[43] FRANCONERI, P., "Selection Factors in Evaluating Large Solid Waste Shredders," *Proc. ASME Natl. Waste Process. Conf.*, Boston, 1976.

[44] ROBINSON, W. D., "Shredding Systems for Mixed Municipal and Industrial Solid Waste," *Proc. ASME Natl. Waste Process. Conf.*, Boston, 1976.

PROBLEMS

3-1. A pneumatic tube is to move wood chips, with a maximum diameter of 1 in. Estimate the air velocity required.

3-2. Using Fig. 3-11, construct a series of bar graphs showing the progression of refuse composition with particle size. Select particle sizes of 0.05, 0.1, 0.2, 0.5, 1.0, 2.0, and 5.0 in.

3-3. Using the data for Middleburg VT shown in Fig. 1-5, calculate the characteristic size, x_0, and the Rosin–Rammler constant n. Using the calculated value of n, estimate x_0 from the expression $x_0 = (x_{90})/(2.3^{1/n})$. Compare and comment.

3-4. Using an appropriate canned computer program, confirm the π selection function for secondary shredding using the Broadbent–Callcott model (Fig. 3-10).

3-5. Calculate the shredder energy requirements for shredding the feed to the product as shown in Fig. 3-9. Use the Bond work index method.

3-6. Calculate the energy requirements for shredding the feed to the product as shown in Fig. 3-9. Use Fig. 3-19.

3-7. It is required to roll aluminum beer cans in a roller mill to a maximum thickness of 0.5 cm. Calculate the size of rollers (steel) required.

3-8. The theoretical maximum capacity of a single lead screw conveyor is to be estimated, and the dimensions are measured as pitch = 20 in., radius to conveyor tip = 18 in., and radius to conveyor hub = 6 in. The speed is measured as 30 rpm. How much will the final answer be sensitive to a 10% error in measurement for each of the variables?

4 MECHANICAL SEPARATION

This chapter is devoted to mechanical (physical) means of separating selected components from mixed municipal refuse. Except for the handpicking operation and some types of screening, it is assumed that the refuse has already been shredded and the materials are in discrete, homogeneous pieces. This is obviously an optimistic assumption, and the application of any of the techniques discussed herein is limited by the fact that shredding does not always successfully break multimaterial products into the individual materials.*

Although the unit processes discussed in this chapter are listed as being mechanical, it should be obvious that chemical considerations are also often involved. Further, the processes can be combined with other separation techniques, such as electromagnetics (Chapter 5), to achieve higher performance. It is, therefore, not suggested that these processes are and should be purely mechanical. They are grouped together in this chapter solely out of a need for logical organization.

All the separation devices included in this chapter (and others that follow) are based on a principle of coding and separation. Some property of the material is used as a recognition code, such as magnetic/nonmagnetic or large/small. At times, the material has to be further coded, such as

---

*The common steel beer can, for example, has an aluminum top, a tin coating, lead/tin solder on the seam, plastic on the inside, and lacquer on the outside. A shredded can is not separate bits of steel, aluminum, tin, and so on.

in froth flotation when conditioning agents are added which allow for the selective attachment of air bubbles to achieve a light/heavy code.

Once coded, the material is separated according to that code. In froth flotation, for example, the separation occurs by simple flotation of some of the materials to the top of the slurry and the sinking of others. In screening, the small pieces fall through the screen and the large ones are retained on the screen. This idea should be kept in mind as an almost philosophical principle of material segregation. First, there must be a recognizable code to differentiate the materials in question, and then this code must be used in a separation device.

Most of the present techniques are binary coding/separation devices, with a few (e.g., jigs) producing more than two classes of product. The objective of research and development in the resource recovery field is to eventually develop coding and separation devices that will distinguish a tomato from an aluminum can full of water and will recognize both as different from a plastic bottle, then separate them accordingly.

GENERAL EXPRESSIONS FOR MATERIALS SEPARATION

In separating various pure materials from a mixture, the separation can be either binary (two output streams) or polynary (more than two output streams). For example, a magnet capturing ferrous material is a binary device, whereas a screen with a series of different sized holes, producing several products, is a polynary separation device.

Binary Separators

Schematics of binary and polynary separators are shown in Figure 4-1. In the case of the binary device, the input stream is composed of a mixture of x and y, and these are to be separated.

The mass per time (e.g., tons/hour) of x and y fed to the separator is x_0 and y_0 respectively, the mass per time of x and y exiting in the first output stream is x_1 and y_1, and the second output stream is x_2 and y_2.

Assume that the device was intended to separate the x into the first output stream and y into the second. If the separator was totally effective, then all of x would go to the first output, and all of y to the second. In practice, this is seldom achievable, and the first stream is contaminated with some y, and the second with some x. The effectiveness of the separation can then be expressed in terms of recovery. The recovery of component x in the first output stream is $R_{(x_1)}$, defined as:

$$\text{Recovery} = R_{(x_1)} = (x_1/x_0)100$$

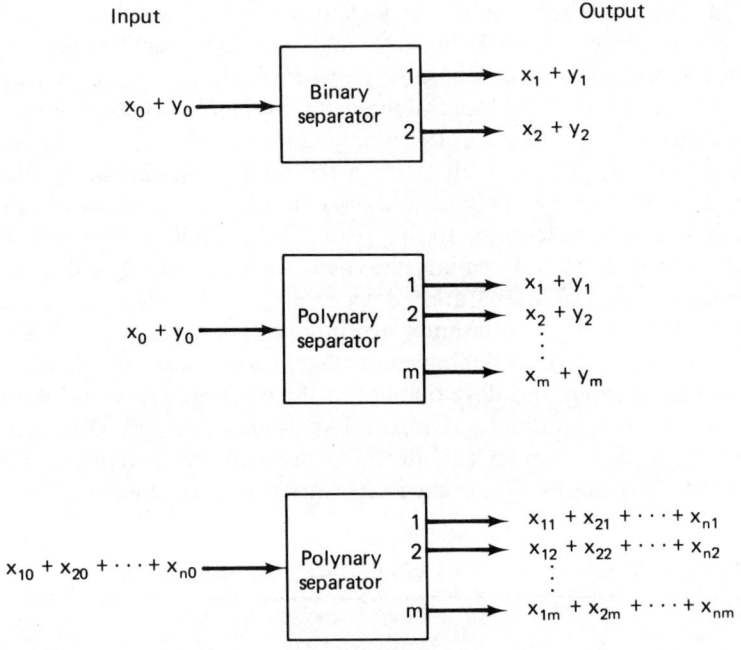

Figure 4-1. Binary and polynary separators.

where recovery is expressed as a percentage, and x_1 and x_0 are in terms of mass/time. Similarly, the recovery of y in the second output stream is expressed as

$$R_{(y_2)} = (y_2/y_0)100$$

Incidentally, since the mass balance holds;

$$x_0 = x_1 + x_2$$

then

$$R_{(x_1)} = [(x_0 - x_2)/(x_1 + x_2)]100$$

It is usually not sufficient to judge a separator only by its recovery. Consider for a moment what would happen if we ran the binary separator in such a way as to achieve $x_2 = y_2 = 0$. In other words, all of the feed is exited as output number one. It is obvious that although the recovery of x was 100%, the device is not performing its desired function, since no separation occurs. A second operational parameter is therefore required, and this is usually an expression of purity, stated as

$$P_{(x_1)} = [x_1/(x_1 + y_1)]100$$

where $P_{(x_1)}$ is the purity of the first output stream in terms of x, expressed as percent. Similarliy, the purity of the second output stream in terms of y is

$$P_{(y_2)} = [y_2/(x_2 + y_2)]100$$

Usually, both purity and recovery are needed for a complete and accurate description of binary separation performance. In some cases, however, this is not true, such as in screening. If the recovery of fines is to be measured, then the recovery term is an adequate description of the process. The purity would have to be 100%, since the product has to fit through the holes in the screen and no oversized particles are possible in the product.

At times, the input and output streams are more conveniently expressed in terms of concentrations instead of mass/time. The equations for calculating the effectiveness of the separation can in this case be shown to be the following:

$$R_{(x_1)} = \frac{[x_1]([x_0] - [x_2])100}{[x_0]([x_1] - [x_2])}$$

where $[x_1]$ is the concentration of x_1 in output stream 1, $[x_0]$ is the concentration of x in the input stream, and $[x_2]$ is the concentration of x in output stream 2, with all concentrations expressed as percentages. A similar expression can be written for component y (with 1 and 2 subscripts reversed, of course). The purity of x would be

$$P_{(x_1)} = \frac{[x_1]\rho_x}{[x_1]\rho_x + [y_1]\rho_y}$$

where ρ_x and ρ_y are the densities of x and y respectively.

Often a binary separator is designed to extract one type of material from a waste stream. For example, a magnet draws off ferrous materials as the desired output. Such an output is often called the *product* or *extract* and the second output is the *reject*. Literally, the magnet extracts the ferrous materials and rejects the rest. In subsequent discussions of materials separation in this text, one output is referred to as the extract and the other the reject.

Polynary separators

Two types of polynary systems are possible, as shown in Figure 4-1. In the first case, x_0 and y_0 are the two components in the feed, and the separator has more than two output streams, with x and y appearing in all of them, but in different amounts. In such a system, the recovery of x in the first output stream is

$$R_{(x_1)} = \left(\frac{x_1}{x_0}\right)100$$

as before, where x is in mass/time units. Similarly, the purity of x in the first output stream is

$$P_{(x_1)} = \left(\frac{x_1}{x_1 + y_1}\right)100$$

The recovery of x in the mth output stream is

$$R_{(x_m)} = \left(\frac{x_m}{x_0}\right)100$$

The second type of polynary separator is the most general case where the feed contains n components ($x_{10}, x_{20}, x_{30} \ldots x_{n0}$) and these are to be separated into m outputs. The notation in Fig. 4-1 is, at output 1, that x_{11} is the x_1 that ended up in the first output, x_{21} is x_2 that ended up in the first output, and so on. The recovery of x_1 in the first output is thus

$$R_{(x_{11})} = \left(\frac{x_{11}}{x_{10}}\right)100$$

where x_{11} is the x_1 that ended up in the first output and x_{10} is the x_1 in the feed. The purity of this stream in terms of x_1 is

$$P_{(x_1)} = \left(\frac{x_{11}}{x_{11} + x_{21} + \ldots x_{n1}}\right)100$$

with all x terms having unit of mass/time.

Another performance parameter sometimes used in literature, but not recommended, is overall recovery, defined as

$$OR_{(x,y)} = \left(\frac{x_1 + y_1}{x_0 + y_0}\right)100$$

This parameter is useful only for process design, such as sizing conveyor belts. Since this term is not a measure of separation effectiveness (i.e. one can attain 100% overall recovery by simply bypassing or turning off the separator), it should not be used in describing the operation of materials separation.

Efficiency of separation

Because of the inconvenience and difficulty in using two measures of separation effectiveness (recovery and purity) in order to adequately define the operation of a materials separator, some effort has been directed

General Expressions for Materials Separation

toward the development of a single-value parameter. Rietema [1] reviewed these efforts and suggested a measure of efficiency. Stated for a binary separation with input of x_0 and y_0, Rietema defined efficiency as

$$E_{(x,y)} = 100 \left| \frac{x_1}{x_0} - \frac{y_1}{y_0} \right| = 100 \left| \frac{x_2}{x_0} - \frac{y_2}{y_0} \right|$$

Another means of obtaining a single value of binary separator performance, developed by Worrell [2], is to multiply the fraction of x in the first output stream by the fraction of y in the second output stream, or

$$E_{(x,y)} = \left(\frac{x_1}{x_0} \right)\left(\frac{y_2}{y_0} \right) 100$$

Both Rietema's and Worrell's definitions of efficiency are rational in that if perfect separation occurs (all of x_0 goes to the first output, so that $x_1 = x_0$, and $y_2 = y_0$), the efficiency is 100%. Likewise, if no separation occurs ($x_0 = x_1$ and $y_0 = y_1$), then both efficiencies are zero.

Example 4-1.

A binary separator has a feed rate of 1 tonne/h. It is operated so that during any 1 hour, 600 kg reports as output 1 and 400 kg as output 2. Of the 600 kg, the x constituent is 550 kg, while 70 kg of x ends up in output 2. Calculate the recoveries and the single valued efficiencies by the methods discussed above.

The recovery of x in the first output is

$$R_{(x_1)} = \left(\frac{x_1}{x_0} \right) 100 = \frac{(550) \times 100}{550 + 70} = 88\%$$

The purity of this output stream is

$$P_{(x_1)} = \left(\frac{x_1}{x_1 + y_1} \right) 100 = \frac{500 \times 100}{600} = 92\%$$

Using Rietema's definition of efficiency,

$$E_{(x,y)} = 100 \left| \frac{x_1}{x_0} - \frac{y_1}{y_0} \right| = 100 \left| \frac{500}{620} - \frac{50}{380} \right| = 76\%$$

and according to Worrell's definition of efficiency

$$E_{(x,y)} = 100 \left(\frac{550}{620} \right)\left(\frac{330}{400} \right) = 76\%$$

Note that the last two statements yield quite similar values and either one seems to be a useful and representative single value parameter for binary separator efficiency.

HAND SORTING

The simplest device for the separation of materials from waste, and historically the first, is the *hand sort*. Ever since civilization began, scavengers have been an integral part of society. Selectively accepting other people's waste, collecting and processing it, and selling it at a profit is a time-honored profession and, in recent times, quite a profitable one.

The first hand-sorting facility in the United States was built by Colonel Waring for New York City in 1898 [4]. The refuse from 116,000 people was sorted, and over 2 1/2 years of operation about 37% of the refuse was recovered. The recovered material yielded an income of about $1 per ton, with a composition as shown in Table 4-1. The income from this and other plants was not sufficient to maintain them, however, and the job of scavengering was given back to private entrepreneurs, who paid the city about $1 per ton for the privilege.

In modern society scavengers are discouraged because of health considerations and the potential for accidents, and the now-common practice of covering collected waste with earth. The trade lives, however, in the person of the "pickers" at resource recovery facilities.

Pickers (or more properly "hand sorters") have two major functions. First, they recover any items of value that need not be processed. Commonly, corrugated cardboard, bundles of newspaper, and large pieces of metal (reinforcing bars, etc.) are recovered by the pickers. Their second function is to remove all those items which could cause damage to the rest

TABLE 4-1
Material Sorted by Hand at the New York City Refuse Processing Facility, 1898–1901

Material	Percentage by Weight
Paper, six grades	74.5
Rags, clothing and twine	12.2
Carpets	3.3
Bottles	2.5
Metals: iron, brass, lead, and zinc	2.1
Tins	1.4
Leather: shoes and scrap	1.9
Rubber: shoes, hose, and mats	0.2
Whole barrels	1.4
Miscellaneous	0.5
Total	100.0

Source: Ref. 4.

of the processing system, such as explosives, as discussed in the previous chapter.

People seldom consider problems of refuse processing when such items are discarded, no more than when all manner of waste, such as pesticides and other toxins, are poured down the drain. Somehow, the public views the disappearance of a waste, whether down the drain or into a rubbish bin, as being the solution of the disposal problem, and no further thought is given to its ultimate fate.

At times the functions of hand pickers (salvage and protection) are combined. For example, one large processing facility recovered a piece of titanium 60 cm in diameter and 10 cm thick off the conveyor belt leading to the shredder. Not only is this a valuable piece of metal, but it would have completely destroyed the inside of the shredder.

The coding and separation functions in hand sorting are simple to define. The material is recognized visually (coding) by such properties as color, reflectivity, and opacity, verified by sensing its density, and removed (separated) by hand picking. Hand sorting is usually done on the conveyor belt leading to the first mechanical process, commonly the shredder. As the belt is loaded and the material is leveled out for the feeding into the shredder, the pickers stand on either side of the conveyor belt and remove the selected materials. Experience has shown that pickers can salvage about 0.5 tonne/h/person. A picking belt should be no more than 60 cm (24 in.) wide for one-sided picking, or 90 to 120 cm (36 to 48 in.) for pickers on both sides, and should not move faster than 9 m/min (30 to 40 ft/min), depending on the number of sorters [5].

If at all possible, the sorting operation should be done in daylight. Artificial light, especially fluorescent bulbs, give off a narrow band of light and this makes identification (coding) of the various components difficult. Large skylights should be installed if outside operation is impractical [6].

Hand sorting is a dirty and dangerous profession, not recommended for the squeamish.

SCREENS

Screening is a process of separation by size. A screen of uniform-sized apertures allows smaller particles to pass while rejecting the larger fraction. A particle can pass a screen if it is smaller than the opening in at least two dimensions.

Screening in resource recovery operations has been used commonly toward the end of a series of unit operations and is intended primarily for glass removal, since glass would by then have been crushed to fine

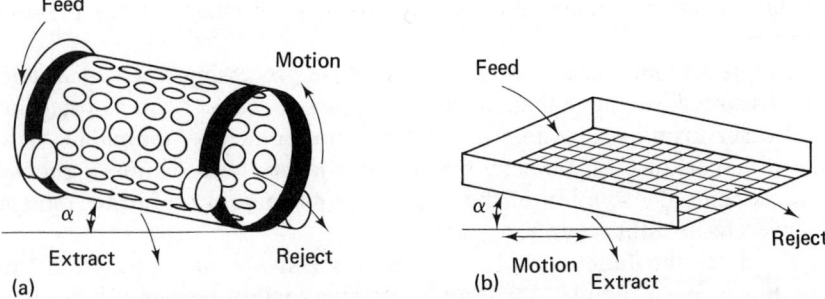

Figure 4-2. Schematics of trommel and reciprocating screens.

particles. Screens have also been used for reclaiming a high-organic (garbage) fraction from shredded waste [7] and as initial rough sorters at the beginning of materials recovery facilities [8]. The breakage and removal of much of the glass in a primary screening has been found to be highly beneficial in reducing the wear on downstream shredders.

By far the most popular screen for processing municipal refuse is the revolving screen or the trommel, which is an inclined cylinder, mounted on rollers, with holes in the side, as shown in Fig. 4-2. The drums roll at slow speeds, 10 to 15 rpm, thus using very little power. Their main advantage is the resistance to clogging. Some of the material within the screen might tend to hang on, but will eventually drop off.

The second type of screen widely used is an inclined or horizontal shaking screen. This screen, however, is readily plugged by rags, paper, and other objects, and is limited in its application to cleaner feeds. One application of such a screen has been the removal of the small pieces of glass so as to produce uniformly sized pieces that might be color-sorted [9]. The fundamental parts of an inclined reciprocating screen are also shown in Fig. 4-2.

Screens, like other separation devices, cannot be expected to attain 100% recovery. In other words, some undersized material (smaller than the screen apertures) will report as reject* and not be removed as extract. The recovery of a screen is expressed as

$$R = \frac{x_1}{x_0} \times 100$$

where R = screen recovery, %
y = amount of undersize in feed (by weight)
x = amount of undersize passing through the apertures and reporting as extract (by weight)

It is assumed that all of the oversize material entering as feed will in fact report as reject, not falling through the apertures, which is only logical,

---

*In screening, a *reject* does not pass through the holes, while the *extract* does pass through the holes.

since "oversize" is defined by screening. By our earlier definition of binary separation, if the extract from output 1 is x_1, then $y_1 = 0$, and the above definition of recovery is adequate to describe screen operation. In other words, the purity of the extract is always 100%.

Theoretically, it is possible to achieve very high recoveries with screening, but only at the cost of limiting the throughput. As a result, most screens are operated between 85% and 95% recovery [10].

Theory of Operation

Almost no data from controlled tests on the screening of shredded or raw MSW have been reported in the literature, and there is no theory that might be applied confidently to refuse screening. The only available theory of trommel screening comes from ore-dressing applications, and a theory of reciprocal screening is available from the chemical process industries. These analyses are discussed below.

Trommel screens

The *trommel screen* works by allowing the refuse in the screen to tumble around until the smaller pieces find themselves next to the apertures and fall through. The tumbling motion may be of two kinds, as shown in Fig. 4-3:

1. *Cascading*: where the charge is lifted up by the circular motion of the screen and then tumbles down on top of the layer heading upward.
2. *Cataracting*: where the speed of the screen is sufficiently great to actually fling the material into the air, where it will drop along a parabolic trajectory back to the bottom of the screen.

Under cataracting, the material is achieving the greatest turbulence and the machine should achieve the greatest efficiency. Obviously, as the drum speed is increased further, some critical speed is attained where a third type of motion is attained—*centrifuging*. In this case, the material adheres to the drum and never drops off, resulting in low recovery.

With reference to Fig. 4-4, consider a particle p in contact with the inside of the screen. The centrifugal force acting to press it against the inside wall is c and the w_1, a component of the gravitational force w acts to pull it away. The angle between the vertical and the line op is α_1, and $w_1 = w \cos \alpha_1$. If $c > w_1$, the particle remains in contact with the screen. However, if $w_1 > c$, the particle will fall off. If c remains greater than w_1 as α_1 decreases to zero (particle at the top), the particle of course never does drop off but remains on the wall through the rotation. At the point of separation, $c = w \cos \alpha$, where α is the angle at which separation occurs.

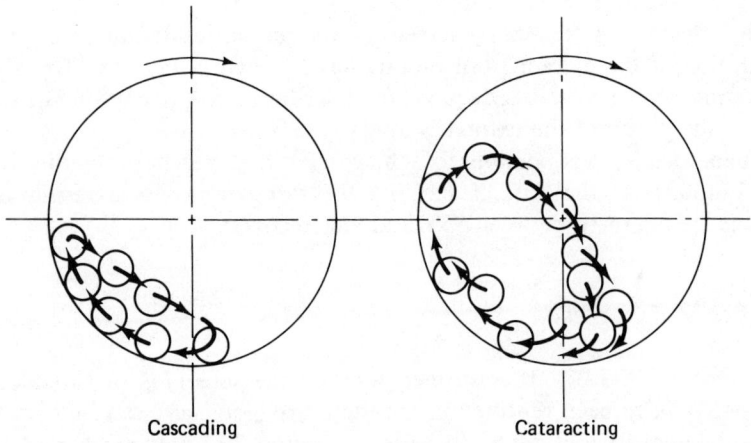

Figure 4-3. Two types of particle paths in a trommel screen.

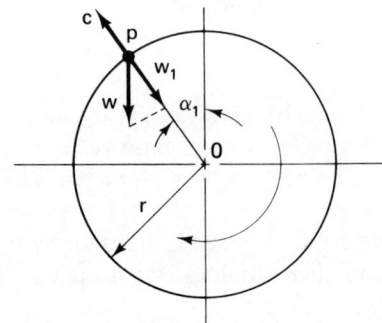

Figure 4-4. Definition of terms for trommel screen analysis.

The centrifugal force is

$$c = \frac{w}{g}(r\omega^2)$$

where ω = rotational velocity, rad/s
r = radius, cm
g = acceleration due to gravity, cm/s^2

Combining the two equations,

$$w \cos \alpha = \frac{w}{g}(r\omega^2)$$

$$\cos \alpha = \frac{r\omega^2}{g}$$

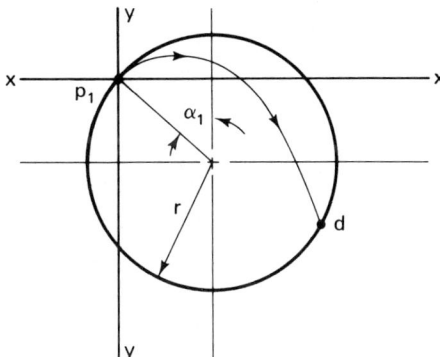

Figure 4-5. Flight path of a particle leaving the inside wall of a trommel screen.

and since $\omega = 2\pi n$, where n is the speed of the screen in revolutions per second,

$$\cos \alpha = \frac{4\pi^2 r n^2}{g}$$

From this relationship, it is clear that the angle α at which the particle leaves the wall of the screen and begins its free flight varies with both r and n. The critical point is of course at $\alpha = 0$, or $\cos \alpha = 1$, so that

$$n_c = \sqrt{\frac{g}{4\pi^2 r}}$$

If $w_1 > c$ at α, the particle will lose contact with the wall and begin its flight in a parabolic path until it once again hits the screen (Fig. 4-5). The equation of the parabola at the origin p_1 is

$$y = x \tan \alpha - \frac{gx^2}{2V_1^2 \cos^2 \alpha}$$

where V_1 = initial velocity of the particle p_1 as it leaves the wall
x, y = coordinates

The equation of the circular path of the screen is

$$x^2 + y^2 = (2r \sin \alpha)x + (2r \cos \alpha)y = 0$$

The simultaneous solution of these equations gives the coordinates of the point d, where these two curves intersect. This point is at the coordinates

$$x = 4r \sin \alpha \cos^2 \alpha$$

$$y = -4r \sin^2 \alpha \cos \alpha$$

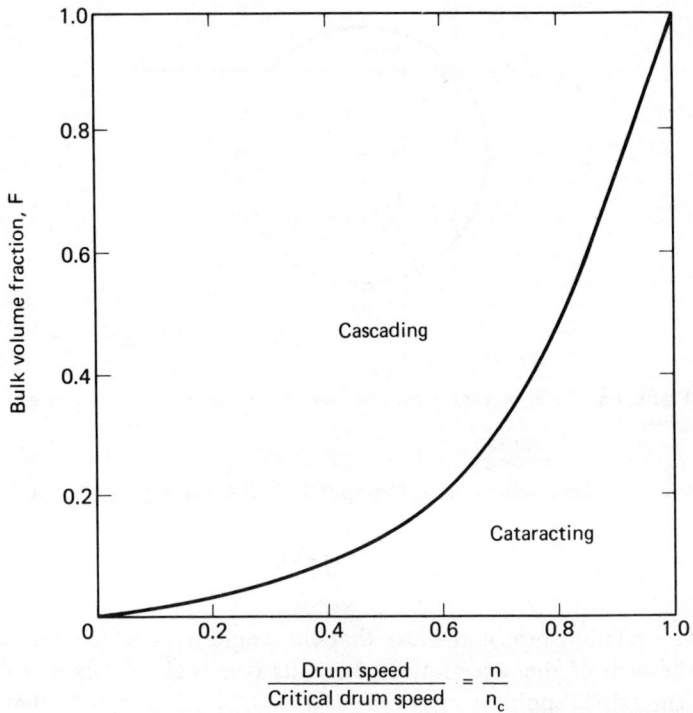

Figure 4-6. Effect of volume occupied by solids on trommel flow characteristics. [*Note*: F = (volume occupied by solids + volume of air between the solids)/(volume of drum).] (From Ref. 11.)

The development above is for a single particle within a screen. This is, of course, an unrealistic assumption. The motion of particles in a rotating drum when hindered by other particles has been analyzed by Rose and Sullivan [11].

The critical speed and the fraction of the screen occupied by the refuse is related as shown in Fig. 4-6. The plot is in terms of bulk volume, defined as the fraction of the drum occupied by the solids and the air spaces between the solids. Note that if the screen is totally full ($F = 1.0$) only cascading is possible for speeds less than critical (there is no space for the particles to fall back through the air). At lower volume fractions, cataracting becomes possible at speeds lower than critical. At the limit, with only one particle in the screen, cataracting occurs at even low speeds since there is no particle–particle interference.

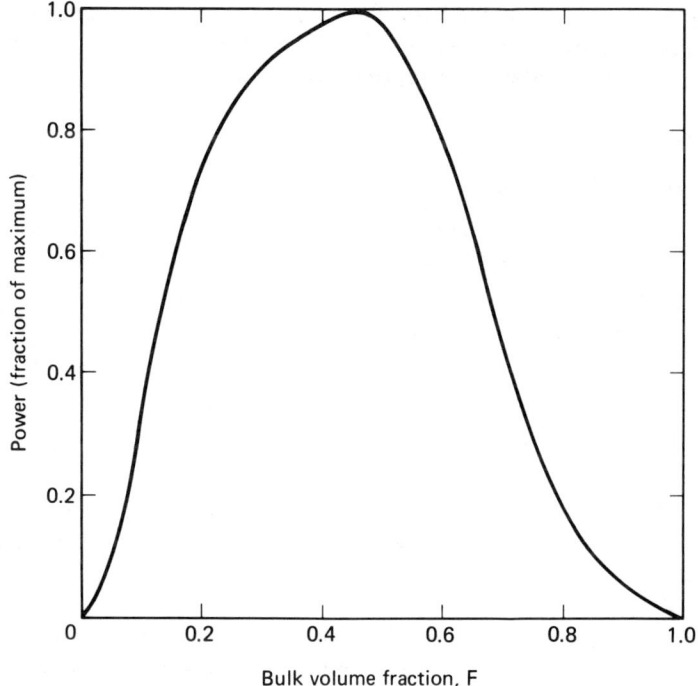

Figure 4-7. Power requirements of a trommel screen as a function of load. (From Ref. 11.)

The horsepower requirements for a screen can be estimated on the basis of ball mill analyses [12] as

$$\text{hp} = Wr^{3/2}\left[0.004467\frac{1-K^3}{(1+K^2)^{1/8}} - 0.0037\frac{1-K^5}{(1+K^2)^{3/8}}\right.$$

$$\left. + 0.00088\frac{1-K^7}{(1+K^2)^{5/8}}\right]$$

where W = weight of the charge in the screen, lb
r = radius of screen, ft
$K \simeq -0.024 + 0.39\sqrt{7-10F}$
F = fractional part of entire screen volume occupied by the charge, defined as bulk volume fraction

This equation was developed for the most efficient ball mill operational speed and must still be proven applicable to MSW. The effect of the bulk fraction on power is illustrated graphically in Fig. 4-7.

Example 4-2

Assume a 2.7-m-diameter trommel. Calculate the critical speed.

$$n_c = \sqrt{\frac{g}{4\pi^2 r}}$$

$$= \left[\frac{980}{4(3.14)^2(270)}\right]^{\frac{1}{2}} = 0.30 \text{ rotation/s}$$

$$= 18 \text{ rpm}$$

The recovery obviously varies with the speed. Alter and Crawford [13], using a 9-ft-diameter (2.7-m) trommel screen found that the best speed was about 45% of the critical. The rule of thumb is that recovery is greatest at a speed where the load rides one-third of the distance to the top of the screen [14].

Another operating variable, not considered above, is the slope of the drum. Within limits, as the slope is increased, the solids retention time is decreased, and the percent of product recovered is decreased because the particles have less chance of finding a hole through which to drop. Punched plate screens used in coal treatment [14] are usually set at 5°. Where trommel screens have different-sized holes along the drum, the slope affects the amount of material within each size range captured. Figure 4-8 shows some laboratory results with shredded MSW. The trommel for these tests had a section of $\frac{1}{2}$-in. holes, followed by 1-in. holes [15]. The recovery was measured as the percent of particles less than 1 in. captured. These data for underloaded conditions show that recovery drops off rapidly with angle of incline. A small slope, however, would have a low throughput, and a proper balance between recovery and throughput must be achieved.

Alter and Crawford [13] suggest that for unshredded MSW, optimum trommel performance can be obtained if the solids retention time is at least 25 to 30 and the material makes 5 to 6 revolutions within the drum. Another author suggests a retention time of between 30 s and 1 min [16].

The Bureau of Mines pilot plant obtained 95 to 100% recovery of $\frac{3}{4}$-in. shredded refuse at a rate of 2 tons/h. At 2.5 tons/h, the recovery dropped to 91%. The 4-ft-diameter, 6-ft-long trommel rotated at 18 rpm [17].

The diameter of a drum is an important design parameter but little information is available on the effect of this variable. For ore processing,

Screens

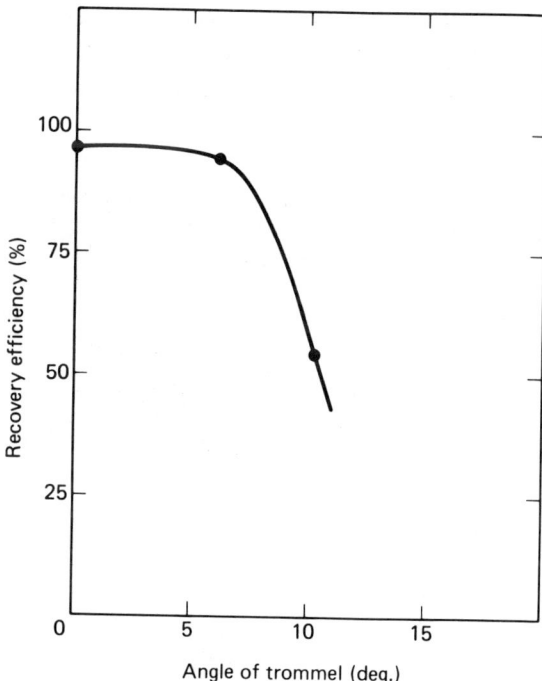

Figure 4-8. Trommel screening of MSW. Angle of the trommel affects efficiency of solids recovery. (From Ref. 15.)

the following formula has been suggested [14]:

$$D = 7.66\sqrt{\frac{Q}{\gamma}}$$

where D = drum diameter, in.
 Q = capacity, tons/h
 γ = specific gravity of material fed

The length of the screen is also important, because a longer screen gives more complete separation. On the other hand, most screening occurs within the first two feet of a screen, and a longer screen requires greater power.

The capacity of a screen is an ambiguous term since it must be related to screen efficiency if it is to have much meaning. For ore processing, a suggested trommel capacity is stated as "0.5 ton per 24 hours per square foot of area per millimeter aperture." Even ignoring the tortured mixture of English and metric units, the capacity has no meaning unless it specifies the efficiency, or the percent of particles accepted of all those which could conceivably fit through the holes, given sufficient time. Trommel capacities for MSW have not been reported.

Reciprocating screens

Experimental work on sieving of particles through standard engineering sieves [18] has shown that the curve of the percentage passing (by weight) vs. residence time on a vibrating screen can be divided into two regions. Region 1 plots as a straight line on log-log paper (Fig. 4-9) and can be described by the simple relationship

$$C = at^b$$

where C = fraction of all the particles which could pass through the sieve that have passed through the sieve, by weight
a = sieving rate constant
b = constant
t = time

The region 2 sieving area is described by the relationship

$$\frac{dN}{dt} = kN$$

where N = number of particles on the sieve which could pass through
k = constant

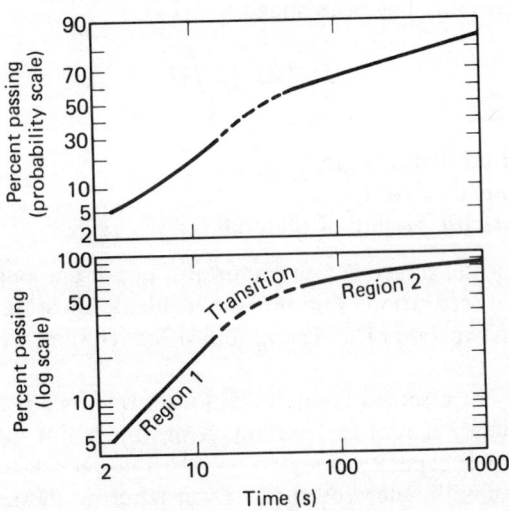

Figure 4-9. Typical curves for vibrating screens. (From Ref. 18.)

Screens

Using empirical relationships and experimental data, Whitby [18] derives relationships for the ideal steady-state reciprocating screen.

An alternative approach [10] to obtaining the required screen area is by using the empirical relationship

$$A = \frac{Q}{C_u} F$$

where A = screen area, ft^2
Q = screen loading, tons/h
C_u = unit capacity (tons/h)/ft^2
F = product correction factor

Obviously, the term F incorporates a number of variables. For some standard materials, it has been found that F can be calculated by multiplying the correction factor for particle-size distribution, moisture content, fraction of screen that is open area, and other practical factors. Table 4-2 is a summary of the factors. The unit capacity is obtained from Fig. 4-10.

TABLE 4-2
Summary of Product Correction Factors for Screens

Factor	Data	Refer to:	Comment
F_f	% half-size	Table 4-3	As % half-size increases, screening becomes more difficult.
F_o	% oversize	Table 4-3	With a larger fraction of oversize, the screen is clogged up and the capacity decreases.
F_e	% efficiency desired	Table 4-3	Obviously, as % efficiency desired increases, required area increases.
F_d	Number of decks	Table 4-4	Lower decks are not as efficient as upper decks.
F_m	Size of opening and moisture	Table 4-5	For smaller openings, a maximum moisture content is permissible.
F_a	% open area	—	—
F_w	Specific weight of material	—	Heavier particles are easier to sieve.

Source: Ref. 10

TABLE 4-3
Fines, Oversize, and Efficiency Factors for Screens

Percent	Fines, F_f	Oversize, F_o	Recovery, F_e
0	0.44		
10	0.55	1.05	
20	0.70	1.01	
30	0.80	0.98	
40	1.00	0.95	
50	1.20	0.90	
60	1.40	0.86	
70	1.80	0.80	
80	2.20	0.70	1.75
85	2.50	0.64	1.50
90	3.00	0.55	1.25
95	3.75	0.40	1.00

Source: Ref. 10.

TABLE 4-4
Deck Factors for Screens

Number of Decks	F_d
Top	1.0
Second	0.9
Third	0.75

Source: Ref. 10.

TABLE 4-5
Limiting Moisture for Screening

Opening size square (in.)	Limiting moisture (%)
$\frac{1}{16}$	1
$\frac{1}{8}$	1
$\frac{3}{16}$	2
$\frac{5}{16}$	4
$\frac{3}{8}$	4
$\frac{1}{2}$	6
$\frac{3}{4}$	6
1	6
$1\frac{1}{2}$ to 2	No limit
Over 2	No limit

Source: Ref. 10.

Screens

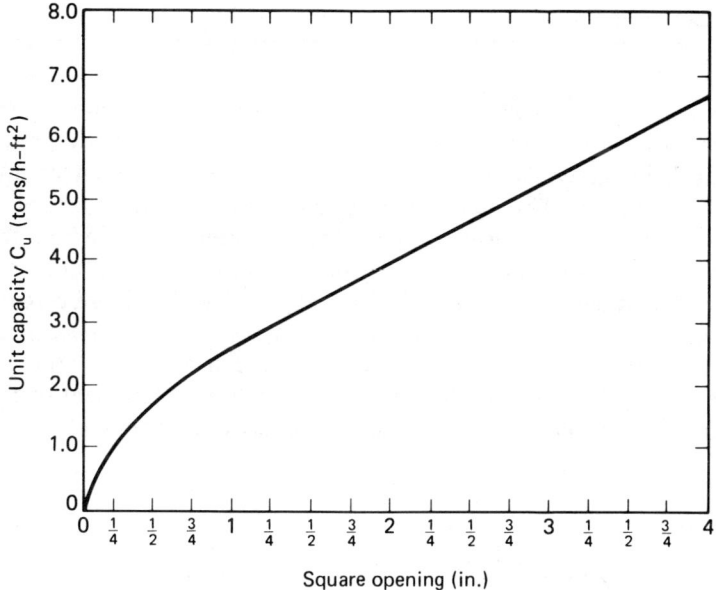

Figure 4-10. Unit capacity of a vibrating screen. (From Ref. 10.)

Example 4-3

A feed rate of 25 tons/h of dry glass is to be screened on a single deck screen at a recovery of 90% through $\frac{1}{2}$-in. screen openings. The percent less than half size is 30% and the percent oversize is 20%. The percent open area is 80%. Find the required area.

$$Q = 25 \text{ tons/h}$$
$$C_u = 2 \text{ tons/h-ft}^2$$
$$F = F_f \times F_o \times F_e \times F_d \times F_a \times F_w$$
$$F_f = 0.80 \text{ from Table 4-3}$$
$$F_o = 1.01 \text{ from Table 4-3}$$
$$F_e = 1.25 \text{ from Table 4-3}$$
$$F_d = 1.0 \text{ from Table 4-4}$$
$$F_a = 0.80$$
$$F_w = \frac{2.7 \times 64}{100} = 1.73$$

where 2.7 = glass specific gravity
 64 = specific weight of water, lb/ft^3

$$F = 0.8 \times 1.01 \times 1.25 \times 1.0 \times 0.8 \times 1.73$$
$$= 1.40$$
$$A = \frac{25}{2} \times 1.40 = 17.4 \text{ ft}^2$$

AIR CLASSIFIERS

The objective of *air classification* is to separate the light, mostly organic materials from the heavy, mostly inorganic fraction. The basic premise is that the light materials will be caught in an upward current of air and carried with the air, while the heavier fraction will drop down, unable to be supported by the air currents. The light fraction entrapped in the air stream must be separated from the air. Commonly, this is done with a cyclone, but it can be accomplished equally well with a large box or bag into which the particles drop while the exit air is filtered and escapes. The air can be either pushed or pulled, and the fan can be placed either before or after the cyclone. Except for smaller installations, placement of the fan so as to suck the material through the blades is not recommended, because

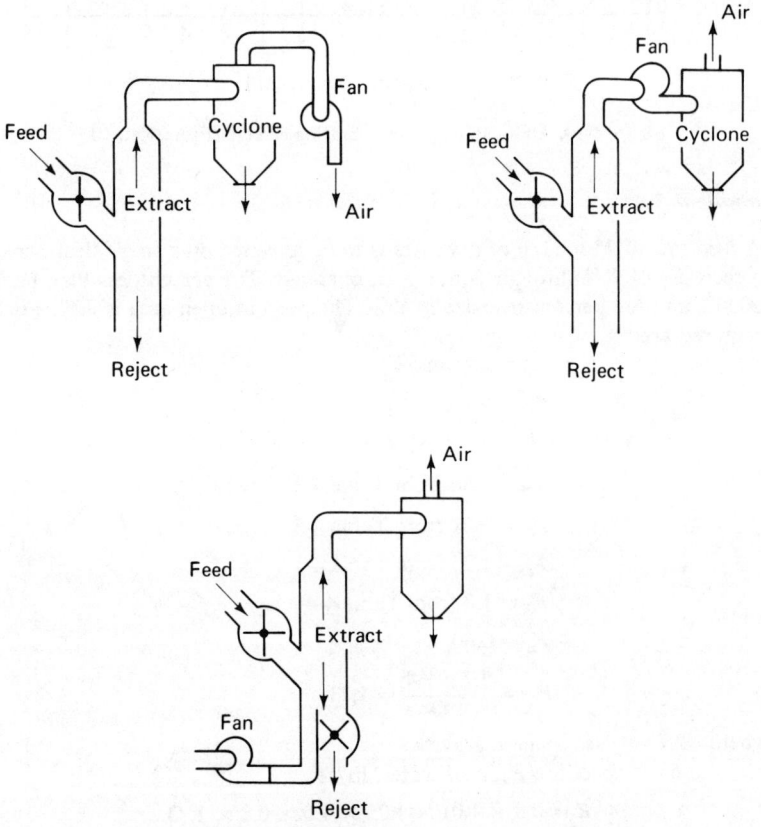

Figure 4-11. Three basic arrangements for air classification.

Air Classifiers

of the wear and tear suffered by the blades. The various arrangements for air classifiers are shown in Fig. 4-11 (see previous page).

Similar in function, but slightly different in construction, is a series of classifiers more properly labeled "air knives," where the air is blown horizontally through a vertically dropping feed. The aerodynamically light particles will be carried with the air stream while the heavy ones will have sufficient inertia to resist a change in direction and drop through the air stream. This technique also allows for separation into more than two categories, as shown by two typical arrangements in Fig. 4-12.

Considerable confusion exists in the terminology for the product (output) streams of air classifiers. Some authors call the material rising with the air stream the "light fraction" and the material falling the "heavy fraction." These terms are misleading, since they imply ideal separation. Better terms for air classifiers are "overflow" and "underflow," but these imply vertical geometry. In keeping with the terminology introduced previously, in this text the material suspended by and removed by the air stream is called the extract, and the material not so removed is the reject.

One of the main requirements for achieving effective separation of light and heavy materials in an air classifier is the creation of turbulence within the column and the development of large shear forces. Both turbulence and shear forces will tend to break up clumps and provide for cleaner separation.

This objective can be achieved in several ways, most notably by what is commonly called a *zigzag classifier*. In this device, the vertical flow column consists of a series of 90° or 60° turns, as illustrated in Fig. 4-13. Laboratory work with smoke tracers has demonstrated that at nominal air velocities (flow/cross-sectional area of tube) of 2.5 m/s (500 ft/min) a central air core is formed, with turbulent vortices at the corners which

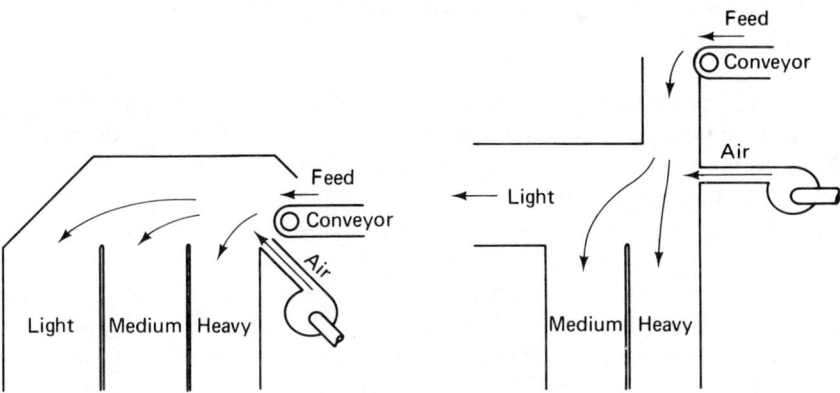

Figure 4-12. Air knife classifiers.

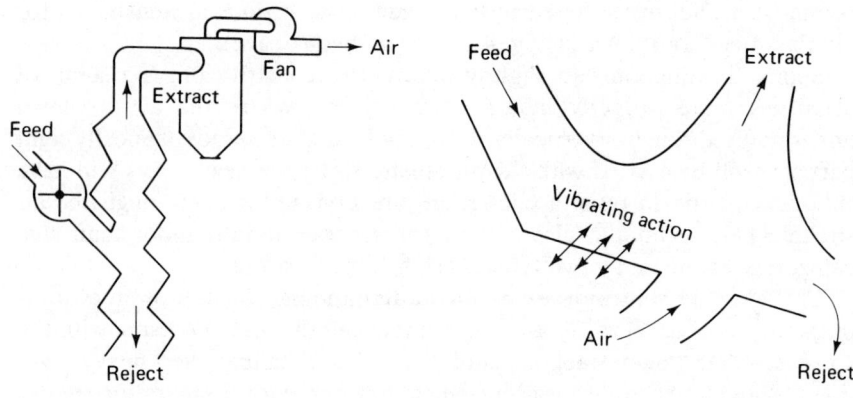

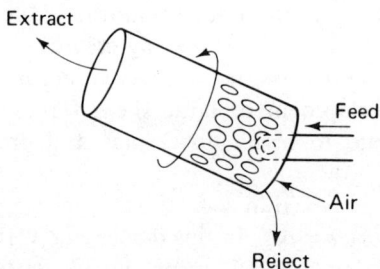

Figure 4-13. Zigzag air classifier, vibrating air classifier, and rotary air classifier.

touch the central core [19]. As the air velocity is increased to 3.5 m/s (750 ft/min), the corner vortices disappear altogether, indicating fully turbulent conditions. The fraction of material reporting as heavies and lights can obviously be altered by changing the flow characteristics in the classifiers.

Within the turbulent vortices, the clumps are broken up and the light particles are transferred to the upward air stream. The heavy particles drop from vortex to vortex until they exit at the bottom. The dropping action also helps break apart any agglomerated particles.

The heavy particles tend to slide down the lower sides of the throat until they are hit by the upward rush of air at the corner (see Fig. 4-14). If the downward velocity of the particle is sufficiently great relative to the air-stream velocity, the particle takes path C, as shown on the figure. This allows it to continue its downward movement on the wall of the next segment. If, however, the air-stream velocity is sufficiently great, the particle may be caught within the air stream and experience trajectory A.

Air Classifiers

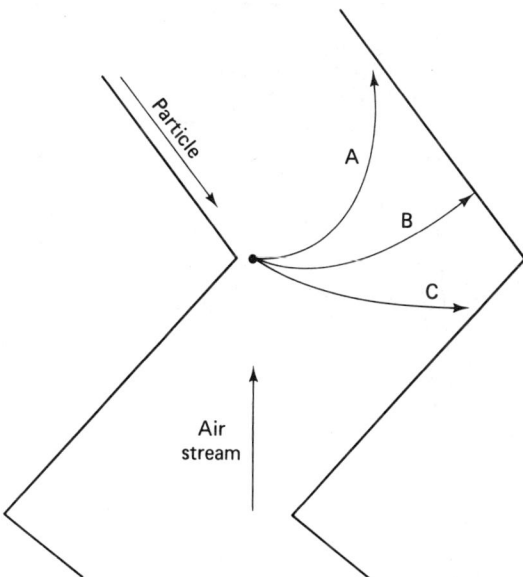

Figure 4-14. Particle motion in a zigzag classifier.

If the particle moves along trajectory B, it has a 50% chance of going up or down [20].

An apparent improvement on the zigzag classifier has been developed at the University of Utah [21]. The pockets for the vortices have been increased by removing the underside of the zigzag, as shown in Fig. 4-15.

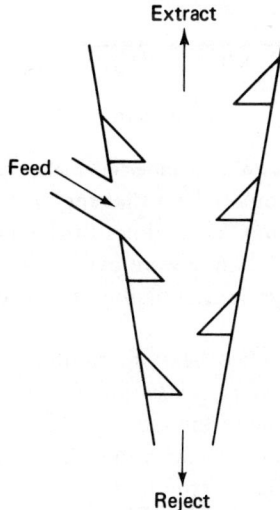

Figure 4-15. University of Utah air classifier.

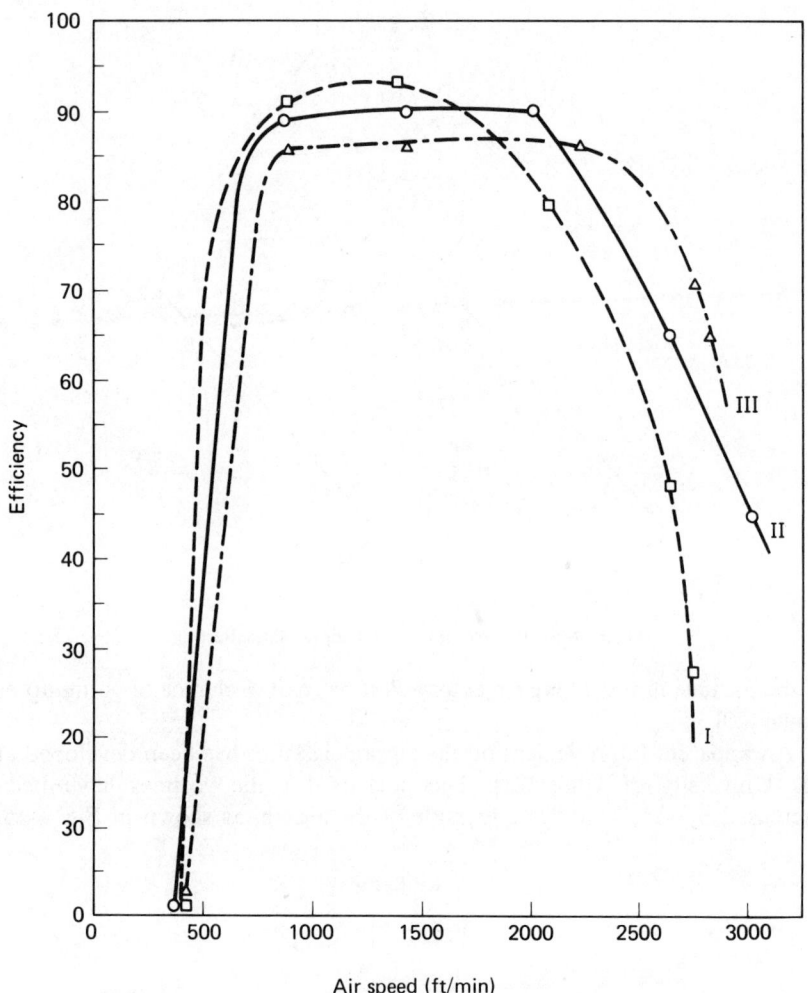

Figure 4-16. Efficiency of three different air classifiers. (From Ref. 22.)

In addition, the throat is tapered, so that the air velocity decreases toward the upper part of the column. This allows the heavy particles, which should have dropped out, but by chance were carried upward, the opportunity to return to the lower stages and possibly be discharged with the heavy fraction.

Two other types of air classifiers are pictured in Fig. 4-13, the vibrating air classifier and the rotary air classifier. *Vibrating air classifiers* combine the separation achieved due to the vibration with air entrainment. The feed vibrates along a sloping surface, with the light materials shaken to the top, where the air stream carries it around a U-shaped curve.

Air Classifiers

The *rotary air classifier* combines the action of a trommel screen (see page 142) and air entrainment. As the drum rotates, the aerodynamically light fraction is suspended in the air stream and carried up into a collection hopper. Small heavy solids fall through the holes, while the large heavy particles exit at the lower end of the drum.

Air classifier geometry can make a substantial difference in performance. Figure 4-16 shows the efficiency (defined here according to Worrell [2] as the product of the fraction of lights reporting to extract times the fraction of heavier reporting of reject) versus the air speed for the different classifiers of equal throat area but different throat geometry [2].

A useful technique for plotting air classifier test results is shown in Fig. 4-16. The data shown are for a simulated refuse, with realistic fractions of shredded paper, plastic, aluminum, and ferrous [22]. These curves can also be used for an economic analysis of air classifiers by estimating the market value of the classifier products at various flow rates [23]. If, for example, the extract is to be used as supplemental fuel, its Btu value is important, but if the total amount of inorganic contaminants is high (at high air flows), the Btu/lb drops, thus reducing the market value. On the other hand, at low air flows, the extract is clean (high Btu/lb), but not all of the organics are captured. The market value of the other recovered materials

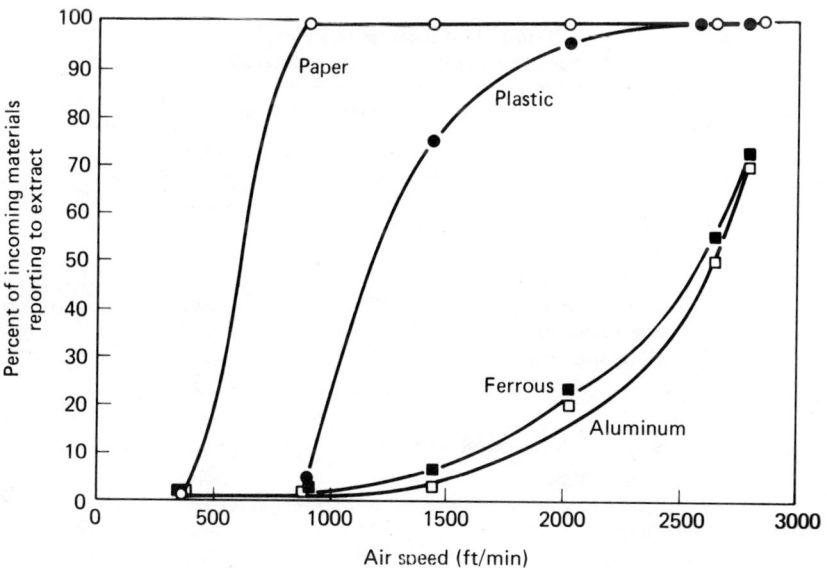

Figure 4-17. Materials reporting to light fraction at various air speeds. (From Ref. 22.)

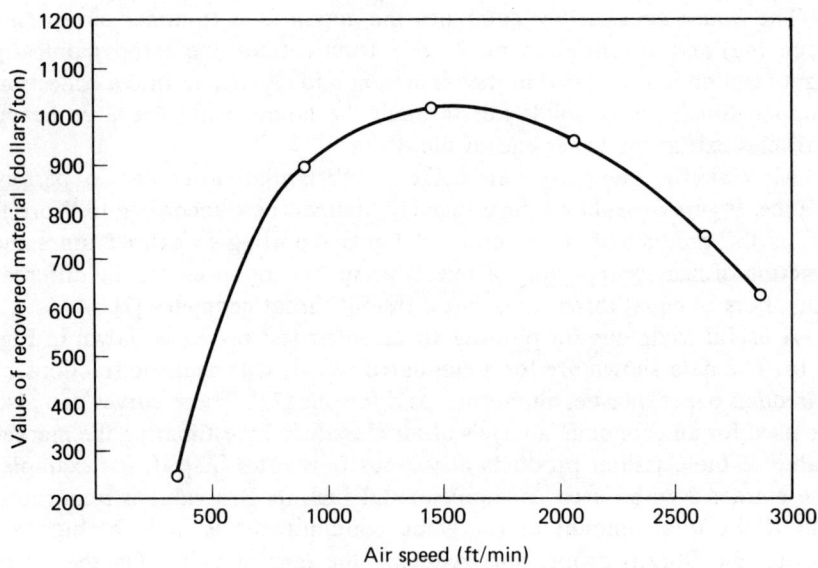

Figure 4-18. Value of light and heavy fractions from air classifiers as a function of air speed. (From Ref. 23.)

TABLE 4-6

Typical Air-Classification Results, for Shredded MSW, at 65% Overall Recovery

	Percent by weight	
Component	Shredded refuse (as received)	Extract (from air classifier)
Noncombustible		
Rocks and dirt	0	0
Ferrous metal	7.8	0.08
Nonferrous metal	1.0	0.05
Glass and other	7.8	1.82
	16.9	1.94
Combustible		
Paper	52.2	78.8
Wet garbage	11.8	0.1
Yard and garden	6.7	8.6
Other	12.2	10.6
	82.9	98.1

Source: Ref. 21.

Air Classifiers

can be similarly analyzed. Figure 4-18 is a plot showing the economics of air-classifier operation using a simulated shredded refuse as before. In real life, it would of course be advisable to operate at the point where the products have maximum value.

As an example of what air classification can do with real municipal refuse, typical data for the Utah University air classifier are shown in Table 4-6.

The fuel characteristics of the extract in Table 4-6 was found to be as shown in Table 4-7. Also tabulated are the fuel properties of two other fuels prepared from MSW by air classification. The first is fuel comprised of cubettes which were prepared from air-classified MSW, and the second is a proprietary product known as Eco-Fuel, which is air classifier extract chemically treated to make it biologically stable [24]. Additional information on the properties of refuse-derived fuel is found in Chapter 8.

TABLE 4-7
Fuel Characteristics of Air-Classified Extract

	(1) Air-Classified MSW	(2) Cubettes	(3) Eco-Fuel
Moisture (%)	15	15	10
Ash (%)	8.3	6	11
Higher heating value, Btu/lb	6930	6800	6900

Source: Column (1), Ref. 21; Column (2), Ref. 21; Column (3), Ref. 24.

TABLE 4-8
Effect of Particle Size on Recovery in an Air Classifier

	Sample		
	1	2	3
Percent moisture	27.0	32.1	24.3
Nominal particle size, in. (90% passing)	1.9	3.1	7.6
Paper and cardboard, percent recovered	92.3	90.4	92.8
Grit, percent recovered	74.8	87.7	86.2
Plastic, percent recovered	73.2	70.6	92.7
Percent heavies in extract	1.5	1.7	3.5
Overall recovery	86.3	85.3	84.0

Source: Ref. 25.

Paper and cardboard seem to be insensitive to particle size, whereas the heavy fraction is somewhat more sensitive, as shown in Table 4-8. Moisture content does not seem to influence greatly the recovery of lights, although a drop of perhaps 5% in recovery is expected when the moisture content doubles. [25]

Air classifier performance can also be expressed in terms of the air/solids ratio, as shown in Fig. 4-19. As the air flow is increased for a given solids feed rate, higher recoveries are experienced. Conversely, lower feed rates for a given air velocity enhances recovery [26].

The capacity of a classifier should be sufficient to withstand sudden surges. A capacity curve such as Fig. 4-20, in this case for a nominal capacity of 4 tons/h, is a valuable description of this attribute [25].

Although the effect of feed rate on air-classifier performance has not been fully studied, it seems reasonable that performance would deteriorate

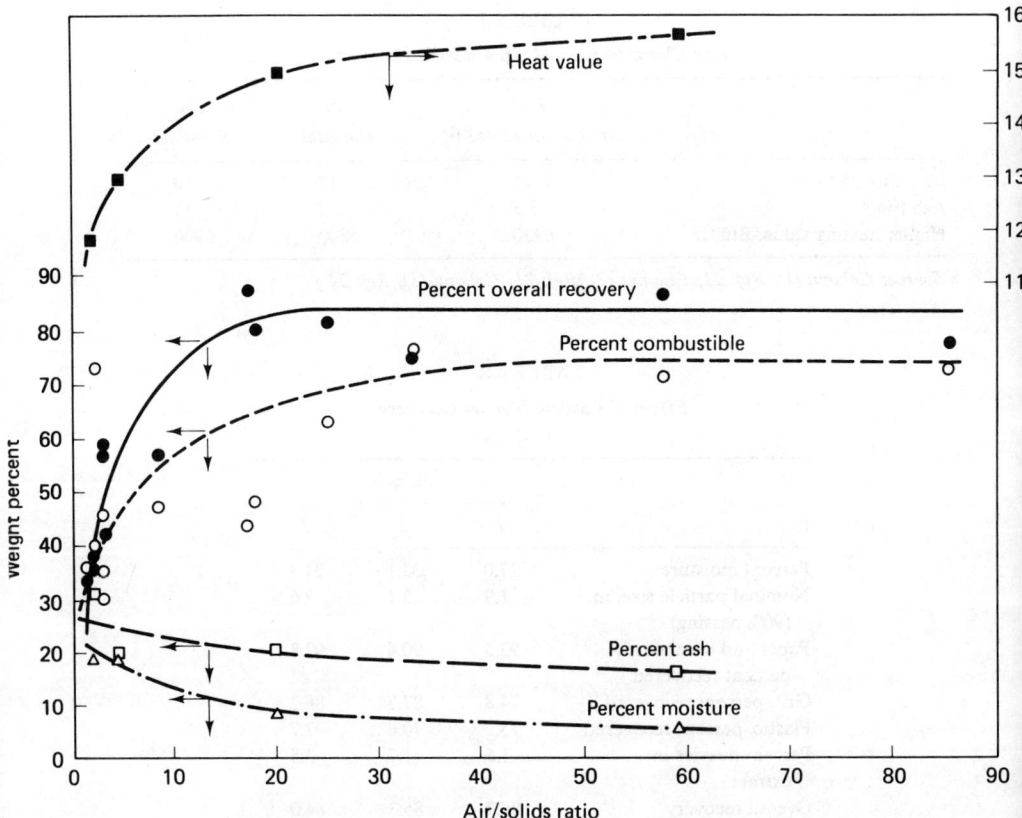

Figure 4-19. Analysis of air classifier overflow. (From Ref. 26.)

Air Classifiers

with feed rate. If a light particle is fed into the throat section, it must accelerate in order to attain the speed of the air stream. If heavier particles are present, these interfere with this acceleration and hence increase further the congestion within the throat. It is likely that the feed rate and average residence time of particles within the throat section are related as

$$t = \left(\frac{V}{Q}\right)^n$$

where t = average residence time of particles
 V = volume within throat section
 Q = feed flow rate
 n = factor that accounts for the congestion within the throat; $n > 1.0$

Theory of Operation

Air classification actually involves two independent unit operations: separation of the aerodynamically light from heavy particles, and the subsequent separation of the light fraction from the air stream. Each of these is considered below.

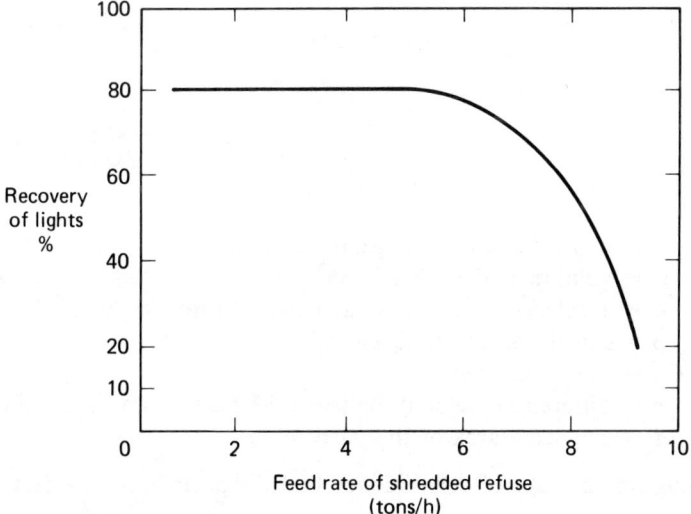

Figure 4-20. Deterioration of the recovery of lights with increased loading of a classifier. (From Ref. 25.)

Air classification

The motion of a particle suspended in a fluid such as an air stream is governed by three forces:

F_E = some external force such as gravity or centrifugal,
F_B = buoyant force,
F_D = drag force,

The motion of the particle is described as

$$\frac{dv}{dt} = (F_E - F_D)\frac{1}{\rho_s V}$$

where V = particle volume, m^3
ρ_s = particle density, kg/m^3
$\frac{dv}{dt}$ = particle acceleration, m/s^2

A particle falling in a fluid under gravity has two phases in its motion: acceleration and terminal velocity. The latter is attained when the three forces F_E, F_D, and F_B are balanced and the acceleration $dv/dt = 0$. Thus, during a steady fall (terminal velocity),

$$F_B = F_D + F_E$$

These three forces can be expressed as

$$F_E = \rho_s V a$$

$$F_B = \rho V a$$

$$F_D = \frac{C_D v^2 \rho A}{2}$$

where ρ_s = density of the solid particle, kg/m^3
V = volume of the particle, m^3
a = acceleration due to some external force, m/s^2
ρ = density of the fluid, kg/m^3
C_D = drag coefficient
v = differential velocity between the particle and the fluid, m/s
A = projected area of the particle, m^2

Assuming for the sake of convenience that the particle is a perfect sphere,

$$A = \frac{\pi d^2}{4} \quad \text{and} \quad v = \frac{\pi d^3}{6}$$

where d is the particle diameter, and assuming that the external force causing the acceleration is gravitational ($a = g$), we can reduce the equation to

$$v = \left[\frac{4(\rho_s - \rho)gd}{3C_D\rho} \right]^{1/2}$$

which is the well-known *Newton's law*. If we now assume laminar flow conditions, $C_D = 24/N_R$, where N_R = the Reynolds number = $vd\rho/\mu$, μ being the fluid viscosity. Substituting, we get the familiar *Stokes's law*:

$$v = \frac{d^2 g(\rho_s - \rho)}{18\mu}$$

This expression is not applicable for air classifiers, however, because the Reynolds number in air classification is about 10,000, placing the flow well into the turbulent flow regime.

Consideration of some of the terms in the Newtonian velocity equation above can yield some clues as to efficiency of air classification. The objective is to have a large variation in the terminal velocities (v) between the organic and inorganic fractions. Recognizing that the difference in densities and the particle diameter vary as the square root of v, it is logical to attempt classification with greater possible density differences and largest particle sizes. The former is, however, tempered by the differential uptake of moisture between organics and inorganics, and wet feed could, in fact, be detrimental to separation. The latter term, however, can be controlled, and it seems that the larger the particle sizes, the greater will be the efficiency of separation (assuming of course equal sizes, and that C_D is not affected).

In the air-classification operation, the difficulty in the application of Newton's law involves the selection of a "diameter" for particles that are irregularly shaped. One solution to this problem is to define an "aerodynamic diameter" which can be back-calculated using known velocities. Using this technique, the aerodynamic diameters for shredded MSW light fraction have been found to be about 40% of actual diameters as defined by screening [22].

Alternatively, it is possible to correct for the diameter by selecting a modified drag coefficient. Although C_D is approximately 1.0 for disk-shaped particles under ideal conditions, $C_D \simeq 2.5$ for refuse particles in air classification.

Example 4-4

Assuming that the drag coefficient is 2.5, calculate the air velocity necessary to suspend 2 cm (screened) particles of shredded aluminum. Note that $\rho_s = 2.70$ and $\rho = 0.0012$.

$$v = \left[\frac{4(2)(2.7 - 0.0012)980}{(3)(2.5)(0.0012)} \right]^{1/2}$$

$$= 1.53 \times 10^3 \text{ cm/s}$$

In some cases, instead of adjusting the drag coefficient or back calculating an aerodynamic diameter, it might be more convenient to define an "effective diameter," which might be the average dimension of the particle as it is presented to the air stream. Since

$$A = \frac{\pi d^2}{4}$$

the effective diameter, d', can be defined as

$$d' = \left(\frac{4A}{\pi}\right)^{1/2}$$

This may be reasonable where flat objects such as pieces of paper are suspended. In other cases, the effective diameter could equally well be defined by the volume as

$$d'' = \left(\frac{6V}{\pi}\right)^{1/3}$$

A further discussion of particle-size analysis is included in the Appendix.

Another problem with applying Newton's law to air classification is that the analysis assumes a single particle settling in an infinite fluid (no boundary conditions). This is obviously not correct, and some accommodation must be made for the problems of interparticle actions and the effect of the walls.

In the turbulent regime, the effect of the wall can be accounted for by a correction factor [27], given as

$$m = 1 - \left(\frac{r}{R}\right)^{3/2}$$

where r = radius of the sphere
R = radius of the tube
m = correction factor

This correction factor can be used to adjust the terminal velocity to take into account the effect of walls, so that

$$v = m\left[\frac{4d(\rho_s - \rho)g}{3C_D\rho}\right]^{1/2}$$

When the particles are sufficiently concentrated, they act as a body, with little interparticle movement. This can occur in air classification when a large slug of feed enters the throat section and is carried upward in mass with the air stream.

One of the earliest attempts (1926) to take the effect of particle concentration into account was to modify the Newton equation as [28]

$$v_c = K\left[\frac{4d(\rho_s - \rho_c)g}{3C_D\rho_c}\right]^{1/2}$$

where v_c = settling rate of a suspension (a mass of particles settling as one body)
ρ_c = density of the suspension
K = constant

Another approach [29] is to express the Newton's equation in terms of the voidage ε, or the fraction of the volume within the settling or flotation column filled by the fluid (in this case, air). The equation would then be

$$v_c = K\left[\frac{4d(\rho_s - \rho_c)g}{3C_D\rho_c}\right]^{1/2} \varepsilon$$

where ε is the voidage.

These preceding two equations are not totally acceptable because of some unsupportable assumptions. For example, the effective buoyancy forces acting on the particle are assumed to be dependent on the density of the suspension. This cannot be true for any single particle, or for a suspension of particles settling at a uniform rate.

A better approach [30] is to express the velocity of the suspension, v_c, relative to the velocity of a single particle, v, as defined by the Newton equation. Under laminar flow conditions (Reynolds number less than about 500, not found in air classification), it can be argued that

$$\frac{v_c}{v} = f_1\left(\varepsilon, \frac{d}{D}\right)$$

where D is the diameter of settling or flotation column. For turbulent conditions,

$$\frac{v_c}{v} = f_2\left(\frac{v d\rho}{\mu}, \varepsilon, \frac{d}{D}\right)$$

Experimental evidence suggests that for laminar flow conditions,

$$v_c = v\varepsilon^{[4.65 + 19.5d/D]}$$

For turbulent conditions, d/D is not significant, and

$$v_c = v\varepsilon^{4.65}$$

As an air-classifier throat section fills up with solids, the effective viscosity of the entire "fluid" changes. This concept has been used to develop a correction factor for situations where the solids concentration within a fluid is sufficiently high to approach "choke" conditions [31]. This factor is given as

$$F = \left[1 - \left(\frac{V_s}{V}\right)^{2/3}\right]\left(1 - \frac{V_s}{V}\right)\left(1 - \frac{2.5V_s}{V}\right)$$

where V_s = volume occupied by the solids in suspension
V = volume of vessel in which suspension takes place
F = correction factor

The terminal velocity of a particle would thus be calculated as

$$v = F\left[\frac{4d(\rho_s - \rho)g}{3C_D\rho}\right]^{1/2}$$

Finally, only rigid shapes have been considered thus far. Obviously, many of the materials in air classifiers are flexible and porous, and this fraction normally makes up the majority of the organic lights that must be suspended by the air stream.

Experimental data with parachutes, wind socks, and flags show that drag coefficients rarely exceed 0.3 for most clothlike materials [32]. It has also been found that drag increases exponentially as a function of fabric weight and linearly as a function of the aspect ratio (horizontal over vertical dimensions) [33].

Experiments with autorotating wings at Reynolds numbers of about 4000 may also provide some insight into the aerodynamics of air classifying shredded refuse [32]. It has been found [34] that drag coefficients of moving objects such as blades increase linearly with the logarithm of the Reynolds number, with the drag coefficient ranging between 1.0 and 1.3.

Recognizing the problems with a theoretical analysis for air classification, a number of empirical models have been suggested as being applicable specifically to air classifiers. Among the earliest is the Dallavalle model for a vertical air classifier:

$$v = \frac{13,300\gamma}{\gamma + 1} d^{0.57}$$

where v = velocity, ft/min
d = particle diameter, in
γ = particle specific weight, lb/ft^3

For horizontal air classifiers, the velocity to carry the entrained particles was proposed by Dallavalle:

$$v_H = \frac{6000\gamma}{\gamma + 1} d^{0.398}$$

where the terms are the same as above, except that v_H is now the necessary velocity for horizontal conveyance [14].

Using pure materials of various shapes, Sweeney [32] found that the terminal velocity could be calculated as

$$v = 1.9 + 0.092\rho_s + 5.8A$$

where v = terminal (falling) velocity, ft/s
  ρ_s = particle density, lb/ft^3
  A = particle area (e.g., for a plate, A = length $\times$ width)

The size limits to this equation are between 0.0625 and 1.00 in.^{2}, which is generally considerably smaller than shredded MSW.

Sweeney also showed that if the area function in the preceding equation were eliminated, there would be little loss in accuracy. In other words, the terminal velocity seems to be most significantly affected by density and only slightly by area and other variables (within the limits used in the experiments). The model, using only density as the independent variable, was found to be

$$v = 1.91\rho_s^{1/2}$$

The model showed reasonable agreement with steel, aluminum, paper, and balsa wood as the materials.

Another model, suggested but unverified [19], is

$$v = \frac{Q}{J\rho_m T}$$

where Q = feed rate to the air classifier, ton/h
  J = dimensionless constant
  ρ_m = bulk density, ton/m^3
  T = throat area of classifier, m^2

Using the dimensions above, v is in m/h. The bulk density of shredded refuse is difficult to evaluate. As defined in Chapter 1, bulk density of the material is its density in bulk, as opposed to particle (material) density. For shredded refuse, this presents a serious problem because refuse compacts readily, and thus the bulk density is changed. However, the advantage of the foregoing model is twofold: it is dimensionally correct, and it is an operational model, relating the air velocity to the feed rate.

It has been possible through the analysis of extensive air-classification experiences [35] to generalize the overall recovery efficiency (percent of feed reporting as light fraction) [36] as

$$\text{overall recovery (\%)} = \frac{v}{20}$$

where v is the air velocity in ft/min.

This model has its obvious weaknesses, such as being able to achieve higher than 100% recovery, but it represents a reasonable approximation for design purposes.

Fan has taken this a step further and developed a series of equations for calculating the moisture ash content and heating value as a function solely of air velocity [37]. His equations are

$$W = 80 + \frac{1520}{65.8 + 0.36v^{1.3}}$$

$$A = 100 - \frac{6270}{65.5 + 0.36v^{1.3}}$$

$$H = \frac{842.4}{65.8 + 0.36v^{1.3}}$$

where W = moisture content of the light fraction, %
A = ash content of the light fraction, %
H = heating value of the light fraction, 10^6 J/kg
v = air velocity, m/s

Henrikson [50], has developed a probabilistic model of air classification, based in part on work by Senden and Tels [20]. In Henrikson's model, the classifier is incrementalized into a number of elements or zones. Since the activity in an air classifier is to a large degree stochastic in nature, particle transition from one element to another can be expressed in terms of probability. The ultimate separation of the particles can then be expressed as a conditional probability; that is, conditional on where the particles have been prior to leaving the air classifier. The bottom element is labeled B and A is the top element, and these two are completely absorbing elements, which is to say that all particles entering them subsequently leave the air classifier. (See Fig. 4-21). Hence, $p_A = 1$, $1 - p_A = 0$, $P_B = 0$, and $1 - p_B = 1$, where p_i is the probability that a particle in element i will rise.

A set of simultaneous equations expressing particle transition in each element can be set up:

$$U_B(A) = 0$$
$$U_B(A - 1) = p_{A-1}U_B(A) + (1 - p_{A-1})U_B(A - 2)$$
$$U_B(A - 2) = p_{A-2}U_B(A - 1) + (1 - p_{A-2})U_B(A - 3)$$
$$\vdots$$
$$U_B(i) = p_i U_B(i + 1) + (1 - p_i)U_B(i - 1)$$
$$\vdots$$
$$U_B(1) = p_i U_B(2) + (1 - p_i)U_B(0)$$
$$U_B(B) = 1$$

Air Classifiers

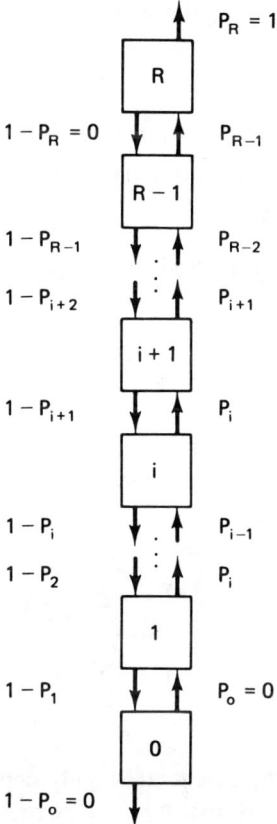

Figure 4-21. Definition of elements in Henrikson's air classification model.

where $U_B(i)$ = the conditional probability that a particle, starting in element i will end up in the bottom product.

Through a series of manipulations, it can be shown that since all particles start in the feed zone, labeled element V, only $U_B(V)$ is of importance, and this can be written as

$$U_B(V) = 1 - \left[\frac{1 - \sum_{m=1}^{V-1} \prod_{i=1}^{m} \left(\frac{1-p_i}{p_i} \right)}{1 + \sum_{m=1}^{A-1} \prod_{i=1}^{m} \left(\frac{1-p_i}{p_i} \right)} \right]$$

The top product of the separation (i.e. the extract) can be developed in a manner similar to that of the bottom product,

$$\begin{bmatrix} U_A(A) = 1 \\ U_A(A-1) = p_{A-1}U_A(A) + (1-p_{A-1})U_A(A-2) \\ U_A(A-2) = p_{A-2}U_A(A-1) + (1-p_{A-2})U_A(A-3) \\ \vdots \\ U_A(i) = p_i U_A(i+1) + (1-p_i)U_A(i-1) \\ \vdots \\ U_A(1) = p_1 U_A(2) + (1-p_1)U_A(0) \\ U_A(B) = 0 \end{bmatrix}$$

and similar manipulations ultimately yield:

$$U_A(V) = 1 - \left[\frac{1 - \sum_{m=1}^{V-1} \prod_{i=1}^{m}\left(\frac{1-p_i}{p_i}\right)}{1 + \sum_{m=1}^{A-1} \prod_{i=1}^{m}\left(\frac{1-p_i}{p_i}\right)} \right]$$

The transitional probabilities, p, are highly dependent on the superficial air velocity, and hence on the type of particles associated with their terminal fall velocities. They are also dependent on the feed rate which helps determine congestion and stochastic disturbances in the throat.

If the superficial air velocity is kept constant, p_i varies with the particle concentration in element i. We can consider each element as a completely mixed reactor vessel with constant concentration throughout, and p_i can be related to concentration as an error function, $p_i = \mathrm{erf}\left(\dfrac{kC_A}{C_i}\right)$

where $\mathrm{erf}(x)$ = error function of x
k = a constant
$\dfrac{C_A}{C_i}$ = the concentration in element i normalized by the concentration in the top (A) element.

The error function is convenient since it can express the probability range of zero to one. The constant k is assumed to be independent of element (stage within the classifier) and the material.

In terms of concentrations, the ultimate separation of the particles is expressed as:

$$U_B(V) = 1 - \frac{1 - \sum_{m=1}^{V-1} \prod_{i=1}^{m} \left[\frac{1 - \mathrm{erf}\left(k\frac{C_A}{C_i}\right)}{\mathrm{erf}\left(k\frac{C_A}{C_i}\right)} \right]}{1 + \sum_{m=1}^{A-1} \prod_{i=1}^{m} \left[\frac{1 - \mathrm{erf}\left(k\frac{C_A}{C_i}\right)}{\mathrm{erf}\left(k\frac{C_A}{C_i}\right)} \right]}$$

$$U_A(V) = 1 - \frac{1 - \sum_{m=1}^{V-1} \prod_{i=1}^{m} \left[\frac{1 - \mathrm{erf}\left(k\frac{C_A}{C_i}\right)}{\mathrm{erf}\left(k\frac{C_A}{C_i}\right)} \right]}{1 + \sum_{m=1}^{A-1} \prod_{i=1}^{m} \left[\frac{1 - \mathrm{erf}\left(k\frac{C_A}{C_i}\right)}{\mathrm{erf}\left(k\frac{C_A}{C_i}\right)} \right]}$$

As noted earlier, several assumptions used in this model are important. The two exit elements, A and B, are completely absorbing (i.e. once a particle reaches those elements, it has a probability of 1 certainty of leaving the classifier). Senden determined that this assumption is good for element A, and not so good for element B. [20] The assumption that k is a constant for all elements was tested by Henrikson, who found this to be true, and further that the value of k seems to be independent of the material processed.

The time history of the particles is not an important parameter in the transitional probability, p. More specifically, whether a particle arrives in the element in a descending manner or an ascending manner has no bearing or preference on whether it will subsequently descend or ascend from that element. Senden determined empirically that this is not necessarily a good assumption for low particle feed rates, whereupon he developed a one-step memory model between particles arriving in a descending or ascending manner. [20]

In conditions where concentrations within each element are high, there will be a large degree of exchange between adjacent elements. The probability that a descending particle will subsequently ascend will be greater than 0.5, and the probability that an ascending particle will continue to ascend will probably be less than 0.5. A descending particle will most likely be knocked back down.

The Henrikson model remains to be experimentally tested, but serves as a first step in the analytical evaluation of air classification.

Air-stream cleaning

The second unit operation in the total air-classification system is to remove the solid particles from the air stream once they have been classified from the heavies. This operation can be accomplished by using a large chamber, but such settling chambers are inadequate both from the standpoint of efficiency and because of the large space requirements. The most popular method of collection is the centrifugal separator or *cyclone*.

The operation of the cyclone is illustrated in Fig. 4-22. The air and solid particles enter the cyclone chamber at a tangent, setting up a high-velocity rotational air movement within the chamber. The solid particles, having greater mass, move outward toward the inside wall, are slowed down on

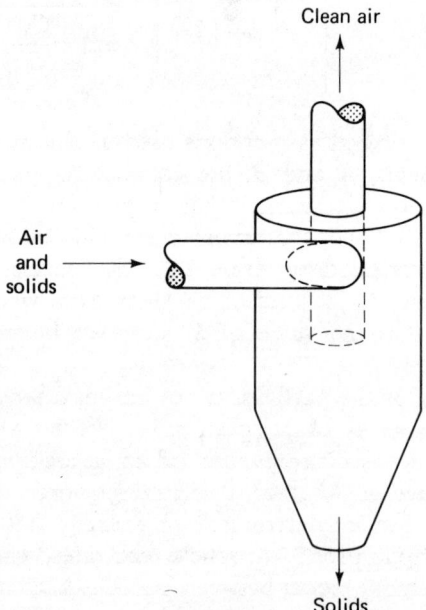

Figure 4-22. Cyclone used for air cleaning.

contact, and eventually drop to the bottom of the chamber under the influence of gravity. The air exits through the central tube, free of solids.

The objective of a cyclone is to move particles to the outside by centrifugal action. The radial velocity of a particle, v_R, is important, as illustrated in Fig. 4-23. Assuming laminar conditions (a bad assumption), the particle terminal radial velocity is governed by Stokes's law, except that the gravitational acceleration must be replaced by centrifugal acceleration, defined as

$$a = r\omega^2$$

where a = centrifugal acceleration, m/s^2
  r = radius, m
  ω = rotational velocity, rad/s

so that

$$v_R = \frac{d^2(\rho_s - \rho)\omega^2 r}{18\mu}$$

where d = particle diameter, m
  ρ = density of air, kg/m^3
  ρ_s = density of particle, kg/m^3
  μ = air viscosity, N-s/m^2

Recognizing that $r\omega^2 = v_{\text{tan}} r$, where v_{tan} is the tangential velocity, the radial velocity can be expressed as

$$v_R = \frac{v}{g} \frac{v_{\text{tan}}^2}{r}$$

which is the basic design equation for cyclones.

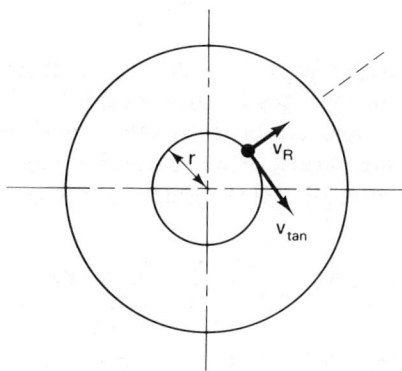

Figure 4-23. Radial movement of a particle within a cyclone.

This analysis may be somewhat simplified by assuming that $\rho_s \gg \rho$, Thus

$$v \simeq \frac{d^2 \rho_s g}{18\mu}$$

and

$$v_R \simeq \frac{d^2 \rho_s \omega^2 r}{18\mu}$$

At any radius r, the tangential velocity is found as

$$v_{\tan} = r\omega$$

$$\omega = \frac{v_{\tan}}{r}$$

where ω is the rotational velocity in rad/s. The time needed for one rotation, t, is thus

$$t = \frac{r}{2\pi v_{\tan}}$$

The distance traveled by a particle during one rotation* is

$$S \simeq v_R t \simeq \frac{d^2 \rho_s \omega^2 r}{18\mu} \frac{r}{2\pi v_{\tan}}$$

where S is the radial distance traveled by a particle during one rotation. Since $v_{\tan} = \omega r$,

$$S \simeq \frac{d^2 \rho_s \omega r}{36\mu\pi}$$

Experience with cyclones for the removal of shredded light fraction has shown that large factors of safety are needed. More than one installation has found itself knee-deep in confetti during initial startup.

While the design of cyclones for shredded MSW is still an art, design formulas have been developed for many other materials. For dust-removal operations, the limiting particle size that can be removed in a cyclone has been estimated by Feifel [38]. This formula, simplified and restated, reads

$$d = 0.7 C_0 \left(\frac{\mu}{\rho}\right)^{1/2} \left(\frac{S_G}{D\Delta P}\right)^{3/4} (Q)^{1/4}$$

*This is necessarily approximate, since the radius and hence the $v_{\tan}$ changes as the particle moves.

Jigs

where d = theoretical particle size, μ
- C_0 = constant relating to cyclone shape, experimentally found to vary between 2.7×10^3 and 4.2×10^3, dimensionless
- D = diameter of cyclone barrel section, ft
- ΔP = pressure drop, in. water
- S_G = specific weight of gas, lb/ft^3
- Q = gas flow rate, ft^3/min
- ρ = density of dust particle, g/cm^3
- μ = viscosity

Substitution of various theoretical limiting particle diameters and back-calculating the pressure drop allows for the comparison of this formula with laboratory results. It should be emphasized that the application of this formula, developed for dust removal, to the removal of large particles extracted by an air classifier is risky at best.

JIGS

A *jig* is a device that achieves the separation of light and heavy particles by using the differences in their abilities to penetrate a shaken bed. One very simple type of jig is the miner's pan used in gold mining days. This pan is filled with dirt and gravel from a stream bottom and shaken until the grains of gold penetrate the pan content and lodge on the bottom, appearing as the remainder of the pan contents is poured off. The modern jig works in essentially the same way.

Figure 4-24 shows a diagram of jig. The feed may be thought of as comprising a light fraction, a heavy fraction, and a middling (an intermediate mixture which contains contaminants as well as some light and heavy fraction, depending on the efficiency of the operation). This mixture is subjected to pulsating forces set up by a plunger (or diaphragm, air, or other mechanism) in the water medium so that the entire bed is lifted up and settles back. As the bed expands, the heavy particles of sufficiently large size and proper shape will crash through the bed since the bed is in the "quick" condition—it offers little resistance to such settlement. As the pulsations continue, the large, heavy pieces end up at the bottom and the light fraction at the top of the bed. A screen allows these heavies to fall into the bottom hopper while the lights (with or without the middling) are drawn off the top.

In addition to the common plunger jig shown in Fig. 4-24, which is called the Harz jig, many other variations on the same theme are available commercially. Most of these vary the motion of the plunger so that the

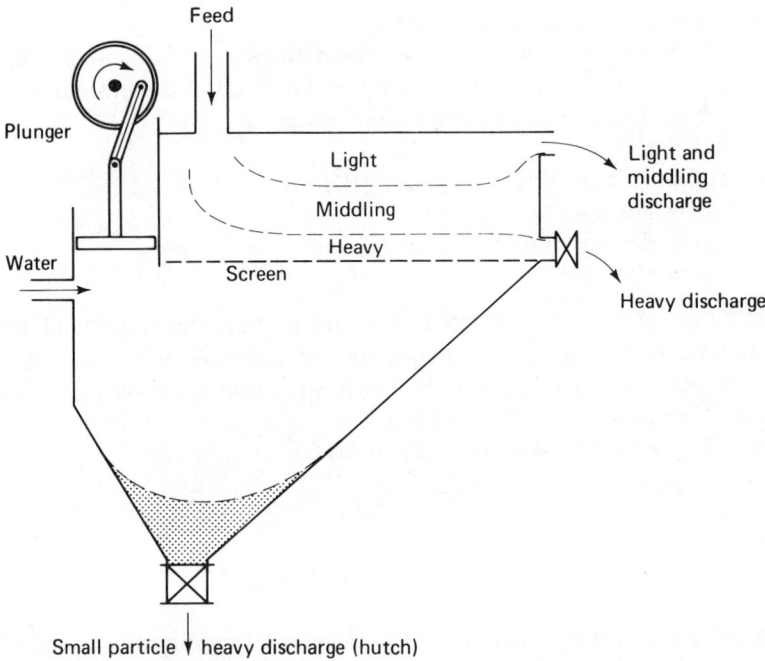

Figure 4-24. Schematic of a common plunger jig.

pressures imparted on the bed have different frequency diagrams, thus achieving specific bed movement.

Instead of pulsating water, the entire screen can of course be moved up and down to achieve separation. This effect can be demonstrated easily by placing mixed material in a standard engineering sieve, submerging this in a sink full of water, and shaking it up and down.

Theory of Operation

Stratification in a jig is accomplished primarily by the relative specific gravity of the material, although particle size and shape are of some importance. As a result, mixtures of different materials can be separated regardless of the particle size, which is difficult if simple particle settling is employed.

To illustrate this point, consider the settling column shown in Fig. 4-25. In such a simple clarifier, vertical stratification occurs because of differences in specific gravity and particle size. If Stokes's law holds, the

Jigs

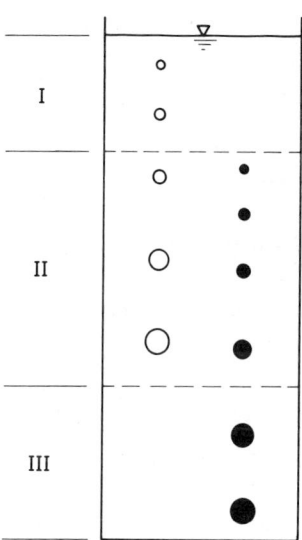

Figure 4-25. Vertical stratification of settling particles, based on differences in diameter and specific gravity. (From Ref. 46.)

terminal settling velocity of either heavy or light material is

$$v \propto (\rho_s - \rho_m)d$$

where v = terminal settling velocity
  d = particle diameter
  ρ_s = particle density
  ρ_m = bulk density of the bed, at any layer, if the particles are crowded together, or the fluid density if the individual particles are not influenced by their neighbors

Note that the terminal velocity is governed by both the size and density difference, and thus the separation achieved is into three phases: the small lights, the large heavies, and a mixture of large lights and small heavies.

The advantage of jigging is that separation can be attained by specific gravity only, irrespective of particle size (within limits). This can be demonstrated by the following argument.

Starting from rest, a particle in a still fluid accelerates until it attains its terminal velocity. This velocity is reached by having the drag force increase from zero (at rest) until it becomes equal to the net gravitational force. As noted in the discussion on air classification, at any instant during the acceleration phase the motion of the particle can be described as

$$F_E - F_B - F_D = \frac{V\rho_s}{g}\frac{dv}{dt}$$

where F_E = force due to gravity
 F_B = buoyant force
 F_D = drag force
 V = volume of particle
 ρ_s = density of particle
 g = acceleration due to gravity
 v = particle velocity
 t = time

At time zero, with $F_D = 0$, and recalling that $F_W = V\rho_s$ and $F_B = V\rho$, where ρ = fluid density, the initial acceleration is

$$a_0 = \frac{dv}{dt} = \left(1 - \frac{\rho}{\rho_s}\right)g$$

Thus particles of the same density (or material), regardless of their size, have the same initial acceleration. Of course, the moment resistance enters the picture, this no longer holds, because particle size is a factor in drag forces. It is also clear from the preceding equation that for any two particles of different density, the heavy particle has a greater initial acceleration than does the light particle. Although this difference is small, and short-lived, it is the basic principle on which the jigging operation is based.

Consider now two particles of the same density but of different size; A = large and B = small. In a still fluid, with both particles starting from rest, the large particles will accelerate faster, but since its terminal velocity is greater, it will still be accelerating when the small particle has attained its terminal velocity. This is shown schematically in Fig. 4-26. The distance traveled by either of the particles in any given time t is simply the area under the curve, or

$$\int_0^t v\, dt$$

Consider now a third particle, C, which is of a greater density but smaller size, so that its terminal velocity is equal to the large particle A. Obviously, it would be difficult to separate these two particles in a simple settling device. It is possible, however, to take advantage of the different accelerations and achieve separation by imposing a series of short accelerations starting from rest. The distance traveled during any one short spurt is small, but the difference in travel, when summed over many accelerations, is sufficient to have the heavy particle A travel further than the light (but larger) particle B, thus achieving stratification and eventual separation.

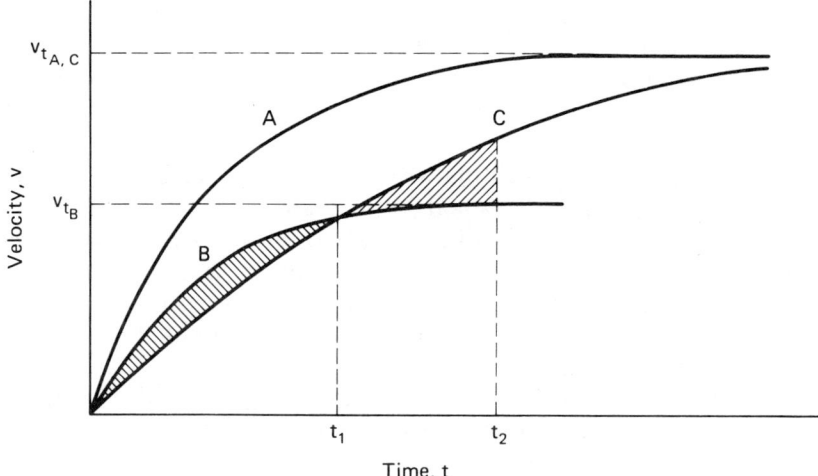

Figure 4-26. Acceleration of particles with different densities and sizes. Particle A is heavy and small, while C is light and large, and they have equal settling velocities. Particle B is also heavy, but smaller than A.

Figure 4-26 illustrates that the denser but smaller particle B has until time t_1 a higher settling acceleration than does the larger but light particle C. After time t_1 the lighter particle has a higher acceleration, and catches up to the smaller particle at time t_2. (The difference between the areas under the curves to time t_1 equals the area from t_1 to t_2.) Thus, if it is intended to separate the heavy but small particles B from the lighter but larger particles C, it is necessary to have the acceleration time well below t_2. This time is thus a design variable which determines the size of particle to be separated as part of the heavy fraction.

In addition to the fact that heavy particles are selectively moved downward in a downward acceleration of the fluid, the reverse of the cycle, the upward pulse, selectively accelerates the light particles so as to move them upward and out of the mixed bed.

Consider two particles of the same size but different density. Each has a terminal velocity with denser particles having the higher v_t. In a situation where the fluid flow is upward at a velocity v_f, and the particle is moving with flow, the net particle velocity is

$$v_f - v_t$$

and the denser particle thus has a lower net velocity. Further, starting from rest, the lighter particle starts faster and continues to accelerate even after the heavy particle has attained its final velocity. An imposition of short accelerations within an upflow bed thus promotes the selective rise of the lighter fraction.

One pulse cycle consists of the bed first being lifted and then compacted. During the lifting (or positive) phases, the light particles are accelerated faster and moved toward the top. During the negative phase, the heavier particles are accelerated faster than the lighter ones and crash through the bed.

Some particles are so small that they travel within the spaces between the larger particles. Such interstitial particles are essentially screened and carried to the top of the bed, where they exit in the overflow layer. The heavy particles fall through the screen into the hutch, thus again achieving separation. As a result of the elimination of fine particles, the jigging operation gives a cleaner light and heavy product than might be obtained by simple gravity separation.

A summary of the action within a continuous jig is as follows [14]:

1. The lowest layer within a bed will exclude from itself all material of specific gravity lower than that of the bed (within reasonable particle-size ranges).
2. Particles with specific gravities greater than that of the lowest layer will always enter, regardless of size. This may not hold true for flat particles, which may need to be properly oriented, and very small (interstitial) particles, which may be caught in the upward flow of the water.
3. The top layer will be penetrated by all particles except those small enough to be caught in the upward flow.
4. The middle layer passes all particles with greater specific gravities and resists penetration by lighter particles.
5. The finest size of particle that can be retained in the bed is determined by the size of the screen apertures and the upward velocity.

Design

The basic objective of jig design is to operate it so as to obtain three distinct layers with a desired purity for the top and bottom layers.

The loading on a bed is in terms of weight per bed area per time. Table 4-9 lists some common loading factors for ores as well as limited experiences on the separation of glass from other shredded and screened refuse.

The actual size of a jig is limited by operational problems such as extraction of the light fraction over an even area and the even motion of the water pulses. The width is seldom over 60 cm (24 in.) and the length is about 1.5 times the width.

A major operational variable is the pulse stroke frequency and acceleration and jigs should be designed to provide this option in operation. Power

TABLE 4-9
Capacities of Jigs

Feed Material	Size of Feed (mm)	Loading (tonnes/m^2/day)	References
Galena-blende-quartz	10 to 5	39–49	14
Galena-limestone	4 to 1	19	14
Blende-chert-dolomite	10 to 0	15–19	14
Minus $\frac{3}{4}$-in. refuse from trommel		24	17

Note: To obtain tons/ft^2/day, multiply by 0.102.

consumption [14] is about 0.1 to 0.15 hp/ft^2 of sieve area and can be estimated as

$$\text{hp} = \frac{Ad^{1/2}}{5000}$$

where A = sieve area, in.^{2}
  d = diameter of feed, mm

Water consumption is fairly low, although it can vary widely, depending on length of stroke, size of screen, materials being separated, and so on. Ore processing jigs use about 800 liters/min/m^2 [20 (gal/min)/ft^2] of jig area [14]. One important consideration is the pollutional characteristics of the wastewater from jigging shredded refuse. The treatment and disposal of this waste must be considered in any economic evaluation of jigs applied to resource recovery processes.

STONERS

Stoners, also called *pneumatic tables*, differ from jigs only in that air is substituted for water as the pulsating fluid. The general principles of operation are the same as for jigs using water. Most commercial models, in addition to pulsating air, use shaking tables, thus providing two forms of the energy input. These devices have been applied for the recovery of aluminum from shredded and screened waste, although operating data have not been made public. Stoners have been used for many years in the agricultural industry for removing stones and other impurities from peanuts, beans, and so on. They are most frequently employed where the two-part separation into light and heavy fractions involves a minor fraction of heavies, with a density difference between the two of at least 1.5 : 1.

Stoners have been employed successfully in cleaning compost of glass and other heavy materials. In one successful compost plant, a stoner operates at about 1.8 to 2.7 tonnes/h (2 to 3 ton/h) processing a screened product. The reject is about 5% of the feed, and has a bulk density of 19 kg/m^3 (1.2 lb/ft^3), whereas the compost has a bulk density of about 9.6 kg/m^3 (0.6 lb/ft^3), easily exceeding the 1.5 : 1 minimum [39].

SINK/FLOAT SEPARATORS

Various fluids, most commonly water, can be used to remove a heavy fraction from a light one by simply allowing the heavies to sink and the lights to float. If water is used, all those materials with specific gravities greater than 1.0 will sink and those with lower specific gravities will float.

When this process is used in resource-recovery operations, many of the organics (wood, textiles) will be recovered from the inorganics. This is not a very advantageous objective, however, since organics are more readily removed by dry operations. Sink/float operations seem to hold most promise in the separation of various types of inorganics, for example aluminum from glass or glass from other metals. This requires an upflow or a fluid specific gravity different from 1.0.

Three basic options exist for achieving this aim:

Heavy liquids.

Heavy media.

Upflow.

Heavy Liquids

The *heavy-liquid* option simply involves the substitution of a denser liquid for water. For example, a mixture of tetrabromoethane and acetone has been used for the separation of aluminum from heavier materials. This liquid has a specific gravity of about 2.4, and the sink fraction of shredded, air-classified, and screened refuse, with the ferrous fraction removed, can produce a fairly high concentration of aluminum. A major problem with the use of such heavy liquids is the inability to readily vary the specific gravity as needed and the cost of the chemicals. A significant fraction can be lost during the operation as a result of adsorption to the waste material.

Another commercially successful heavy liquid is pentachlorethane (C_2HCl_5 : specific gravity 1.67 at 20°C)[14] which has been used for coal cleaning. Some of the more common heavy liquids used in heavy media separation are listed in Table 4-10.

Sink/Float Separators

TABLE 4-10
Compounds Used in Heavy-Liquid Separation

Compound	Specific Gravity
Tetrabromoethane (acetylene tetrabromide)	2.96
Bromoform	2.89
Methyl bromide	2.48
Ethylene dibromide	2.17
Pentachloroethane (pentalin)	1.67
Tetrachloromethane (carbontetrachloride)	1.50
Trichloroethane	1.46

Source: Ref. 42.

Heavy Media

Heavy media differs from heavy liquids in that the specific gravity is varied by adding colloidal solids. For example, a mixture of ferrosilicon and water (85:15), with a surface-active agent can be used to attain specific gravities over 3.0. In one study [40], aluminum was removed by first sinking it in a fluid with a specific gravity of 1.4 and then floating it in a liquid of specific gravity at 3.0.

This method also suffers from the problem of capture and retention of the fluid, resulting in higher operating costs for replacement, as well as potential wastewater treatment problems. But it has advantages.

The principal advantage of heavy-media separation can be demonstrated by the following argument [41]. Suppose that a mixture is comprised of two solids, a and b, a being denser than b. Because of size differences, the larger b particles may have higher settling velocities than the small a particles, and hence separation by settling cannot be complete. The range of sizes that can be separated can be calculated by recognizing that, at equal settling velocities (assuming laminar flow and Stokes's law),

$$v_a = \frac{d_a^2 g(\rho_{s_a} - \rho)}{18\mu} = v_b \frac{d_b^2 g(\rho_{s_b} - \rho)}{18\mu}$$

$$\left(\frac{d_a}{d_b}\right)^2 = \frac{\rho_{s_b} - \rho}{\rho_{s_a} - \rho}$$

Hence separation is possible for any two particles if

$$\frac{d_a}{d_b} > \left(\frac{\rho_{s_b} - \rho}{\rho_{s_a} - \rho}\right)^{1/2}$$

If it is possible to choose a fluid with a density very nearly equal to one of the solid materials, the classification can be made more efficient. For example, if the fluid density approaches that of particle b, the ratio d_a/d_b approaches zero, meaning that very large b particles can be separated from even the very small a particles. Obviously, if the fluid density is greater than the density of b and less than that of a, complete separation is possible.

The settling (or rising) velocity of the heavy or light material within the heavy media can be estimated using a modified Stokes's law,

$$v = \frac{d^2 g(\rho_s - \rho_a)}{18\mu_a}$$

where v = velocity of a particle in the heavy-medium suspension
d = particle diameter
ρ_s = particle density
ρ_a = apparent density of the medium
μ_a = apparent viscosity

The apparent viscosity of a suspension of fine particles can be estimated from a theoretical equation first developed by Einstein [43]:

$$\frac{\mu_a}{\mu} = 1 + 2.5C$$

where μ_a = apparent viscosity of the suspension
μ = viscosity of the fluid
C = concentration, solids fraction by volume

This relationship only holds for dilute suspensions considerably more dilute than the ones used for heavy-media separation. Numerous other equations have been suggested for estimating the influence of suspended particles on viscosity, all based in part on empirical evidence [44]. For example, one equation that seems to hold over a wide range of solids concentrations is

$$\frac{\mu_a}{\mu} = \frac{10^{1.82(1-x)}}{x}$$

where μ_a = apparent or bulk viscosity of the slurry
μ = viscosity of the fluid fraction
x = volume fraction of the liquid in the slurry

Alternatively, it is possible to develop a correction factor R for the Stoke's equation which incorporates both the change in viscosity and density:

$$v_t = \frac{d^2 g(\rho_s - \rho)}{18\mu} R$$

Sink / Float Separators

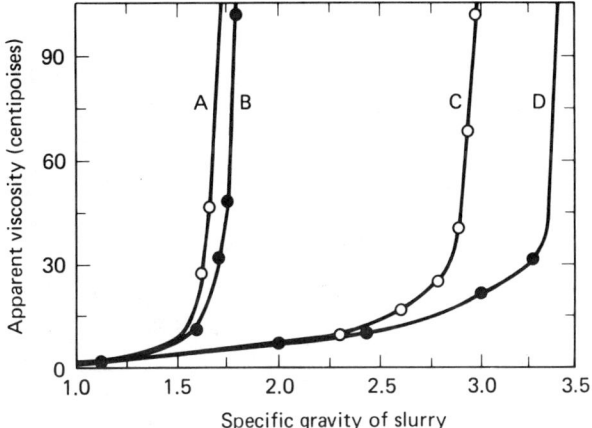

Figure 4-27. Apparent viscosity of heavy-media suspensions. Curve A, quartz; B, magnetite; C, new ferrosilicon; D, abraded ferrosilicon. (From Ref. 14.)

DeVaney and Shelton [14] found that the apparent viscosity of suspensions increases at a relatively slow rate with the density of a suspension until a critical point is reached at between 17 and 30 volumetric percent, whereupon the viscosity changes rapidly until the mixture ceases to behave like a liquid, as shown in Fig. 4-27.

The shape of the particles also affects viscosity. For example, Fig. 4-27 shows two curves, one for new ferrosilicon, one for abraded ferrosilicon.

Another variable that can be a serious problem in the application of heavy-media separation processes to resource recovery operations is contamination. The viscosity–specific gravity curve is moved to the left with increased contamination, producing high viscosities at lower specific gravities. Running at a specific gravity of 3.0 with ferrosilicon, for example, can be difficult, because contamination can substantially increase the viscosity at that specific gravity. Figure 4-28 shows the change in -200-mesh ferrosilicon with bentonite contamination.

The apparent viscosity, as previously noted, must be obtained from experimental evidence. The apparent density of the suspension can be calculated as

$$\rho_a = \frac{w_1 + SGw_2}{g(w_1 + w_2/SG)}$$

where SG = specific gravity of the solid
  w_1 = weight of fluid
  w_2 = weight of solids

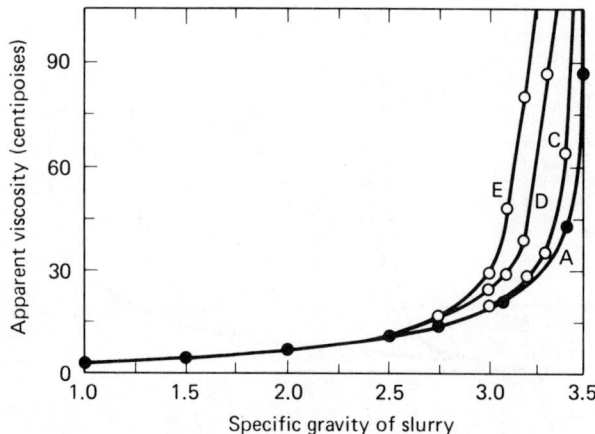

Figure 4-28. Apparent viscosity changes in a −200-mesh ferrosilicon with bentonite contamination. Curve A, no bentonite; C, 0.3%; D, 0.6%; and E, 1% bentonite. (From Ref. 14.)

TABLE 4-11
Common Heavy-Media Materials

	Material Specific Gravity	Maximum Practical Suspension Specific Gravity
Magnetite	5.0–5.2	2.5
Galena	7.4–7.6	3.3
Ferrosilicon	6.7–6.8	3.5
Lead (atomized)	11.3	6.2
Quartz sand	2.7	1.5

Source: Ref. 14.

The higher the value of SG, the greater can be the apparent density of the suspension without producing intolerably high viscosities. The specific gravities of some common heavy-media materials are shown in Table 4-11, together with some practical limits of the apparent specific gravity.

It should be noted that ferrosilicon, an alloy of iron and silicon with a small amount of carbon, can vary from 10 to 25% Si. Alloys with over 22% Si are substantially nonmagnetic, while those with less than 15% Si tend to rust. The most common forms of ferrosilicon contain about 15% Si.

Ferrosilicon is also difficult to wet, and surfactants are generally used to produce acceptable suspensions.

Upflow Separators

The third method of achieving float/sink separation is to use an upflow device with water. Effective specific gravities of between 1.1 and 2.0 have been used in commercial devices. Such upflow devices have been used to separate heavy organics (leather, plastics, textiles, etc.) from metals and glass in the heavy fraction of air-classified refuse [45].

The effective specific gravity is calculated by recognizing that the net velocity of a particle (relative to the walls of the container) is

$$v_t - v_r$$

where v_t = terminal velocity of the particle
v_r = rise velocity of fluid

Using Stokes's equation

$$v_t - v_r = \frac{d^2 g(\rho_s - \rho_e)}{18\mu}$$

where ρ_e is the effective fluid density, and recognizing that specific gravity is related to density as SG = ρ/ρ_{water}, or

$$v_t - v_r = \frac{d^2(SG_s - SG_e)\rho_{water}}{18\mu}$$

$$SG_e = SG_s - \frac{18\mu(v_t - v_r)}{d^2 \rho_{water}}$$

Example 4-5

Water is to be used to float glass particles, SG_s = 2.7, d = 1 cm. Calculate the effective specific gravity if the particle rise velocity is 1 cm/s (viscosity of water, μ = 0.01 poise).

$$SG_e = SG_s - \frac{18\mu(v_t - v_r)}{D^2 \rho_{water}}$$

$$= 2.7 - \frac{(18)(0.01)(-1)}{(1)^2(1)}$$

$$= 2.7 + 0.18 = 2.88$$

INCLINED TABLES

The differential movement of particles on an inclined table can be used to separate particles of various densities and sizes. For example, coal can be washed by separating slate and other heavy materials from the raw ore

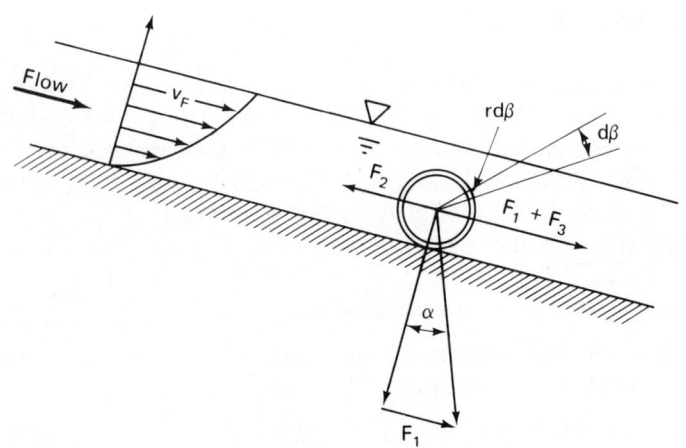

Figure 4-29. Submerged particle moving on an inclined plane. (From Ref. 46.)

by washing the mixture down an inclined table and removing the heavy contaminants through ports located along the incline.

A particle on an inclined table, completely submerged by the fluid, moves as a result of both the pull of gravity and the flow of the fluid. The forces on the particle are shown diagrammatically in Fig. 4-29.

The parallel component of the gravitational force is F_1. The frictional force between the particle and surface is F_2, and the force exerted by the flowing liquid is F_3.

Since the gravitational force is the difference between the force due to gravity and buoyancy, F_E and F_B,

$$F_1 = (F_E - F_B) \sin \alpha = V(\rho_s - \rho)g \sin \alpha$$

Similarly, the drag force is

$$F_2 = k(F_E - F_B) \cos \alpha = kV(\rho_s - \rho)g \cos \alpha$$

where α = angle of incline
 V = particle volume
 ρ_s = particle density
 ρ = fluid density
 k = coefficient of friction between the particle and the surface

The force exerted by the flowing fluid is

$$F_3 = 3\psi\mu \int_0^{2\pi} (v_F - v)r \, d\beta$$

Shaking Tables 191

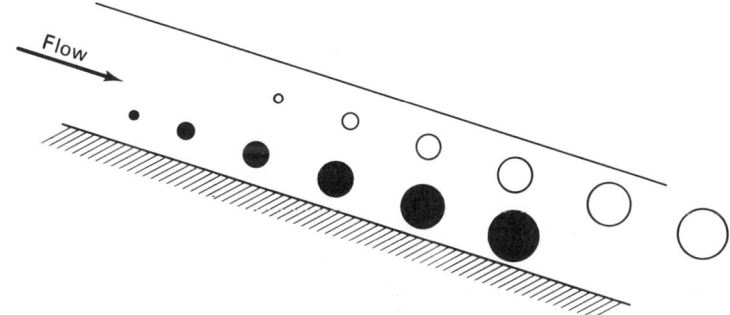

Figure 4-30. Stratification on a submerged inclined bed. (From Ref. 25.)

where r = radius of the sphere equal to the volume of the particle
v = velocity at any distance relative to the inclined plane
v_F = velocity at any distance from the inclined plane (see Fig. 4-29)
ψ = shape or sphericity factor (see the Appendix)
β = angle of particle roll (see Fig. 4-29)

If the velocity distribution is known (or assumed linear, or calculated from average velocity), the shape factor determined, and the friction coefficient determined from published data, the three forces can be equated and angle α determined. This is a minimum slope for the movement of the particles on the inclined table.

The force exerted by the fluid is an important variable for segregating particles of different sizes, as is readily apparent from Fig. 4-28. Thus large particles would have higher velocities than the smaller particles of equal density. For any two particles of equal size, but different densities, the lighter particle will have a higher velocity due to the greater effect of the frictional resistance on the heavy particle. Thus for a binary mix, the stratification on an inclined bed is as shown in Fig. 4-30.

SHAKING TABLES

Shaking tables differ from simple inclined tables only in that the table is shaken with a differential movement in the direction perpendicular to fluid flow. In addition, all modern shaking tables, such as the Wilfrey Table, are equipped with riffles, which are long slots in the table, also perpendicular to the flow. A schematic of a shaker table is shown in Fig. 4-31.

The shaking mechanism consists of an eccentric wheel with a flexible connecting rod. The motion is suddenly stopped by the bumping block. The feed is introduced at the top of the inclined table (see Fig. 4-30). The

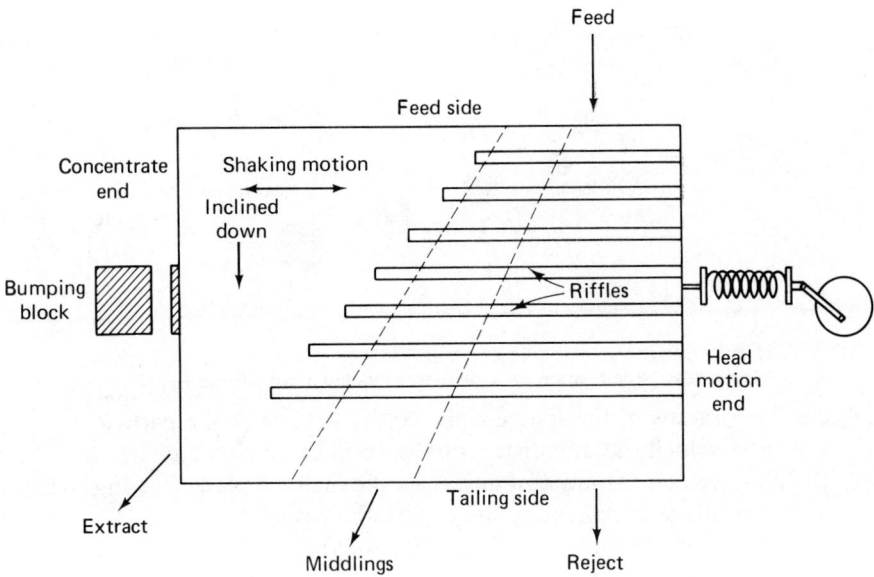

Figure 4-31. Schematic of a shaking table. (From Ref. 14.)

slime flows directly across the table and exits, while the granular material is moved toward the concentrate end by the bumping action which at the start of a stroke gives the particles sideways velocity by means of the moving table. When the motion of the table is suddenly stopped by the bumping block, the momentum of the granular materials carries them further sideways. The flexible connection between the drive shaft and table absorbs the eccentric motion, slowly withdraws the table from the bumping block, and thus completes a cycle.

The stratification within the riffles is as shown in Fig. 4-32. The small heavy fraction (concentration) is on the bottom, the large heavies on the top of that, followed by the small lights and finally the large lights. This is due first to the difference in density (heavy from light) as the particles are rearranged during the shaking action, and then to the movement of the small fraction through the intersticies of the larger particles of equal density. Further action within the riffle trough is the flow of fluid above the ridges, causing small eddies to occur within the troughs.

The height of the riffles is reduced from the head-motion end to the concentrate end, thus allowing the large lights to flow down the table first, since their support was withdrawn. These are followed by the light small particles, and so on, until the small concentrate (heavy) fraction is removed at the far end. The ideal exit of the various products is then as shown in Fig. 4-33.

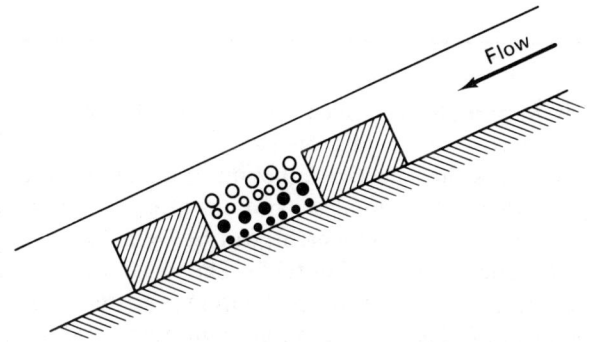

Figure 4-32. Material stratification in a riffle trough on a shaking table.

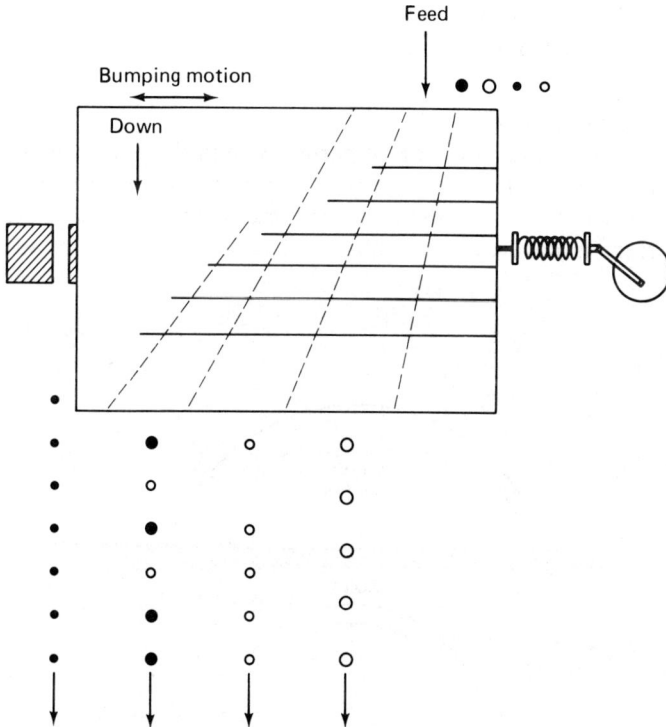

Figure 4-33. Products produced from a shaking table.

193

FLOTATION

Flotation is a process by which solids are selectively floated to the surface of the slurry by means of attached gas bubbles. The key to successful flotation is the selective adhesion of air bubbles to the material that is to be floated. The actual separation, after the material has been made lighter by air-bubble attachment, can be by frothing or by a simple gravity separator such as a shaking table. The usual separation method in resource recovery operations is froth flotation, and the common application is the removal of glass from ceramics and other contaminants.

The basic steps in the froth flotation method are:

    Control of particle size.

    Conditioning.

    Addition of collection agent.

    Addition of frothing agent.

    Aeration.

    Separation of froth.

The first step is the reduction of particle size to the point where flotation can occur. The smaller the particle, the larger the surface area/volume ratio and hence the greater the reduction in effective density of the particle as its surface is covered with bubbles. The second step, the addition of

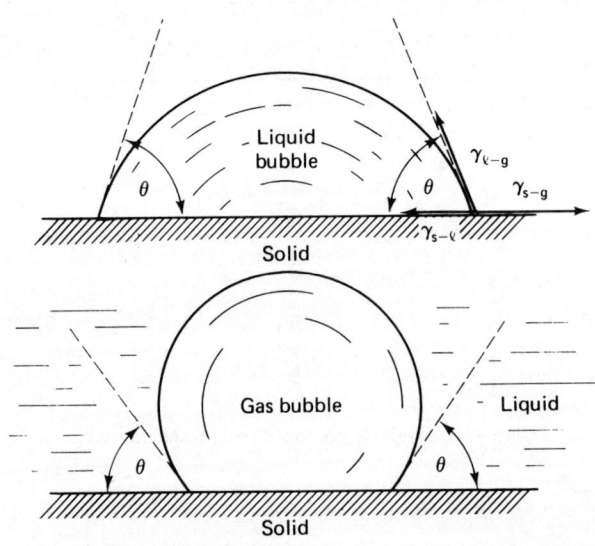

Figure 4-34. Contact angles and surfaces tensions of bubbles.

conditioning agents, assists the selective adhesion of the collection agent. Conditioning is necessary because of the dirty nature of most slurries, in which various materials can react with and adversely affect the action of the collection agent. This may be a simple pH adjustment [14], since the contact angle of the bubble can vary significantly with pH or the addition of detergent.

The interaction of air bubble and particle surface can be characterized by the contact angle, which may be measured by viewing a liquid droplet on the surface through a microscope [46]. The usual assumption is that the contact angle of water/solid droplet is the same as that of an air/solid bubble, as shown in Fig. 4-34.

The significance of contact angles in froth flotation arises from the fact that surface energy represents the work of breaking atomic bonds when a surface is formed. Contacting phases partially satisfy these broken bonds, lowering the energy of the surface somewhat. The surface energy of a solid–liquid interface will depend in part on the relative polarities of liquid and solid. A polar liquid will wet (completely coat) a polar solid, because their polar bonds result in surfaces that are mutually attractive. For example, water is a polar liquid and will wet a polar solid, such as clean glass; but water will not wet a nonpolar hydrocarbon, such as polyethylene or paraffin. A nonpolar liquid, such as benzene, will wet polyethylene but not glass. When wetting occurs, the sum of solid–liquid and liquid–gas surface energies must be less than, or equal to, the solid–gas surface energy.

$$\gamma_{s-l} + \gamma_{l-g} \leq \gamma_{s-g}$$

Since surface energies represent work per unit area, they also may be interpreted as force per unit length (in Fig. 4-34, the length being perpendicular to the plane of the figure).

$$\gamma = \frac{dE}{dA} = \frac{dF}{dl}$$

where E = energy
A = area
F = force
l = length

That is, the surface energy represents a surface tension, or resistance of a particular interface to being expanded by force. When wetting is incomplete (Fig. 4-34), force equilibrium in any particular plane requires that

$$\gamma_{s-g} - \gamma_{s-l} = \gamma_{l-g} \cos \theta$$

where γ_{s-g} = surface tension between solid and gas
γ_{s-l} = surface tension between solid and liquid
γ_{l-g} = surface tension between liquid and gas

For pure water and air, $\gamma_{l-g} = 72$ dyn/cm, depending on the temperature. The contact angle θ thus depends on the difference between γ_{s-g} and γ_{s-l}. The objective of flotation is to achieve a large contact angle between gas and solid so as to firmly attach the bubble to the solid. Unfortunately, γ_{s-g} and γ_{s-l} can be measured only indirectly.

After conditioning, the materials to be floated are covered with a water-repellent film which will allow for the ready attachment of bubbles. The collectors are selective, and do not coat the other materials in the slurry. For example, experiments at the U. S. Bureau of Mines laboratory at College Park, Maryland, have shown cocoa amine to be an effective collection agent for glass, selectively coating it to the exclusion of ceramics, stones, and other porous materials [47].

A common series of collectors used in ore processing are xanthates, which are produced from alcohols according to the following reaction:

$$ROH + CS_2 + NaOH \rightarrow H_2O + ROCSSNa$$

The structural formula for sodium ethyl xanthate is

$$C \begin{cases} O-C_2H_5 \\ =S \\ S-Na \end{cases}$$

Higher xanthates are also widely used.

When sodium xanthate is dissolved in water, it ionizes into Na^+ ions and negatively charges xanthate ions. The latter are attached to the surface of selected materials in such a way that the hydrocarbon chains (e.g., ethyl radical) point outward, thus producing the hydrophobic characteristic necessary for bubble attachment.

Once the particles have been treated so that the desired ones will adhere to air bubbles, a frothing agent is added and the slurry is aerated or otherwise agitated to produce the froth. The agents that produce frothing vary greatly, but a typical agent is simple soap. Various types of commercial froth flotation separators are described in reference 14.

In the recovery of glass, the contaminants that would most likely be screened with the glass are aluminum, iron oxide, and ceramics. Froth flotation has been shown to cleanse the cullet of these impurities [48]. Prior to flotation, the contaminated glass is first screened, which produces a product consisting of 63% glass. This is then rod-milled and further screened to produce a glass-rich product with a size range of $-32, +200$ mesh. A proprietary conditioner and collector is then added before frothing.

The chemical composition of three different samples of recovered glass is shown in Table 4-12, along with the composition of typical amber, flint, and green glass. The sample from Franklin, Ohio, resembles an amber

TABLE 4-12
Composition of Glass

Oxide	Typical Amber	Typical Flint	Typical Green	Franklin, Ohio	Palo Alto, Calif.	San Francisco, Calif.
S_iO_2	72.3	73.0	71.6	72.4	72.4	72.3
Al_2O_3	2.3	1.7	1.5	1.5	2.5	2.2
MnO	0.5	0.01	0.01	0.03	0.03	0.008
CaO	10.6	10.8	10.8	10.6	9.9	9.8
MgO	1.5	0.8	0.7	2.6	0.61	0.79
Na_2O	13.4	13.2	14.5	13.7	13.5	13.8
K_2O	0.5	0.3	0.3	0.25	0.88	0.84
Fe_2O_3	0.27	0.043	0.19	0.23	0.137	0.106
PbO	0.003	0.003	0.003	0.012	0.022	0.031
Cr_2O_3	Nil	Nil	0.23	0.008	0.011	0.048

Source: Ref. 48.

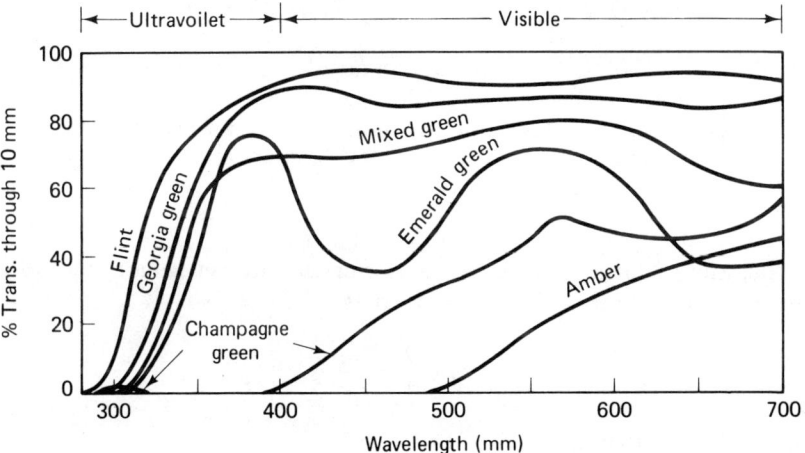

Figure 4-35. Transmission spectra for recovered mixed green glass and other types of glass made from virgin raw material. (From Ref. 48.)

glass, while the two California samples produced a glass quite similar to a typical green glass. The transmission spectra for the mixed green glass is shown in Fig. 4-35, together with other types of glass.

A promising application of flotation is the separation of various types of plastics. It is well known that although most plastics are hydrophobic (difficult to wet), the wetting characteristics can be selectively adjusted [49]. Using a "goniometer" to measure the contact angle of a water droplet on variously wetted plastic surfaces, it was possible to estimate the segregation possibility of different plastics. The measurement involves dropping exactly 0.80 mg of water through a microburet onto a piece of treated

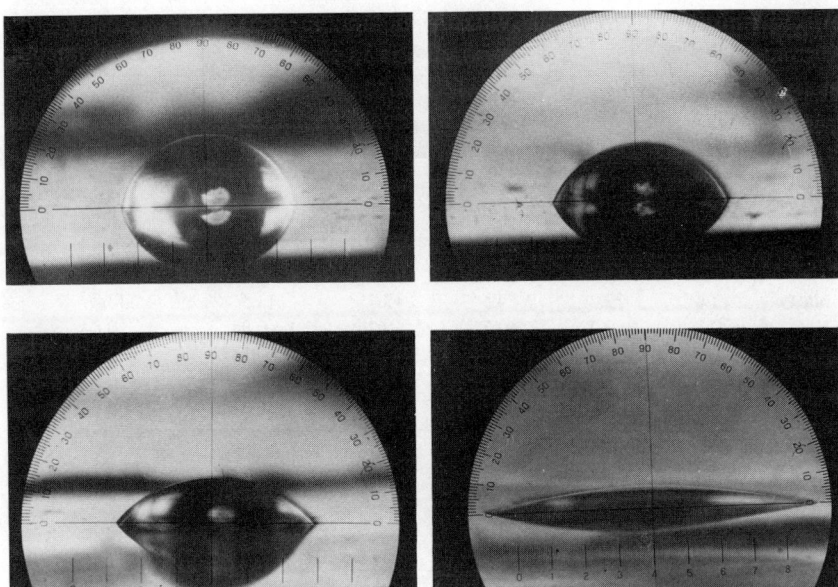

Figure 4-36. Water drops on a PVC plate: (a) untreated PVC; (b) 0.2% wetting agent; (c) 0.3% wetting agent; (d) 0.5% wetting agent. (Courtesy of Mesco, Inc., Tokoyo.)

plastic, and under a microscope measuring the droplet height and radius. The contact angle may then be defined as

$$\tan \frac{\theta}{2} = \frac{h}{r}$$

where θ = contact angle
h = height of droplet
r = radius of droplet

Figure 4-36 shows the shape of a water droplet on a PVC plate. As the plastic becomes more hydrophylic, the drop spreads out, decreasing the contact angle as defined above.

This wetting characteristic is related to the ease of air-bubble attachment, since a plastic that is highly hydrophobic will selectively hold on to air bubbles. By changing the wetting characteristics, it is possible to float some plastics while allowing others (with perhaps equal initial densities) to sink. A commercial separator has proven to be over 98% effective in separating out various types of plastics, and this method shows great promise for the future [45].

COLOR SORTING

One of the major problems with the recovery of glass is that the waste glass is of many colors, and such a mixture has low market value. Clear glass alone has substantial value, while contamination by even 5% amber glass makes it essentially unmarketable, since it cannot be used to produce clear glass products.

At this time, the only technique for color-sorting glass seems to be by the use of the wavelength of transmitted or reflected light from the glass.

The most popular color sorter, which has in the past been used for the separation of diamonds in diamond mines and rotten corn and peanuts in agriculture, is shown in Fig. 4-37, which schematically illustrates the basic principle. The individual particles are moved by means of a high-speed

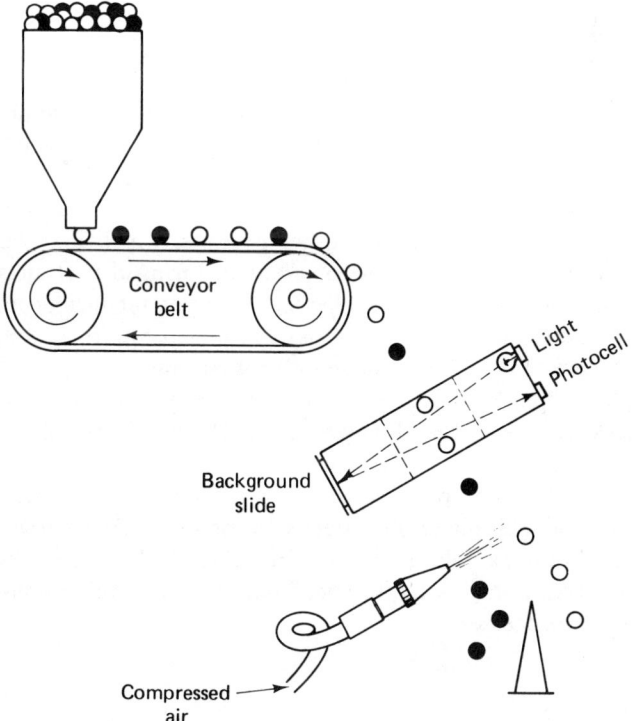

Figure 4-37. Schematic of an optical color-sorting device.

pulley from a bin and flung through a light detection box or sensor. Within this detector are three lights and three photocells. Each light is bounced off a background reference slide into a photocell, and the current produced is sensed electronically. Should the particle falling through the sensor be of the same color (some average of reflected and transmitted light) as the background slide, the photocell will not detect any difference and nothing happens. These particles are thus passed through the device as "accepts." Should a particle be lighter or darker than the background slide, however, the current change will trigger a compressed-air ejector located immediately below the sensor box and the "reject" particle is blown into a different bin. Dividing a feed of glass particles into three categories—flint (clear), green, and other—requires two passes, because the color sorters on the market are inherently binary devices. In such a case, the first sorting would be based on light intensity reaching the photocells, passing the flint particles, and ejecting the colored particles. In the second sorting of the colored particles, green filters might be placed over the photocells and the electronics tuned so that green particles are perceived as "light" and are passed; amber (and other colors on the red end of the spectrum) particles are perceived as "dark" and are ejected.

This device works well when the particles are uniform in size and the electronics can be so set as to make the device reject even suspicious particles. For example, of 100 peanuts entering the sensor, 2 might be rotten. The device can be adjusted to sense these two, but since it is so finely tuned, may also eject two other peanuts, which may be perfectly good but which may have been so oriented during their trajectory as to produce a shadow and thus be rejected. The peanut farmer accepts this margin of safety, and gladly discards two good peanuts as long as he is reasonably sure that he has removed all rotten ones.

With glass separation, however, the mix is not 2 colored pieces and 98 clear pieces out of a 100. Instead, a 50/50 mix is common, and this presents problems of efficiency of operation.

To date, it has not been possible to achieve the required purity of product that would make this device a practical alternative for glass sorting. In addition, a color sorter requires a relatively narrow size distribution in the feed, and this requirement places considerable constraints on size-reduction processes.

PARTING SHOTS

This chapter is an outline of some significant mechanical methods of materials separation techniques. No attempt has been made to present an exhaustive list of the possibilities. Indeed, handbooks such as Taggart's *Handbook of Mineral Dressing* [14] present an impressive compilation of machines and processes and include some techniques that will find use in future resource recovery facilities. An emerging field such as resource recovery has to borrow from many fields and select those devices which show most promise.

REFERENCES

[1] RIETEMA, K., "On the Efficiency in Separating Mixtures of Two Components," *Chem. Eng. Science*, 7, 89 (1957).

[2] WORRELL, W. A., AND P. A. VESILIND, "Evaluation of Air Classifier Performance," *Resour. Recovery Conserv.*, 4 (1980).

[3] RIMER, A. E., Ph.D Dissertation, Duke University, Durham NC pending.

[4] HERING, R., AND S. A. GREELEY, *Collection and Disposal of Municipal Refuse*, McGraw-Hill Book Company, New York, 1921.

[5] ENGDAHL, R. B., *Solid Waste Processing*, U.S. EPA OSWMP, Washington, D.C., 1969.

[6] HENSTOCK, M., Discussion at Engineering Foundation Conference, Rindge, N.H., July 1978.

[7] TREZEK, G. J., and G. SAVAGE, "MSW Component Size Distribution Obtained from the Cal Resource Recovery System," *Resour. Recovery Conserv.*, 2, 67 (1976).

[8] National Center for Resource Recovery, *Bulletin*, Washington, D.C.

[9] CUMMINGS, J. P., "Glass and Non-Ferrous Metal Recovery Subsystems at Franklin, Ohio—Final Report," *Proc. 5th Miner. Waste Util. Symp. U.S. Bur. Mines*, IIT Research Institute, Chicago, 1976.

[10] MATTHEWS, C. W., "Screening," *Chem. Eng.*, July 10, 1972.

[11] ROSE, H. E., and R. M. E. SULLIVAN, *A Treatise on the Internal Mechanics of Ball, Tube and Rod Mills*, Constable and Co., London, 1957.

[12] DAVIS, E. W., "Fine Crushing in Ball Mills," *Trans. AIMME*, 61, 250 (1920).

[13] ALTER, H., and B. CRAWFORD, *Materials Recovery Processing Research*, U.S. EPA OSWMP, Washington, D.C., in press.

[14] TAGGART, A. F., *Handbook of Mineral Dressing*, John Wiley & Sons, Inc., New York, 1974.

[15] NELSON, W., A. KRUGLAK, and M. OVERTON, "Trommel Screening," Duke Environmental Center, Duke University, Unpublished, 1977 and 1978.

[16] HILL, R. M., "Rotary Screens for Solid Waste—Ring in the Old," *Waste Age*, Sept. 1977, p. 33.

[17] MAKAR, H. V., and R. S. DeCESARE, "Unit Operations for Nonferrous Metals Recovery," in *Resource Recovery and Utilization*, ASTM STP 592, 1975, pp. 71–88.

[18] WHITBY, K. T., "The Mechanics of Fine Sieving," *Symposium on Particle Size Measurement*, ASTM Spec. Tech. Publ. 234, Philadelphia, 1958.

[19] BOETTCHER, R. A., *Air Classification of Solid Waste*, U.S. EPA OSWMP SW-30c, Washington, D.C., 1972.

[20] SENDEN, M. M. G., and M. TELS, "Mathematical Model of Air Classifiers" *Resour. Recovery Conserv.*, 2, 129 (1978).

[21] ECKHOFF, D., University of Utah, Personal communication.

[22] WORRELL, W. A., *Testing and Evaluation of Three Air Classifier Throat Designs*, Duke Environ. Center Publ., Duke University, 1978.

[23] POLKOWSKY, B. V., "Economic Considerations of Air Classifier Design," Unpublished, Duke Environmental Center, Duke University, 1978.

[24] Arthur D. Little, Inc., Personal communication.

[25] MURRAY, D. L., and C. L. LIDDELL, "The Dynamics, Operation and Evaluation of an Air Classifier," *Waste Age*, March 1977.

[26] Midwest Research Institute, "Study of Processing Equipment for Resource Recovery Systems," U.S. EPA Contract 68-03-2387 Final Report (1979).

[27] MONROE, H. S., "The English vs. the Continental System of Zigging—Is Close Sizing Advantageous?" *Trans. AIME*, 17, 637 (1888–1889).

[28] ROBINSON, C. B., "Some Factors Influencing Sedimentation," *Ind. Eng. Chem.*, 18 (18), 869 (1926).

[29] STEINOUR, H. H., "Rate of Sedimentation," *Ind. Eng. Chem.*, 36 (10), 902 (1944).

[30] RICHARDSON, J. F., and W. N. ZAKI, "Sedimentation and Fluidisation," *Trans. Inst. Chem. Eng. (Br.)*, 32, 35 (1954).

[31] KERMACK, W. O., A. G. MCKENDRICK, and E. PONDER, "The Stability of Suspensions," *Proc. R. Soc. Edinburgh*, 49 (1929).

[32] SWEENEY, P. J., *An Investigation of the Effects of Density, Size and Shape Upon the Air Classification of Municipal Type Solid Waste*, Rep. CEEDO-TR-77-25, Civil and Environmental Engineering Development Office, Tyndall Air Force Base, Fl., 1977.

[33] HOERNER, S. F., "Fluid-Dynamic Drag," Midland Park, N.J., 1965 (as reported in ref. 32).

[34] SMITH, E. H., "Autorotating Wings: An Experimental Investigation," *J. Fluid Mech.*, **50**, pt. 3 (1971).

[35] National Center for Resource Recovery, Inc., *Materials Recovery Systems*, Washington, D.C., 1972.

[36] VESILIND, P. A., "Materials and Energy Recovery," in *Solid Waste Engineering*, N. Pereira and L. Wang (Eds.), in press.

[37] FAN, D. N., "On the Air Classified Light Fraction of Shredded Municipal Solid Waste: I. Composition and Physical Characteristics," *Resour. Recovery Conserv.*, **1**, 3 (1977).

[38] FEIFEL, E., *Forschung*, **9**, 68 (1938), as reported by C. Doerschlag and G. Miczek, *Chem. Eng.* Feb. 14, 1977.

[39] DROBNEY, N. L., H. E. HULL, and R. F. TESTIN, *Recovery and Utilization of Municipal Solid Waste*, U.S. EPA OSWMP SW-10c, Washington, D.C., 1971.

[40] MICHAELS, E. L., K. L. WOODRUFF, W. L. FRYBERGER, and H. ALTER, "Heavy Media Separation of Aluminum from Municipal Solid Waste," *Trans. Soc. Min. Eng.*, **258**, 34 (Mar. 1975).

[41] FOUST, A. S., et al., *Principles of Unit Operations*, John Wiley & Sons, Inc., New York, 1960.

[42] Oil, Paint, and Drug, *Reporter*, May 6 and 13, 1967, as reported in Ref. 39.

[43] ROSCOE, R., "Suspensions," in *Flow Properties of Disperse Systems*, J. Hermans (Ed.), North-Holland Publishing Co., Amsterdam, 1953.

[44] RUTGERS, R., "Relative Viscosity and Concentration," *Rheol. Acta*, **2** (4), 305 (1962).

[45] SALTOH, K., I. NAGANO, and S. IZUMI, "New Separation Technique for Waste Plastics," *Resour. Recovery Conserv.*, **2**, 127 (1976).

[46] LARIAN, M. G., *Fundamentals of Chemical Engineering Operations*, Prentice-Hall, Inc., Englewood Cliffs, N. J., 1958.

[47] U.S. Bureau of Mines, Unpublished data.

[48] MOREY, B., and J. P. CUMMINGS, "Glass Recovery from Municipal Trash by Froth Flotation," *Proc. 3rd Miner. Waste Util. Symp., U.S. Bur. Mines*, ITT Research Institute, Chicago, 1972.

[49] SALTOH, K., I. NAGANO, and S. IZUMI, "New Separation Technique for Waste Plastics," *Proc. 5th Miner. Waste Symp.*, ITT Research Institute, Chicago, 1976.

[50] HENRIKSON, R. "Analytical Evaluation of Air Classification," Duke Envir. Center Publ., Duke University, 1979.

PROBLEMS

4-1. Suppose that a proposal is made by some well-meaning people to hand-sort all the refuse from a town of 100,000. Estimate the manpower required, the cost, and describe the problems involved in implementing such a program.

4-2. Estimate the critical speed and horsepower requirements for a trommel screen, 3 m in diameter, processing refuse so as to run 25% full.

4-3. If the angle of a trommel screen was increased (steeper), how would this affect the critical speed, horsepower, and efficiency?

4-4. Aluminum chips of uniform diameter 1.0 in. are to be separated from glass of 100% less than 0.5 in. by use of a single-deck reciprocating screen. The total feed rate is 1 ton/h, with 0.1 ton/h of aluminum and 0.9 ton/h of glass. It is required to produce of an aluminum fraction that is 99% pure (by weight). Find the area and size of screen required.

4-5. Using the aluminum/glass mixture specified above (and assuming that the smallest glass particle of significance is 0.05 in. in diameter) estimate the air-classifier size and throat velocity required to separate the glass from the aluminum.

4-6. If the pressure drop through a cyclone is assumed to be 1 in. of water, estimate the diameter of cyclone required for various air-flow rates to remove 1-in.-diameter aluminum chips.

4-7. Estimate the air velocity required in an air classifier to just suspend a piece of glass of 0.5 in. in diameter. Suppose that a 0.4 volume fraction of the throat section was occupied by glass, what would the required air velocity be to suspend this glass?

4-8. What density of fluids is needed to separate glass, aluminum, and plastics into the three individual components? What specific fluids might be used for this purpose?

4-9. Two air-classifier manufacturers report the following performance for their units:

    Manufacturer A:
        Recovery of organics = 80%
        Recovery of inorganics = 80%

    Manufacturer B:
        Overall recovery = 60%
        Purity of light fraction = 95%

Assume a feed that consists of 80% organics. Compare the performance of the two units.

5 MAGNETIC AND ELECTROMECHANICAL SEPARATION PROCESSES

This chapter is devoted to magnetic and electromechanical means of separating selected components from mixed municipal refuse, as contrasted with the purely mechanical means of separation described in Chapter 4.

The devices described in this chapter use a property such as magnetic permeability as a recognition code. In the case of a ferromagnetic material, coding occurs when the material interacts with a magnetic field, and separation is effected by magnetic attraction, while the nonferrous materials are separated by gravity prior to the release of the magnetic field. Certain electromagnetic processes can produce more than two products by a continuous coding/separation scheme. One such process incorporates the use of a magnetic fluid to separate a wide range of nonferrous materials on the basis of their differences in density. Such sophisticated processes may facilitate the economic recovery of even small fractions of various materials in future waste streams.

MAGNETIC SEPARATION

The use of the force associated with a magnetic field gradient for the extraction of ferrous materials is probably one of the simplest, most effective, and economical processes used to separate components of municipal solid waste. Magnetic separation is a well-known technique used first for the removal of tramp iron from scrap and for the concentration of iron ores. Since 1849, numerous U.S. patents for magnetic separation processes have been issued, and even before 1910, a number of texts described a variety of magnetic devices for mineral processing [1, 2.]

The conventional use of magnetic separation falls into two general categories: the purification of feed streams containing unwanted magnetic impurities, and the concentration of magnetic materials. The desired product in the first case is the nonmagnetic material, while in the second case the magnetic product is desired. Such separations are performed for the removal of tramp iron from a variety of feeds, for the benefication of several ferromagnetic ores, and for the removal of ferrous materials from municipal solid waste.

Conventional magnetic separation devices are generally restricted to separating ferromagnetic materials, such as iron and magnetite. A number of different types of devices have been developed, resulting in the tendency for a more refined mechanical design and higher recovery rates.

Theory of Operation

A magnetic field may be defined in terms of a solenoid comprising a number of turns of wire around a core and conducting a current I. The *magnetic field intensity* H in the interior of a long solenoid (i.e. sufficiently far away from the end faces) is given by

$$H = \frac{nI}{l}$$

where H = magnetic field intensity, A/m
 N = number of turns
 l = length of solenoid, cm
 I = current, A

Magnetic field intensity is the analog of mechanical stress. The magnetic analog of mechanical strain is the *magnetic induction B* (also called the magnetic flux density). B is defined by Faraday's law of electromagnetic induction as [3]

$$-\frac{dB}{dt} = \frac{V}{A} \tag{5-1}$$

where B = magnetic flux density
 V = voltage
 A = cross-sectional area normal to the lines of magnetic flux

According to Faraday's law, a voltage V will be generated in a conductor that is "cut" by lines of magnetic flux (i.e., in a conductor that experiences an increasing or decreasing magnetic flux). If Φ is defined as

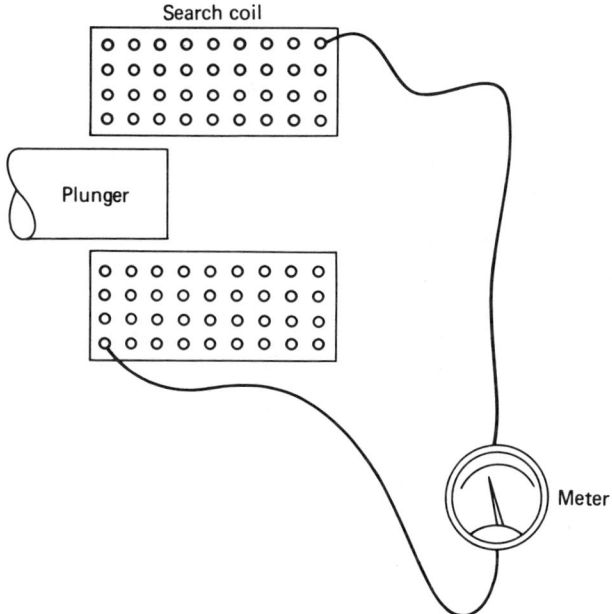

Figure 5-1. Solenoid arrangement to measure magnetic flux density.

the magnetic flux, then the flux density is $B = \Phi/A$. If the magnetic flux is defined in terms of Webers (W), where 1W = 1 Volt second, then B is in W/m^2, which is also known as teslas (T). A typical experimental arrangement for measuring B is shown schematically in Fig. 5-1; the search coil has essentially the same cross-sectional area A as the solenoid, so when the solenoid is filled with a material, the voltage generated in the search coil can be used to calculate the magnetic induction of the material in the solenoid for that magnetic field intensity:

$$B(H) - B_0(H) = \int_0^t \frac{V(H)\,dt}{A}$$

where B = magnetic flux density, T
  B_0 = magnetic flux density at time zero, T
  H = magnetic field intensity, A/m
  V = voltage
  A = cross-sectional area, m²

The solenoid is therefore used as an experimental volume for determining magnetic properties of any material filling that volume. The functional dependence of B on H describes the magnetic behavior of the material filling the solenoid. When the behavior is linear (the mechanical analog would be an elastic solid),

$$B = \mu H$$

where B = magnetic flux density, T
  μ = magnetic permeability (constant)
  H = magnetic field intensity, A/m

A material for which μ is a positive constant is called *paramagnetic*; it tends to concentrate the magnetic flux lines slightly more than would empty space. A material for which μ is a negative constant is called *diamagnetic*; it tends to spread the magnetic flux lines slightly farther apart than would empty space. Ferromagnetic materials exhibit large positive values of μ and therefore are associated with a high concentration of magnetic flux lines. Their behavior is distinctly nonlinear and for ferromagnetic materials:

$$dB = \mu\, dH$$

The work of magnetizing a volume $\overline{V}$ containing a material of permeability μ is

$$\text{Work} = \overline{V} \int H\, dB = \overline{V} \int \mu H\, dH$$

Since work may also be expressed as

$$\text{Work} = F\, dl$$

The force with which a magnet attracts a ferromagnetic material can be expressed in one dimension as

$$F = \left(\frac{dW}{dl}\right) = \overline{V} H \left(\frac{dB}{dl}\right) = \overline{V} H \mu \left(\frac{dH}{dl}\right)$$

where F is the force of attraction and l is length.

In other words, the force on a material in a magnetic field is increased by a high field intensity, a high field gradient, and a high magnetic permeability. When the magnetic permeability of a substance exceeds that of the medium in which it is immersed, the force is attractive. Similarly, if the magnetic permeability of the substance is less than that of the medium in which it is immersed, the force is repulsive. With respect to resource recovery applications, air is the general medium in which the material to be coded and separated is immersed; thus ferromagnetic materials move in the direction of increasing field intensity.

The magnetic field of a typical bar magnet is depicted in Fig. 5-2, where the dashed lines represent the lines of magnetic flux. Typical magnetic fields can be mapped by these so-called "lines of force," which all elementary physics students have created with a magnet and iron filings. Magnetic fields are often classed as uniform if the lines of force are

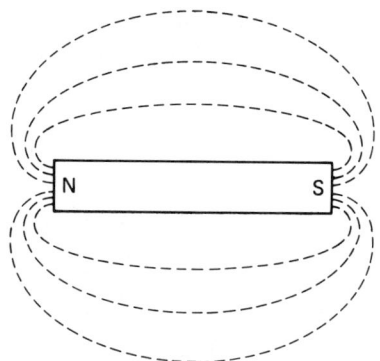

Figure 5-2. Typical magnetic field surrounding a bar magnet.

parallel and equally spaced, or nonuniform if they are not. Only nonuniform fields are of use in attracting ferromagnetic materials in resource recovery separation processes.

The forces in magnetic separators which compete with the magnetic forces and act on all particles that travel through the separator are those of gravity, hydrodynamic drag, friction, electrostatic attraction, and inertia. Since these forces also act on the nonferromagnetic components of the feed, they can be utilized to effect separation. For separators that treat large particles, the gravitational, friction, and mechanical forces clearly predominate, whereas in those that treat very small particles, electrostatic forces and hydrodynamic drag are of primary significance [4, 5].

Recovery is generally less dependent on small differences in permeability between substances than it is on the particle size. Thus a device that must process particles of different sizes cannot be very selective between materials whose magnetic properties differ only slightly. This difficulty is inherent in most types of magnetic separators. In practice, the materials to be separated often have widely different magnetic properties. Typically, ferromagnetic materials are separated from nonferromagnetic materials and one accepts that certain nonferromagnetic steels (e.g., austenitic and ferritic stainless steels) will be lost in the reject fraction.

The total combined effect of the interaction of magnetic, competing, and interparticle forces on the characteristics of a magnetic separator may be described quantitatively for a given feed. The important characteristics include both the ability of the separator to recover magnetic particles, (known colloquially as *mags*) as the extract from the feed and the purity of that extract. The purity P and the recovery R are both used as quantitative measures of separator performance. As previously defined, the recovery is the ratio of mass of the magnetic material in the extract relative to that in the feed, and the purity is the fraction of the mass of magnetic materials in the extract stream. The materials which should enter the reject stream, incidentally, are often referred to especially in mining technology as *tails*, or *tailings*.

Example 5-1

Assuming that the material from the heavy fraction of air classified MSW has the following characteristics, compute the recovery R and the purity P.

$$
\begin{aligned}
\text{Feed:} \quad & \text{Mags} = 3000 \text{ kg/h} \\
& \text{Tails} = 8000 \text{ kg/h} \\
\text{Extract:} \quad & \text{Mags} = 2700 \text{ kg/h} \\
& \text{Tails} = 200 \text{ kg/h} \\
\text{Reject:} \quad & \text{Mags} = 300 \text{ kg/h} \\
& \text{Tails} = 7800 \text{ kg/h}
\end{aligned}
$$

The recovery is

$$R = 100 \left(\frac{2700}{3000} \right) = 90\%$$

The purity is

$$P = 100 \left(\frac{2700}{2700 + 200} \right) = 93\%$$

Magnetic Separation Equipment

Magnetic separators are used primarily to remove ferrous materials from municipal solid waste. For example, in waste-to-energy systems, the magnetic separation of materials has two objectives: (1) to increase the heat content of the refuse derived fuel and (2) to recover marketable materials. Furthermore, the removal of metal reduces wear on subsequent processing and handling equipment and also reduces the amount of ash when the fuel is burned.

The recovery of magnetic separators is dependent on the degree of size reduction provided and the extent to which ferrous materials are physically liberated from other materials. Almost without exception, MSW is shredded prior to magnetic separation in a waste-to-energy system. Magnetic separators are commonly located after the primary shredder; in some cases an air classifier is inserted between the two to concentrate the ferrous and other heavy materials.

Two types of magnetic separators are generally used to remove ferrous materials from shredded wastes. These are (1) the holding-type separator, where the waste is fed directly onto the collecting surface; and (2) the suspended type, where the collecting surface must pick up the ferrous material from among the particles of shredded refuse. Design parameters include the field-intensity and field-gradient system, the mechanical means

Magnetic Separation

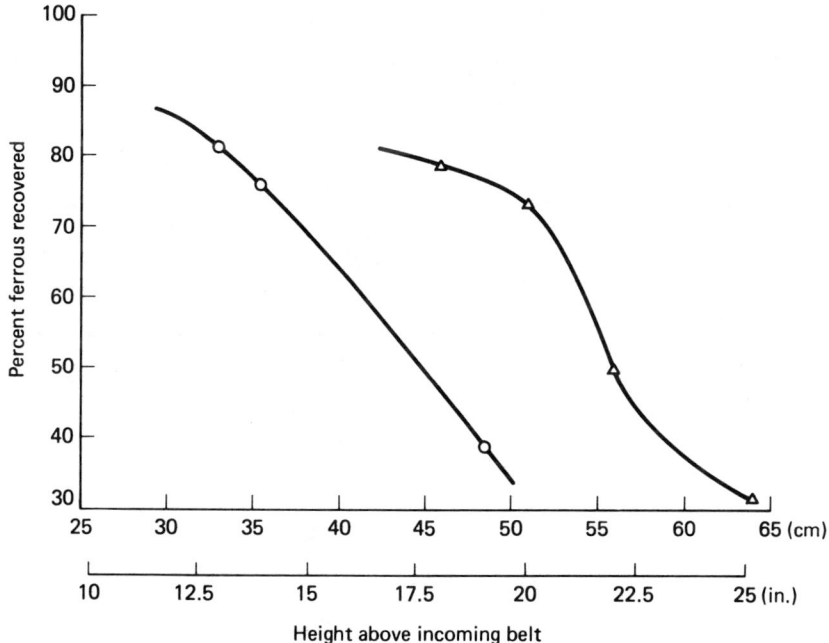

Figure 5-3. Height of magnet above incoming refuse affects ferrous recovery. Source: Ref. [28].

used to bring the waste to the magnet, the type of magnet (permanent or electromagnet) used to create the magnetic field, whether or not the magnets are stationary or moving and the height of the magnet above the feed belt. Fig. 5-3 illustrates the effect of the last design variable.

Holding-type separators

Two configurations of holding-type separators are the drum and belt separators, as shown in Fig. 5-4 [6].

For the *drum* holding-type separator, the rotating horizontal cylinder is fabricated of some nonmagnetic material such as brass or stainless steel. This rotating drum surrounds a magnetic bank which is formed in the same shape as the drum itself. As the material flows onto the drum, the magnetic material is collected and carried to a point below a doctoring blade, where it moves out of the influence of the magnetic field and is scraped off by the blade. The other nonferrous materials drop directly off the drum.

Drum separators range in size from about 30 cm (12 in.) to about 150 cm (60 in.) in width [5]. The drums are often about 30 to 75 cm (12 to 30

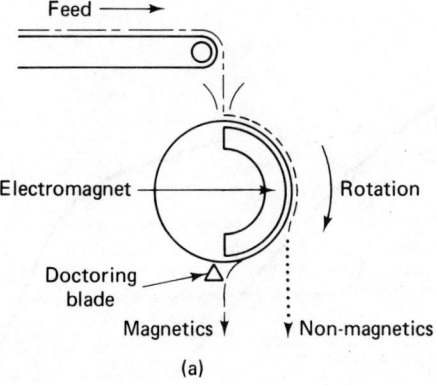

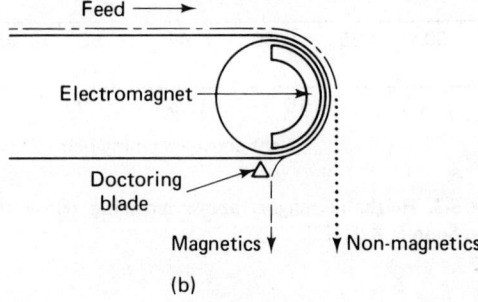

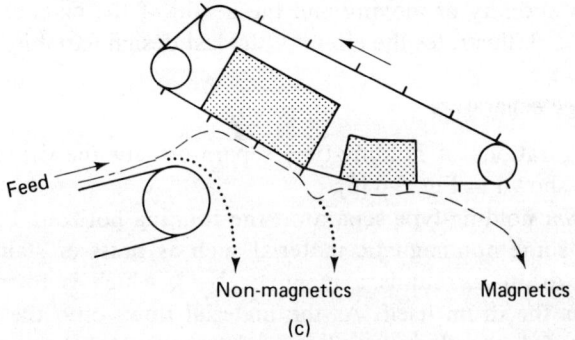

Figure 5-4. Holding- and suspended-type magnetic separators: (a) drum-type holding separator; (b) belt-type holding separator; (c) suspended-type separator. Source: Ref. 6.

in.) in diameter. Design speeds of the units may range from about 50 m/min (2.7 ft/s) to 250 m/min (14 ft/s). Power requirements generally do not exceed 3.7 to 7.5 kW (5 to 10 hp) for the mechanical operation of the unit, and a similar amount of power is required to create the magnetic field. In general, the design criteria for the magnets must be established after the size, range, amount, and composition of the shredded refuse has been determined for the particular waste to be handled.

In the *belt* holding-type separator, a thin layer of MSW is fed directly onto the collecting surface. Gravity holds both the magnetic and nonmagnetic materials to the top of the belt. As the belt moves, the ferrous material is attracted to the head pulley, and the divided receiving hopper receives both ferrous and nonferrous materials in separate compartments. The magnetic forces in this system are generated in much the same way as in the drum-type separator except in this case the drum itself is covered by the moving belt.

In the belt holding-type separator, the pulley diameter may range from 30–60 cm (12 to 24 in.) on a belt that may range up to nearly 2.50 m (8 ft) in width. Belt speeds are more moderate with this unit than the speeds that can be achieved with the drum-type separator and generally range from about 2.5 to 7.5 m/min (0.14 to 0.42 ft/s) [5]. Power requirements for a belt-type magnet may be somewhat greater than for a drum-type magnet because of the increased energy utilized with the long belt to overcome the frictional resistance of the rollers and so on. Power requirements may range from less than 1 kW for the very small systems to 7 to 15 kW (10 to 20 hp) for the larger conveyor-type.

Suspended separators

The other common type of separator used in MSW applications is the suspended separator. In this operation, none of the feed comes into direct contact with the collecting surface, as in the holding-type separator. Instead, the feed is conveyed through a relatively strong magnetic field gradient which lifts the ferrous material from the surrounding medium and attaches it to a belt by the strength of the magnetic field and the magnitude of the field gradient. Fig 5-4c illustrates the details of this separator.

Gravity usually opposes the attractive forces, but since the particles which are magnetically attracted are moved through a relatively short distance to the collecting surface, gravitational interference generally does not pose much of a problem. In fact, gravity is used to enhance the purity of the ferrous extract by allowing the particle to momentarily drop off the belt, only to be attracted by a second magnet which is located downstream on the conveying belt.

Separator design

The design of a magnetic separator involves a number of options. The magnet itself can be an electromagnet or a permanent magnet. An electromagnet has the advantage that it can be turned off, but it has the disadvantage of consuming power. Permanent magnets require no power and can be either metallic (typically Alnico-type alloys, which contain aluminum, nickel, and cobalt, in addition to iron), or they can be nonmetallic ferrites. Ferrites usually do not exhibit such high field intensities as do Alnico-type alloys, but they are much lighter and thus do not require such heavy supporting structures. Properly designed ferrite magnets are usually less expensive than Alnico magnets, but their magnetic properties may deteriorate if exposed to very high or very low temperatures [6].

In summary, the design of any magnetic separator depends on the following factors: (1) a means of controlling the material flow through the magnetic field so as to permit easy movement of the particles; (2) a magnetic field that is sufficiently strong and has a sufficiently high field gradient to remove the size and type of material in the waste stream; and (3) an adequate means of separating both the magnetic and the nonmagnetic materials from the magnetic field [5,7].

Although magnetic separators have been used for numerous industrial applications, their use to code MSW presents some unusual problems. The amount of ferrous metals in MSW is higher than in many other "cleaning" operations. There is also a tendency for nonmagnetic materials to be entrained within the ferrous metals. The sharp metal edges on shredded ferrous products also shorten the life of the nonmetallic belts. Some caution must be exercised in selecting a unit, but the wide adaptability of the process and a history of successful experience permit some latitude in choice in applying the criteria employed in process design.

EDDY CURRENT SEPARATION

Theory of Operation

Eddy currents represent another manifestation of Faraday's law of electromagnetic induction. If the magnetic induction B in a material changes with time, a voltage is generated in that material, according to Faraday's law of magnetic induction:

$$-\frac{dB}{dt} = \frac{V}{A}$$

where B = magnetic flux density, T
  V = Voltage
  A = cross section of enclosed area normal to the lines of magnetic flux, m^2

In an electrical conductor, the induced voltage will produce a current, called an *eddy current* (named because an axial magnetic field that increases or decreases in intensity produces circular current loops reminiscent of the eddies created in a turbulent-fluid-flow regime). The direction of the current loop is determined by *Lenz's law*: If B is increasing, the current direction will be such as to create a magnetic field that opposes the applied magnetic field; if B is decreasing, the current direction will be such as to create a magnetic field that reinforces the applied field.

Using an aluminum beverage can as an example (Fig. 5-5), the eddy current shown in the illustration would be produced when the magnetic field is increasing (Lenz's law coupled with the right-hand rule). The aluminum can becomes in effect a solenoid with its north pole pointing toward the north pole of the applied field. Therefore, it will be repelled by the applied field, affecting a coding (recognition of its magnetic polarity) and a separation (by the forces of magnetic repulsion). If the magnetic field had been decreasing, the eddy current in Fig. 5-5 would have been in the opposite direction, and separation would have been accomplished by magnetic attraction.

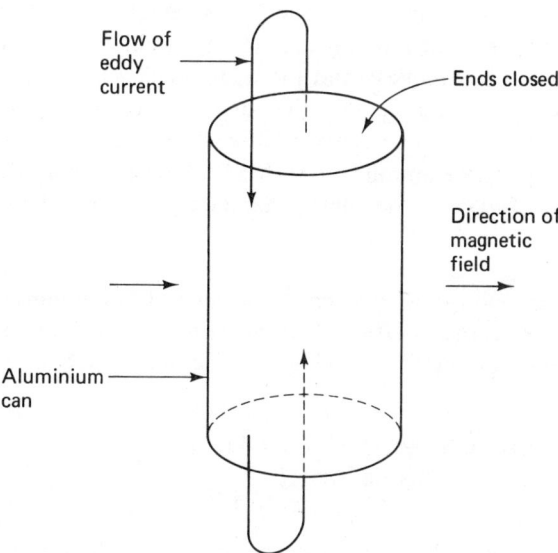

Figure 5-5. Eddy currents in an aluminum can.

The force that orients a current loop or bar magnet in a stationary magnetic field, and that moves a current loop or bar magnet in a moving magnet field is called the *Lorentz force*. The direction and magnitude of the Lorentz force is given by the vector equation

$$F = j \times B$$

where F = force on current element j
 B = magnetic flux density, T

For the aluminum can in Fig. 5-5, the net Lorentz force on the can would be zero (right-hand convention for vector product, applied at different points around the current loop) if the field were uniform over the can. But if the field were stronger on one side of the can than on the other, the can would be propelled in the direction in which **B** was increasing. (The maximum Lorentz force can be expected when **B** is in one direction on one side of the can and in the opposite direction on the other side of the can.) When the magnetic field is moved (as a traveling **B** sine wave) by electrically phasing the current to the motor windings, any conductors containing eddy currents will experience a net force in the direction of field motion. Similarly, if conducting materials move through a stationary, but alternating (–N–S–N–S– ...) magnetic field, they will experience a Lorentz force in the direction connecting adjacent north and south poles.

Sommer and Kenny describe four basic methods of inducing eddy currents in metals [8, 9]; (1) physically moving a sample through a magnetic field; (2) moving a magnetic field through a sample by physically moving the magnet; (3) moving a magnetic field through a sample via electrical phasing techniques; and (4) temporarily changing the magnetic field intensity in a sample. Studies of eddy current separation have demonstrated that eddy currents produced by such methods could be employed to separate aluminum from MSW. However, considerations such as the induced force on the metal, size effects, electrical resistivity, and other factors affect the efficiency of the separation processes to varying degrees.

The field geometry of the magnetic field is an important factor in determining the recovery affected by an eddy current unit. In one set of studies, force on the metal was related to the magnetic field as follows [9]:

$$F = aB^2$$

where F = force on the electrical conductor, N
 B = magnetic flux density, T
 a = constant

This relationship may be interpreted to mean that the force exerted on a metal (the major selection process) is proportional to the product of the applied magnetic field and the induced magnetic field (proportional to the

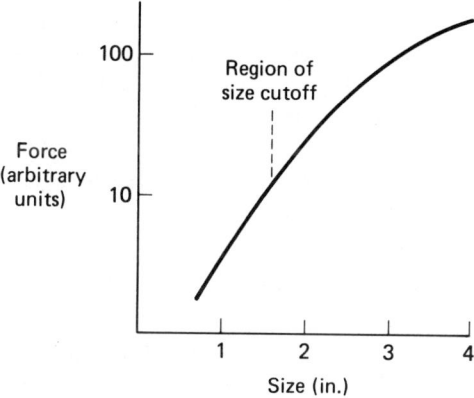

Figure 5-6. Relationship between eddy current force and particle size. Source: Ref. [9].

TABLE 5-1
Electrical Conductivity/Mass Density Ratio for Various Nonferrous Metals

Metal	Electrical Conductivity, σ (10^8 mho/m)	Mass Density, ρ (10^3 kg/m^3)	σ/ρ 10^3 mho-m/kg
Aluminum	0.35	2.7	13.0
Copper	0.59	8.9	6.7
Silver	0.63	10.5	6.0
Zinc	0.17	7.1	2.4
Brass	0.14	8.5	1.7
Tin	0.09	7.3	1.2
Lead	0.05	11.3	0.4

Source: Ref. 9.

eddy current) in the sample. The varying configurations of the applied field and the approach of the metal to that applied field result in substantial differences among eddy current separators.

Size effects also were found to affect all eddy current processes [9]. The factor most affecting the maximum size of the particle removed (size cutoff) is the method of eddy current generation. Fig. 5-6 illustrates the functional dependence of the force and size for the removal of aluminum.

Since eddy current separation depends on the ability of the magnetic field to levitate the conductor (by magnetic repulsion), the ratio of conductivity to density provides some measure of discrimination among materials. Table 5-1 illustrates the fact that aluminum exhibits a ratio of conductivity almost twice as great as copper [9]. Thus the eddy current separation process permits a highly selective coding and separation process for aluminum. Presumably, eddy current separators could also be tuned to discriminate between good conductors (Al, Cu, Ag) and other materials, but the principal market now is specifically for separating aluminum. Since aluminum has such a high scrap value, it may justify the relatively high cost of an eddy current separator.

Eddy Current Separation Equipment

Three basic systems of eddy current separation are described below. Two involve the principle of the linear induction motor with different configurations of the magnetic poles, and one involves the use of a falling stream of refuse through a permanent magnetic field. The feed to an eddy current separator generally consists of the "heavies" or reject component from an air classifier from which the ferromagnetic component has been removed. Removal of ferromagnetic materials prior to eddy current separators is primarily to avoid the problems of blocking the flow and shielding the conductors from the applied magnetic field.

Linear induction motor

Eddy current units that employ a traveling field intensity appear to be relatively efficient units. One such device is a basic modification of a *linear induction motor* (LIM). A typical linear induction motor is similar in principle to that of a stator in a rotary induction motor, the significant difference being that the configuration is flat rather than cylindrical [10]. Figure 5-7 illustrates a typical linear induction motor and the conceptual modification of that motor to accommodate the removal of the metal utilizing such a configuration.

The linear induction motor generates a sine wave of magnetic field intensity which travels down the length of the motor with alternating north-and south-pole components. As the metal-rich concentrate passes over the LIM, eddy currents are induced in any electrical conductor. The induced magnetic fields associated with the eddy currents in the metals interact with the moving field generated by the motor, which pushes the nonferrous metals along the linear motor. All that is necessary to achieve removal is to orient the motor transverse to the direction of the feed, so as to repel the metals away from the main direction of travel of the refuse along the belt.

Figure 5-8 illustrates a typical LIM configuration. Material is fed to the end of a nonmagnetic belt which travels over linear induction motors positioned on the underside of the belt. Recovery of the metal concentrate is to the top side of the belt, where the material to be removed is ejected by

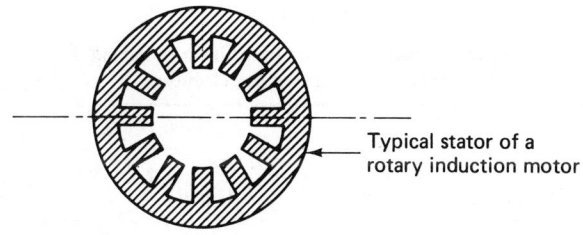

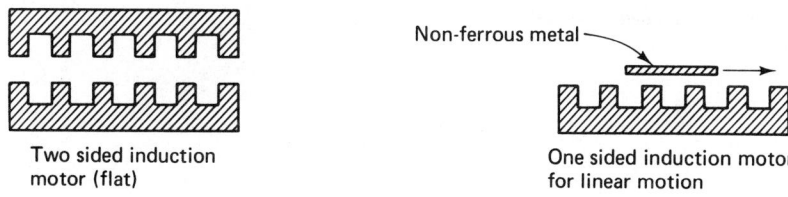

Figure 5-7. Principle of linear induction motor. Source: Ref. 29.

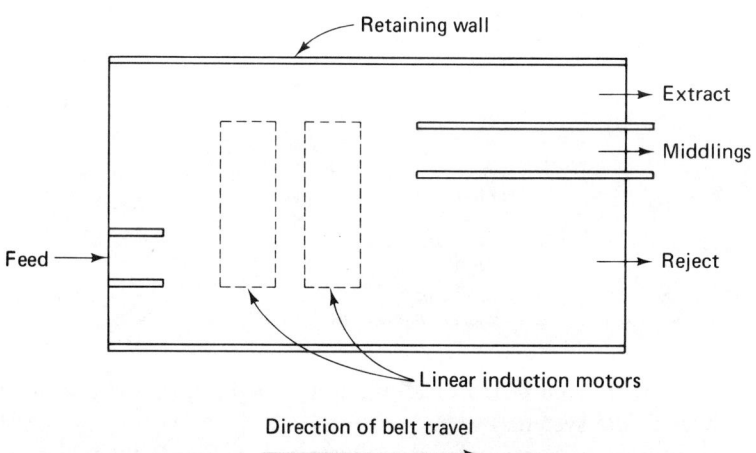

Figure 5-8. Typical plan view of linear induction motor system for aluminum recovery. Source: Ref. 29.

the LIM against the retaining wall into the extract area. Waste materials that are not affected by the eddy currents flow along the lower portion of the belt area.

This system yields an overall recovery of aluminum described in Tables 5-2 and 5-3. Foil recovery in this system is low. This type of separator does result in a relatively low recovery of other metals, which means that the

TABLE 5-2

Eddy Current Separation by a Linear Induction Motor of Aluminum from Refuse[a]

Fraction	Weight (kg)	Grade (wt %)			
		Al can and other Al	Al Foil	Other Metals	Organics
Extract	14.5	88.8	0.2	5.5	5.5
Middlings	7.4	60.2	2.1	11.6	25.1
Reject	85.7	7.0	3.4	3.0	86.6
Feed	107.5	21.9	2.6	3.9	71.5

[a]After one pass through LIM unit.
Source: Ref. 9.

TABLE 5-3

LIM System Performance on San Diego Refuse[a]

Element	Weight (%)
Si	0.6
Fe	0.5
Cu	0.6
Mn	0.7
Mg	0.3
Zn	2.6
Pb	<0.1
Others	0.1
Aluminum	94.5

[a]After two passes through LIM unit.
Source: Ref. 11, 30.

purity of the extract in terms of aluminum is high, particularly after two passes through the system.

A second type of configuration is illustrated in Fig. 5-9 which has two pairs of magnets with each pair having a magnet both above and below the belt. In this process, four water-cooled, two-sided electromagnets are arranged about a 0.5-m (1 ft 7 in.)-wide conveyor as illustrated. The unit operates on the principle that the first two pairs of magnets extract material from one-half of the belt and eject it to one side. The other pair is downstream of the waste flow, and eject the remainder of the material from the other half of the belt to the other side of the unit. In this configuration, the top of each double-sided magnet has five poles, while the bottom has four poles. The air gap between the magnets in this

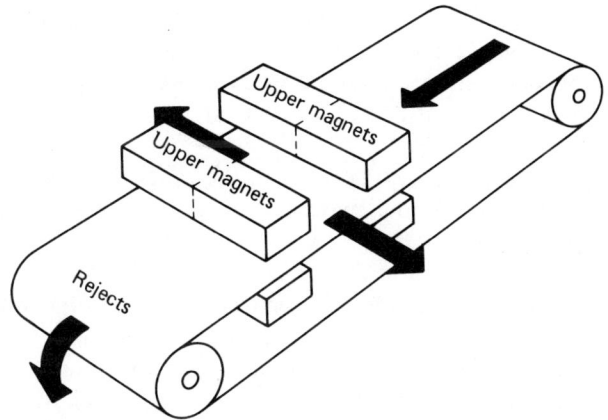

Figure 5-9. Typical cross section of double magnet system for aluminum recovery. Source: Ref. 10.

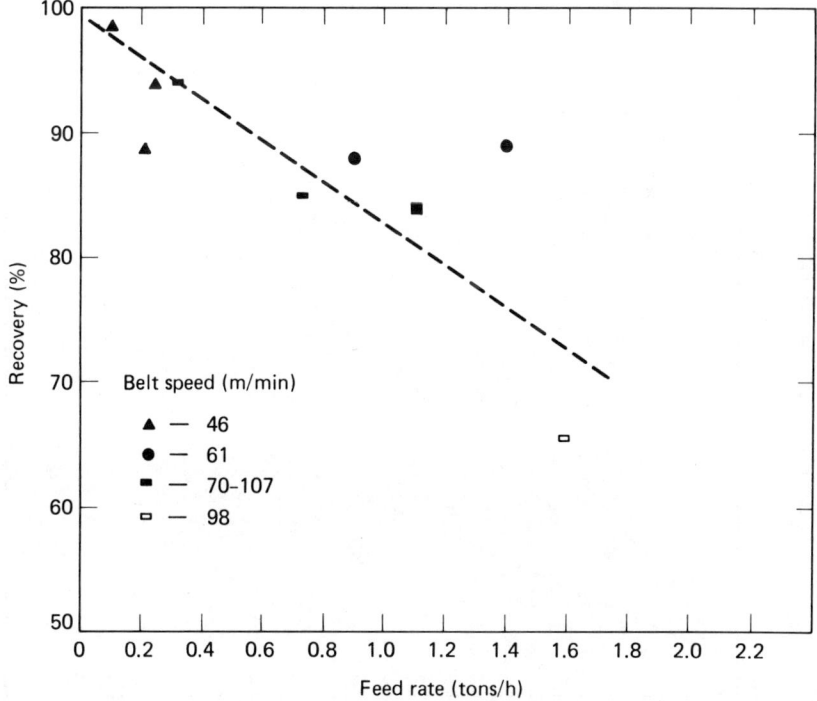

Figure 5-10. Effect of loading and belt speed on recovery. Source: Ref. 11.

configuration is approximately 9 cm ($3\frac{1}{2}$ in.) and the electrical energy consumption is about 22 kW. The process works by switching the electrical currents on and off at intervals (about every 45 s), thereby preventing any ferrous materials that might be in the waste stream from blocking the flow on the belt [10].

The purity of the extract is generally in the range 75 to 80% aluminum and 10 to 15% other nonferrous materials, including heavy aluminum, and 10% loose organics [11, 12].

Aluminum separation in this system is a direct function of the feed rate to the unit. Figure 5-10 illustrates the recovery rate as a function of speed, or the feed rate of the belt system. Figure 5-11 illustrates the effect of loading and speed on loose organics in the aluminum extract. It should be noted that the ordinate in this figure is not (1 − Purity), since contaminants other than loose organics may be in the extract. A conservative design factor of 0.9 tonne/h (1 ton/h) of input material consisting of roughly 40% nonferrous metal is considered typical [11].

It is evident that the feed rate has a much more significant effect on efficiency than does the belt speed. It is expected that in the design of this

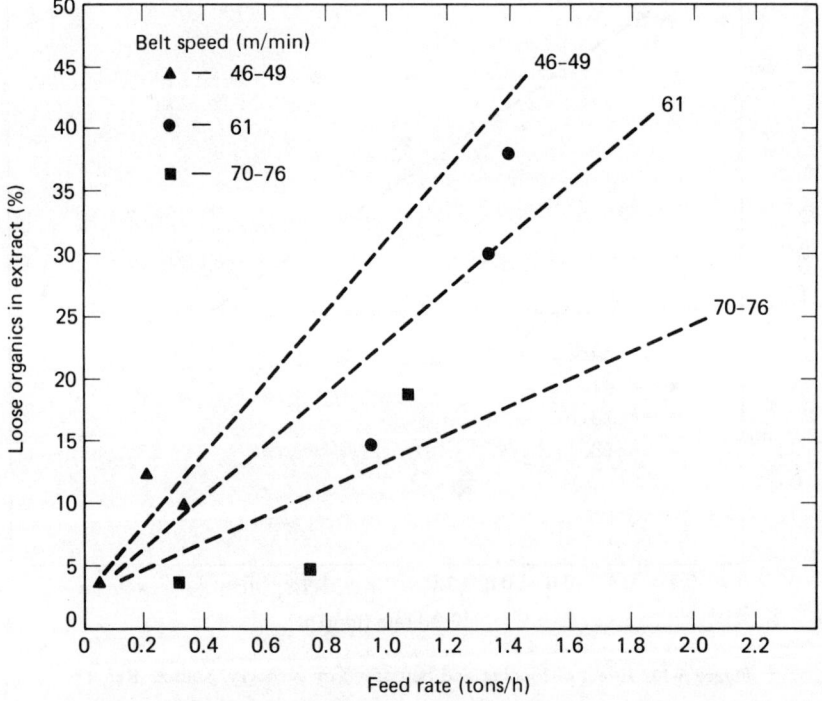

Figure 5-11. Effect of loading and belt speed on loose organics. Source: Ref. 11.

type of system, the feed rate will be at a rate of 0.54 to 0.72 tonne/h (0.6 to 0.8 ton/h) and that the recovery of aluminum will be in the range 85 to 90%. As belt loading increases, the extract purity will drop significantly.

Example 5-2

An eddy current separator (using two sets of magnets) must be chosen to achieve 80% recovery of aluminum. How fast can the material be fed (tons/h) to achieve this recovery.

From Fig. 5-10, the feed rate would be about 1.2 tons/h.

Vertical eddy current separators

In the vertical system, the separation of aluminum and some other nonferrous metals takes place during a vertical fall of the heavy fraction from previous separation processes. The refuse is fed down a vibrating chute which enhances density classification, and is then passed directly through the magnetic field of the eddy current module as illustrated in Fig. 5-12. The depth of the stream bed at the point of discharge into the eddy current field should not be much larger than the nominal characteristic particle size of the nonferrous metal products [9]. As the feed flows downward, the linear induction motor is phased to move the magnetic field upward; interaction with the magnetic field lifts the metal out of the refuse, after which it is deflected out of the system.

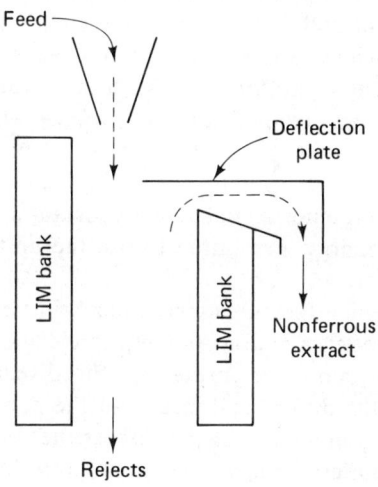

Figure 5-12. Vertical eddy current separator.

The eddy current induction module used in this process has fairly low power requirements, on the order of 1.5 to 7 kW for each tonne/h (2 to 10 hp for each ton/h) of material input, at a bulk density of about 480 kg/m^3 (30 lb/ft^3). Tests conducted on the vertical system indicate that particle size and feed rate are important factors in the systems performance [9].

To increase the efficiency of the separation processes, it has been suggested that the most efficient way to separate the aluminum fraction from other nonferrous metals is to screen the waste first [9]. Particles in the +1.3 to −5.1 cm (+$\frac{1}{2}$ to −2 in.) size could then be directed to one eddy current module and particles greater than 5.1 cm in size could be sent to a second module. Although both of these units selectively sort out aluminum, in the first module, while aluminum predominates, it is intermixed with other nonferrous materials. In the second module, the product normally is aluminum can stock and there is little contamination of the aluminum with other nonferrous materials.

Tests conducted on the vertical eddy current unit indicate that aluminum can stock recovery rates above 80% are typical and in some experimental runs have exceeded 95% [9, 10]. On the other hand, selectivity toward the recovery of aluminum foil has been much lower, but this appears to be the case for all eddy current separators. Recovery rates for foil ranged between 20 and 50% for this unit.

Inclined table separator

There has been considerable interest in developing methods for the recovery of aluminum and other nonferrous metals from municipal solid wastes by devices which are less energy intensive than the previously described eddy current separators. One such unit induces eddy currents in conductors by using an inclined table (or ramp) which has an array of alternate-polarity permanent magnets aligned just below a surface skin of austenitic stainless steel. In practice the table is inclined at an angle of approximately 45° to the horizontal. The important attribute of this system is that no electrical energy is required to run the unit. Fig. 5-13 illustrates such a metal separator.

The waste from which ferrous material has been removed enters the top part of the ramp through a chute. As the particles begin to slide down the ramp in a relatively thin layer formed by the discharge from the chute, they accelerate and the differential masses of the particles tend to thin out the waste stream even more as it moves across the slick stainless skin. This thinning out of the material helps to avoid any interference with the lateral movement of the metallic particles. The metal particles tend to be laterally displaced by the eddy current force. The resultant forces that are induced

Figure 5-13. Permanent magnet separator. Source: Ref. 14, 32.

by the magnets are perpendicular to the magnetic strips in the panel (i.e., parallel to the flux lines between alternate rows of opposite polarity magnets), while the nonmetallic particles move straight down the ramp and are not deflected. Figure 5-14 illustrates the influence of the induced eddy currents on the metallic particles which are deflected by the permanent magnetic field from the stream of nonmetallic particles. The lateral distance through which a metallic particle moves from its entrance at the top of the chute to the bottom, where it is discharged into the collection hopper, depends on many parameters, such as the electrical conductivity of the particle, its size and shape, the mass density, the length, surface conditions and angle of inclination of the ramp, and the strength and period of the magnetic field at the ramp surface.

The operation of this nonferrous metal separation process can be modeled by a thin metal ring which is allowed to slide down the ramp [13–16]. Figure 5-15(a) illustrates the magnetic flux distribution near the ramp's surface. The component of flux density B perpendicular to the ramp surface is labeled B_z. The voltage V induced in the metal ring can be

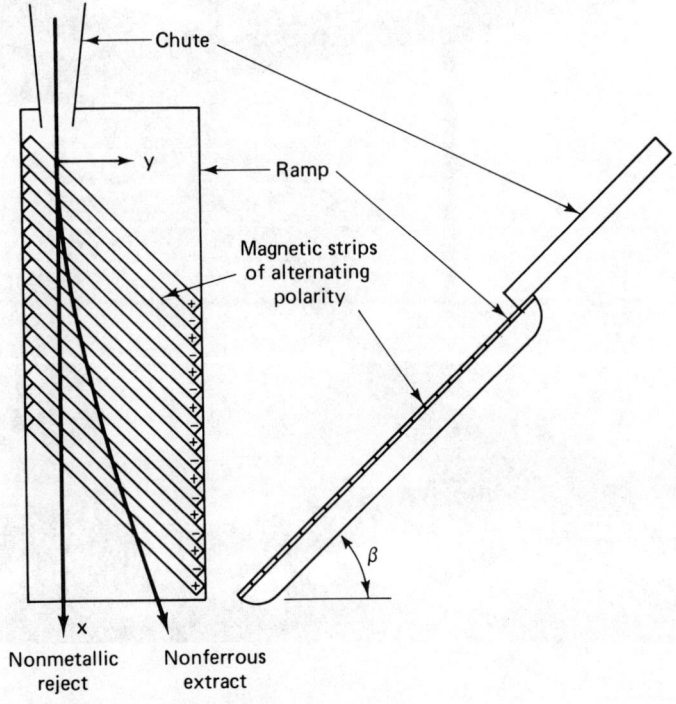

Figure 5-14. Schematic diagram of permanent magnet separator: (left) front view; (right) side view. Source: Ref. 15.

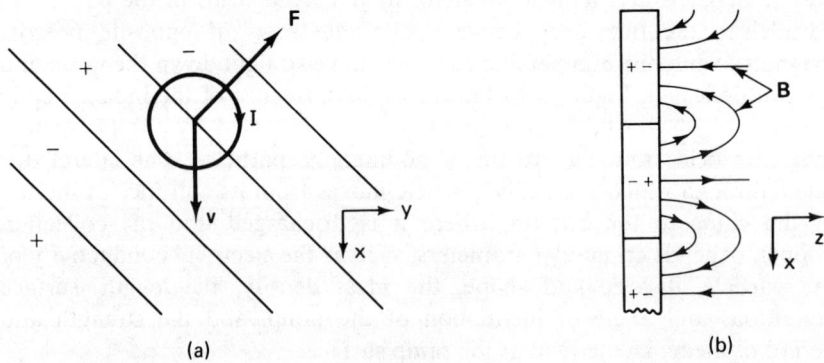

Figure 5-15. (a) Magnetic flux distribution near ramp surface; (b) induced forces in the ring-shaped conductor. Source: Ref. 16.

explained electrodynamically as being equal to the rate of change of the magnetic flux enclosed by that ring and is equivalent to equation on 5-1.

$$V = -\frac{d}{dt}\int B_z \, da$$

where V = voltage induced in ring, V
  A = area enclosed by ring, m²
  B_z = flux density, T

In Fig. 5-15(a), the center of the ring is just crossing the border between two magnetic strips of opposite polarity, and it is at this moment that the rate of change of the magnetic flux enclosed within that ring is at its maximum.

Since the metal ring is now carrying an induced electric current I, the Lorentz force **F** on the ring carrying that current is equal to the vector product of $q\mathbf{v}$ and the magnetic flux density **B**, where q is the charge on the electron and **v** is the average drift velocity of the electrons.

$$\mathbf{F} = \mathbf{I} \times \mathbf{B}$$

where **F** = induced force per unit length, on a conductor, N
  I = current, vector A
  B = magnetic flux density, T

In Fig. 5-15(a), it is evident that the flux is in the downward direction in the upper-right-hand portion of the ring and in the upward direction in the lower-left-hand portion of the ring. The resulting force **F** is thus perpendicular to the orientation of the permanent magnetic strips. The separator is capable of moving metallic particles primarily because there are no adjacent particles that would impede the lateral movement created by the induced force **F** [13, 15, 16].

The lateral component of the induced force **F** is proportional to the x component of the particle velocity v. Thus:

$$F_y = m\alpha v_x$$

where F_y = induced force on a conductor, N
  m = particle mass, kg
  v_x = particle velocity, cm/min
  α = constant

The constant has a dimension of inverse time, which depends on the size, shape, and electrical conductivity of the particle [13].

One design parameter that affects the movement of the particle on the ramp is the inclination of the ramp itself. Using a lead disk 3.8 cm in diameter, it was determined experimentally that the critical ramp inclination is 21° from the horizontal, as shown in Fig. 5-16, and that the

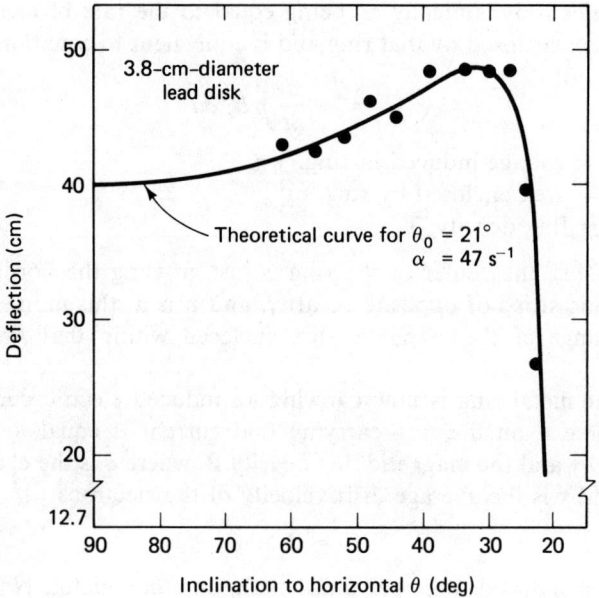

Figure 5-16. Measured and calculated deflection of a 3.8-cm diameter lead disk as a function of ramp angle. Source: Ref. 16.

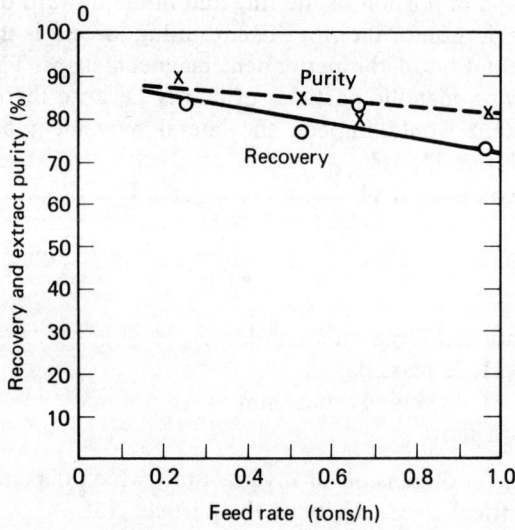

Figure 5-17. Recovery and purity as a function of feed rate. Ramp angle 30° to horizontal. Source: Ref. 13.

Eddy Current Separation

deflection of the particle is at a maximum when the ramp is inclined approximately 30 to 45° [16, 17].

Figure 5-17 illustrates the effect of feed rate on recovery [16]. While the feed rate increases by a factor of 5, the recovery of the separator decreases by only 10% and the purity drops to 80%. In this case, the results were reported in terms of all nonferrous metals, such as aluminum, copper, zinc, lead, and certain stainless steels, and thus the inclusion of the other metals tends to reduce the efficiency of aluminum removal.

The recovery of aluminum foil is poor, since complex and small eddy currents are generally established within foil material. It has been reported that at a feed rate of 0.45 tonne/h (0.5 ton/h), only a 38% of the aluminum foil was recovered. This problem of the recovery of aluminum foil is not unique to the permanent magnet system, and has plagued researchers using other types of eddy current systems previously described.

Fig. 5-18 illustrates the effect of moisture on recovery and extract purity. As would be expected, when the moisture content begins to rise, the material tends to stick to the inclined ramp, impeding the performance of the separator. At moisture contents in excess of 30% recovery efficiency drops off markedly.

Usually, a substantial portion of the feed to the permanent magnet separator is organic (some 85% of the feed material) and a substantial

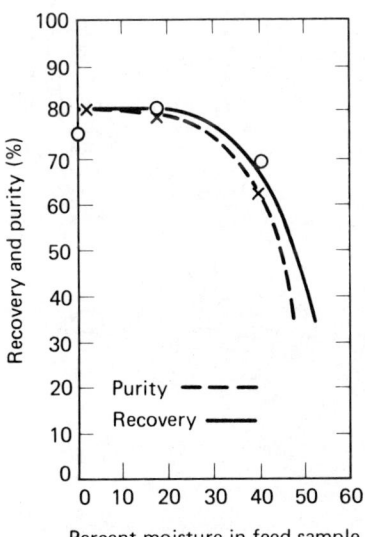

Figure 5-18. Moisture effects on recovery; a permanent magnet system. Source: Ref. 12.

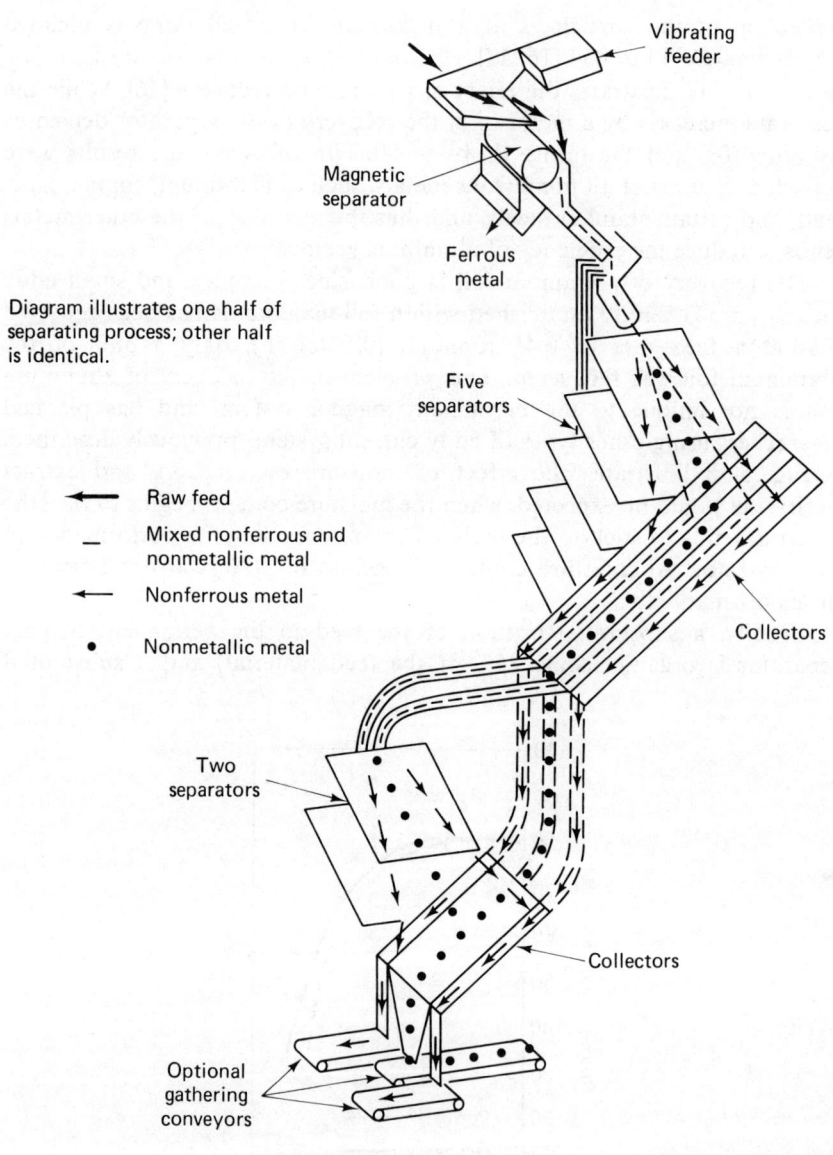

Figure 5-19. Schematic of prototype permanent magnet separation device. Diagram illustrates one half of separation process; other half is identical. (Courtesy of Iowa Manufacturing Co., Cedar Rapids, Ia.)

portion that reports to the tails or rejects. In prototype design, however, by the use of a cascade-type system (ramps in series) the amount of organics and other metals that report to the concentrate is nominal [18].

Figure 5-19 illustrates a suggested configuration of the permanent magnet concept in a prototype design unit. Presuming that the nonferrous metal separator is placed in the resource recovery system after the material has been shredded, air-classified, and magnetically separated, five 6- by 8-ft inclined separators are mounted in a cascade pattern in such a way as to permit varying the inclination of the panels from 35 to 55°.

The process appears to be cost-effective, is reliable, and is relatively simple to operate. Potential drawbacks to the system include a need for a fairly dry waste and a feed relatively free of ferrous materials. Advantages of the system are that it requires no energy (other than that to raise the feed to the top of the ramp) and should develop very few, if any, maintenance problems because there are essentially no moving parts in the separator.

ELECTROSTATIC SEPARATION

Theory of Operation

Charged particles under the influence of electrostatic forces obey laws of attraction and repulsion similar to those describing the interaction of permanent magnets. The fundamental equation of electrostatic separation is *Coulomb's law*, which describes the force of attraction between point charges of opposite electric polarity (or, equivalently, the repulsion between like charge particles):

$$F = \frac{q_1 q_2}{4\pi\varepsilon_0 r^2}$$

where q_1, q_2 = charges on two particles
r = distance between particles, m
ε_0 = permittivity constant = 8.854×10^{-12} C^2/N-m^2

Point charges are charged objects whose size is considerably smaller than the distance between them.

The premise of electrostatic separation is that certain materials can be coded by being electrically charged, then separated by being attracted to an opposite-charged electrode (or by being repelled from a like-charged electrode). An electrical conductor can be charged by adding electrons to it, or withdrawing electrons from it; the conductor then becomes an electrode, as for example when a conductor is attached to one terminal of a battery or a voltage generator. Two surfaces of a nonconductor can be oppositely charged by placing the nonconductor in an electric field. The simplest geometry for measuring *dielectric* properties (i.e., the capacity of insulating materials to store electric charge) is that of a parallel-plate capacitor. For dielectric materials, a capacitor represents an "experimental volume" in much the same way that a solenoid permits the measurement of magnetic properties for magnetic materials. The equation describing the behavior of a parallel-plate capacitor can be used to introduce the important parameters in dielectric behavior:

$$D = \varepsilon \times E$$

where D is called the *electric displacement* and equals the charge stored on one electrode divided by the area of that electrode:

$$D = \frac{q}{A}$$

E represents the *electric field intensity*, or *voltage gradient* between the electrodes:

$$E = \frac{dV}{dx}$$

where V = voltage
x = distance

The dielectric behavior of an insulating material is defined by its permittivity ε. Since dielectric behavior is usually measured in a circuit that compares the capacitance of a capacitor filled with the material being measured with the capacitance of an identical capacitor containing air, or a vacuum, dielectric behavior is more commonly expressed in terms of a *dielectric constant* K:

$$\kappa = \frac{\varepsilon}{\varepsilon_0}$$

If an insulator is subjected to very high field intensities, it may exhibit nonlinear dielectric behavior, or it may break down. The mechanical analog of linear dielectric behavior is linear elasticity, so the analog of nonlinear behavior is plastic deformation. Just as a low-carbon steel deforms plastically past the yield point, a nonlinear dielectric (called a *ferroelectric*) will exhibit a decreasing instantaneous permittivity

$$\varepsilon = \frac{dD}{dE}$$

Electrostatic Separation

at very high field intensities. The mechanical analog of dielectric breakdown is fracture. Just as tensile strength measures the stress at which mechanical breakdown occurs, the *dielectric strength* measures the electric field intensity at which electric breakdown occurs. When this electric field intensity (usually measured in volts/mil, where 1 mil = 0.001 in.) is reached, electrons are figuratively "torn" from their atoms, and internal arcing occurs in the insulator.

In addition to coding on differences in dielectric behavior, electrostatic separation processes also code particles on the basis of a difference in electrical conductivity. Electrical conductivity is defined by

$$\frac{1}{A}\frac{dq}{dt} = \sigma E \tag{5-2}$$

where A = cross sectional area of the conductor
dq/dt = current (usually measured in amperes)
σ = electrical conductivity of the material
E = electric field intensity or voltage gradient dV/dx (where x is the length of the conductor)

For a linear voltage gradient $E = V/x$, Eq. 5-2 can be combined with Ohm's law

$$V = IR$$

when $I = dq/dt$, to yield

$$\sigma = \frac{1}{R}\frac{x}{A}$$

Conductors are often characterized by their electrical resistivity, e', which is the reciprocal of conductivity

$$e' = R \times \frac{A}{x}$$

Taggart developed a useful description of the electrostatic forces and how they are utilized in the mining separation process [5]. The principle was illustrated by a negatively charged insulated sphere which is brought into the proximity of an uncharged particle of an electrical conductor suspended by an insulated string. Polarization is induced in the particle as electrons are repelled from its surface and the particle is attracted toward the sphere, since a sphere possesses a converging field of force. When the particle which is attracted to the sphere touches the surface, it becomes negatively charged immediately by conduction and is then repelled by the Coulomb forces. However, should either the sphere or the particle be a poor conductor, there may be a relatively high interfacial resistance, which causes a delay before the particle becomes charged by conduction and is repelled.

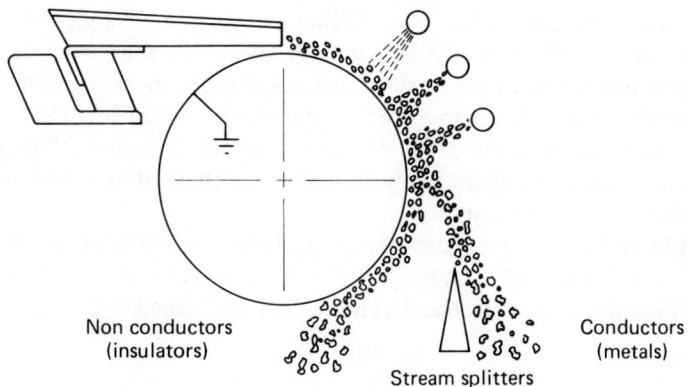

Figure 5-20. Electrostatic separation. Source: Ref. 31.

One way of using electrostatic forces is to code and separate on the basis of electrical behavior [20]. In this type of electrostatic separator, a mixture of conductors and nonconductors is fed onto the charged, conducting surface of a drum. Conducting particles become charged by induction, attracting them to the drum surface. Once contact is established between the particle and drum, however, the particle loses some of its charge from one surface, resulting in a net charge which is the same as that on the drum. Electrostatic repulsion then rejects the conductors from the drum. The electric field intensity in this case is not high enough to polarize nonconducting particles, so they just fall off the drum, essentially unaffected by the electric field (Fig. 5-20).

Another way of using electrostatic forces to separate conductors from nonconductors is illustrated in Fig. 5-21. In this case the roll is a grounded conductor, and a fine wire is used to "sprinkle" electrical charges on the particles passing between it and the roll. The particles thus possess a net charge opposite to that of the roll, before contacting the roll, and are attracted to it. When a conducting particle subsequently contacts the roll, it loses its excess charge by conduction, becomes neutral, and just drops off the roll. Nonconducting particles are also attracted to the roll; however, when a charged nonconducting particle contacts the roll, it retains its charge and remains attracted to the roll. These nonconducting particles can then be doctored off the roll into a separate hopper.

Another approach to electrostatically coding and separating materials relies on differences in dielectric constants. This method codes and separates nonconductors from one another rather than separating them from conductors. It involves suspending particles of different dielectric constant in a liquid medium with a dielectric constant intermediate between the two components to be separated; a converging electric field is then established within the suspension. Particles of higher dielectric constant than the

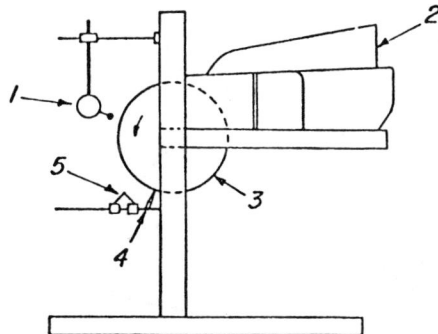

1	Combination 4-inch aluminum electrode with wire electrode
2	Vibratory feeder
3	Grounded rotating drum
4	Brush
5	Adjustable stream splitter

Figure 5-21. Electrostatic separation. Source: Ref. 20.

medium will travel in the direction of increasing electrical field intensity. Particles of lower dielectric constant than the medium will travel in the direction of decreasing electric field intensity. Although this method has not been used commercially for processing MSW, it offers interesting possibilities for the future.

Electrostatic Separation Equipment

Recoveries of up to 99.4% of the plastic and 100% of the paper in separate fractions have been achieved in pilot scale electrostatic separators [20]. The moisture content of the paper has a significant impact on the separation, as illustrated in Fig. 5-22, showing that the recovery goes up with an increase in moisture content. Interestingly, this is often contrary to the efficiencies of other unit operations, which show a marked decrease in the extract fraction as the moisture content increases. Voltages ranging from 35 to 50 kV with electrodes spaced on about 15-cm (6-in.) centers have been used effectively. The removal of plastics from MSW is possible by means of electrostatic separation if the feed has a moisture content of at least 20 to 25% in order to permit adequate charging of the particles.

Another promising application of the electrostatic separation process is the concentrating of metals that come from the air classifier rejects. Of primary interest in this metals separation process is aluminum, and experimental data indicate that the electrostatic separation process can upgrade

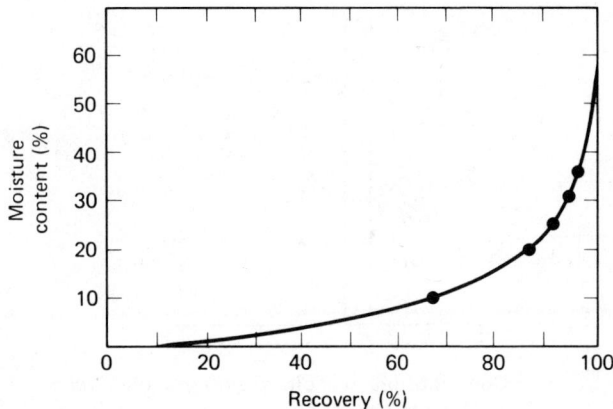

Figure 5-22. Effect of moisture on Electrostatic Separator recovery. Source: Ref. 19.

TABLE 5-4
Electrostatic Separation on Air-Classifier Underflow Products

	Sample A			Sample B		
Product	% Weight	% Al	Al Distribution	% Weight	% Al	Al Distribution
Feed	100.0	7.7	100.0	100.0	13.5	100.0
Nonconductors (Reject)	90.4	1.4	16.4	88.1	3.4	22.2
Conductors (Extract)	9.6	67.2	83.6	11.9	88.2	77.8

Source: Ref. 21.

the feedstock from a metal purity of 40% to a purity of approximately 82%, with 96% recoveries possible [20]. It has been found that a multiple electrode separator unit (with two, three, and four stages) can develop good recovery of materials at relatively high feed rates; 36 kg/h/cm (200 lb/h/in.) [21]. Table 5-4 illustrates the products of separation on an air-classifier underflow product.

MAGNETIC FLUIDS

Theory of Operation

The development of magnetic fluids is a relatively recent technological advancement inspired by NASA research. When the colloidal and magnetic states of matter are combined, a magnetic colloid results, the be-

Magnetic Fluids

havior of which may be modified by the influence of a magnetic field. Such a characteristic enables this magnetic colloid, generally called a magnetic fluid, to exhibit properties which potentially permit its use in separation of nonferrous components of the solid waste stream.

A magnetic fluid generally contains ferromagnetic particles, but ferrimagnetic particles such as ferrites may also be used as the colloid. The ferro- or ferrimagnetic materials are suspended as submicron particles (about 100 Å in diameter) in a hydrocarbon liquid base, such as kerosene or heptane. The fluids are generally quite stable and exhibit slight viscosity changes while retaining their fluid properties in the presence of a magnetic field [22].

The magnetic fluid exhibits an attraction to a magnet, which appears to be due to the linking of the magnetic particles with the solvent [23], but it behaves as a homogeneous suspension. In effect, magnetic levitation forces (sometimes known as "antigravity" forces) are created within the magnetic fluid. It is this magnetic response that is exploited to separate materials of varying specific gravities.

The properties of magnetic fluids can be demonstrated by placing alumina beads in a magnetic fluid [24] (Figure 5-23). If the magnetic fluid uses a hydrocarbon-base solution (kerosene and oleic acid), as the solvent, it has a density of about 0.8 g/ml and contains water as a contaminant. When the alumina pellets with a density of 3.8 g/ml are placed in this suspension, both the water and alumina beads are initially submerged below the magnetic fluid. Fig. 5-24 illustrates the effect of magnetic forces created by the placement of a permanent magnet (a laboratory horseshoe magnet) beneath the solution. This magnet, in effect, creates a magnetic field in which the magnetic fluid appears to have a density greater than 3.8 g/ml, and the water and alumina "float" or levitate on top of the magnetic fluid. In effect, the magnetic particles in the magnetic fluid are attracted to the magnet, and the solvent moves along with the particles in such a way that both appear to be a continuous one-phase system.

The apparent density of the magnetic fluid is increased by an increasing magnetic field gradient, and separation of the material on the basis of density is possible. This property suggests that selective separation of particles in a solid waste stream at any particular "density cutoff" is possible. While this simple experiment illustrates the principles of such a separation scheme, it does not illustrate a practical process.

Nonvertical levitation forces, generated by the placement of the magnetic field in such a location as to permit the multiple separation of particles in the fluid, is possible. The process can be described as a density spectrograph, whereby particles are diverted in different trajectories according to their specific densities [24]. Figure 5-25 illustrates diagrammatically the force orientations in material separation using the concept of nonvertical levitation. The gravitation and levitation forces acting on a

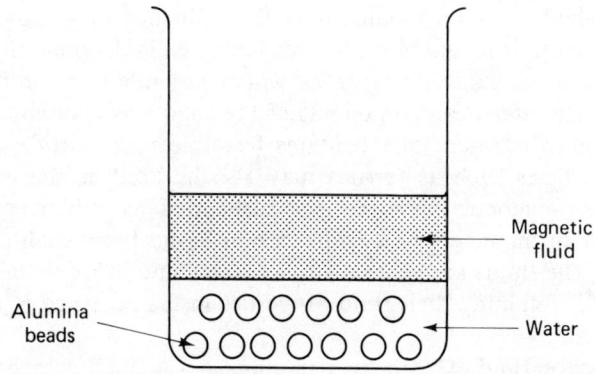

Figure 5-23. Magnetic fluid contaminated with water and alumina beads. Source: Ref. 22.

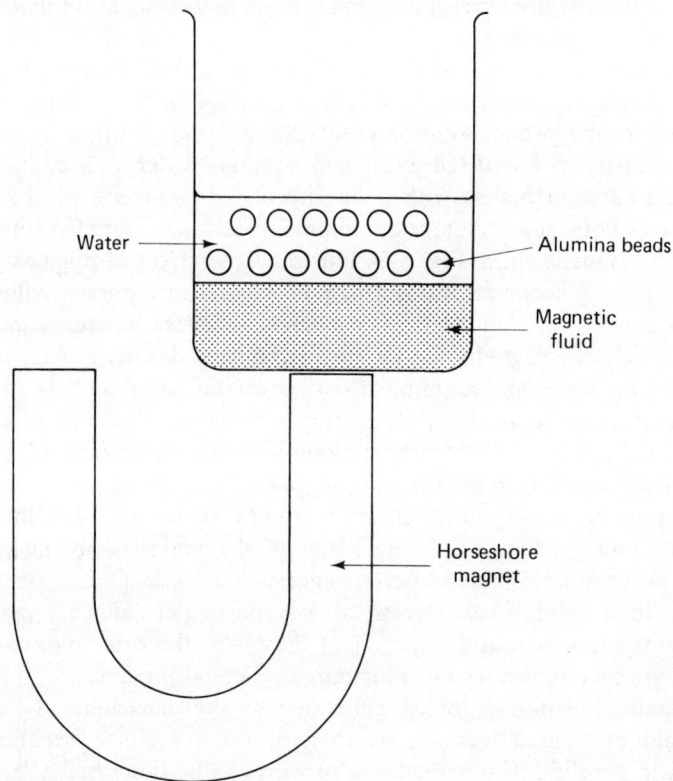

Figure 5-24. Magnetic fluid contaminated with water and alumina beads under the influence of applied magnetic field. Source: Ref. 22.

Magnetic Fluids

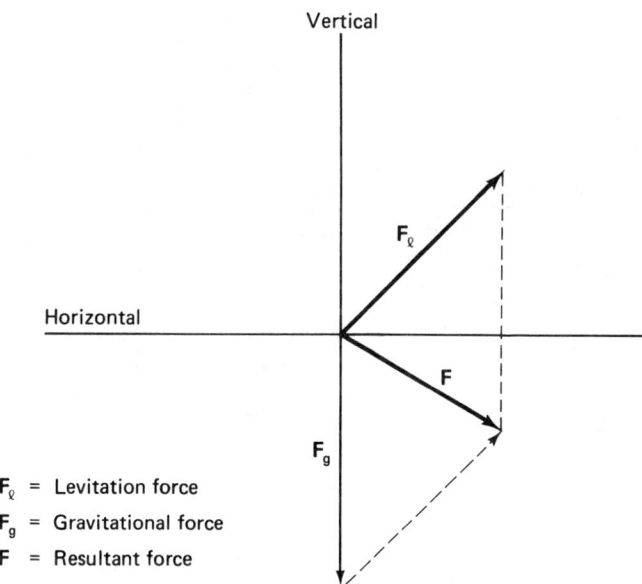

Figure 5-25. Diagram of force orientations in material magnetic fluid separation. Source: Ref. 26.

specific particle have a resultant force that is sufficient to permit selective separation. Obviously, a heavier particle will have a larger deflection angle from the horizontal, while a lighter object will have a smaller deflection angle. By the proper placement of collection devices to separate the particle, the varying deflections of these particles within the magnetic fluid can be exploited and the materials removed.

Magnetic Fluid Separators

The use of magnetic fluids may be attractive for large-scale operations to separate the denser materials and minerals found in municipal solid wastes. Figure 5-26 illustrates the separation of a "typical" nonferrous incinerator residue containing aluminum, zinc, copper, and other miscellaneous products [25].

Although heavy liquids and dense-media slurries can also be used to remove most of these same particles (see Chapter 4) the distinct advantage of a magnetic fluid is that many more materials with a broad range of

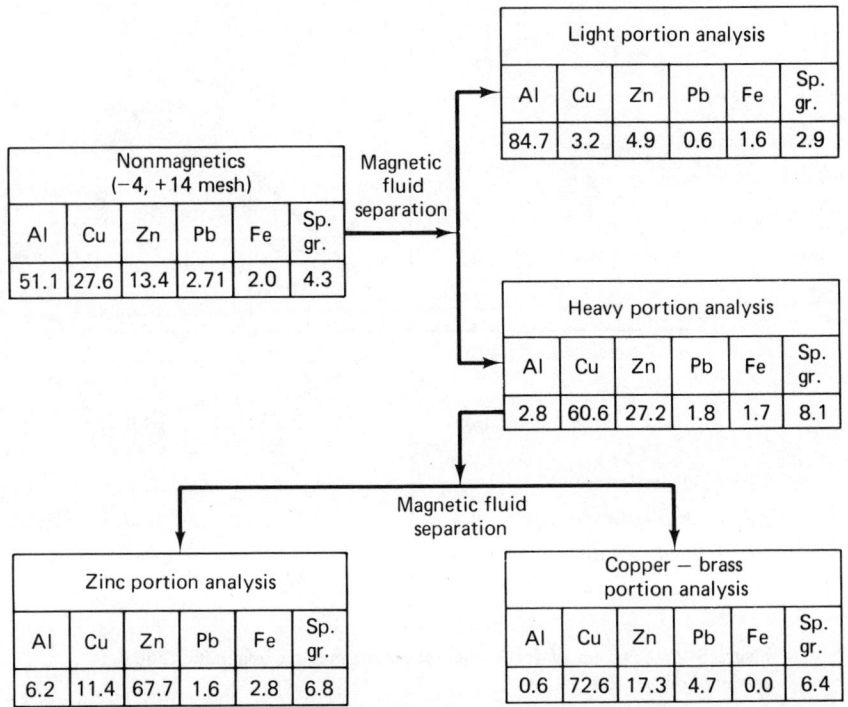

Figure 5-26. Chemical analysis of incinerator residue fractions after separation by magnetic fluids. Source: Ref. 25.

densities can be removed. Figure 5-27 illustrates the density of various nonferrous metals and the range of those metals which heavy liquids, dense-media slurries, and the magnetic fluids, will separate. For example, dense-media slurries are really only effective in separating aluminum from other materials. The proper application of a magnetic fluid may permit the separation of such diverse items as platinum and aluminum.

Figure 5-28 illustrates a typical prototype magnetic fluid separation process. In the operation, incoming shredded mixed metals are placed into a static pool of magnetic fluid suspended between the poles of an electromagnet. Depending on the apparent density of the material being separated, the material must either float to the top of the pool or sink to the bottom. Floating pieces and sinking pieces are conveyed to separate locations for further washing and removal of the magnetic fluid. In this operation, mixtures of three or more components, such as aluminum, copper, and lead, may be separated in a multipass operation. Aluminum

Magnetic Fluids

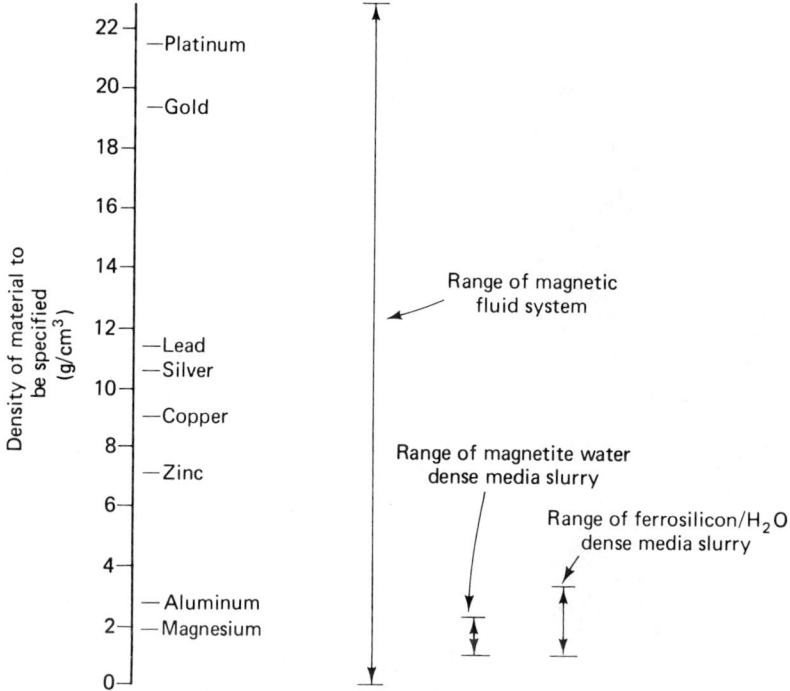

Figure 5-27. Capability of various separation devices to separate metal components from MSW.

would be separated in the first pass, and then the zinc and the lead separated in the second pass by changing the apparent density of the magnetic fluid (increasing the magnetic field strength). To a large extent, design of such systems must rely on information from the various manufacturers of the magnetic fluids. Few empirical design data are available in the literature.

PARTING SHOTS

Although we can "make gold" out of all the components of solid waste, the only true reason one separates materials is because there is a market for the recovered product. Simple (sometimes not so simple?) economics are the true motivating force to develop a particular separation process.

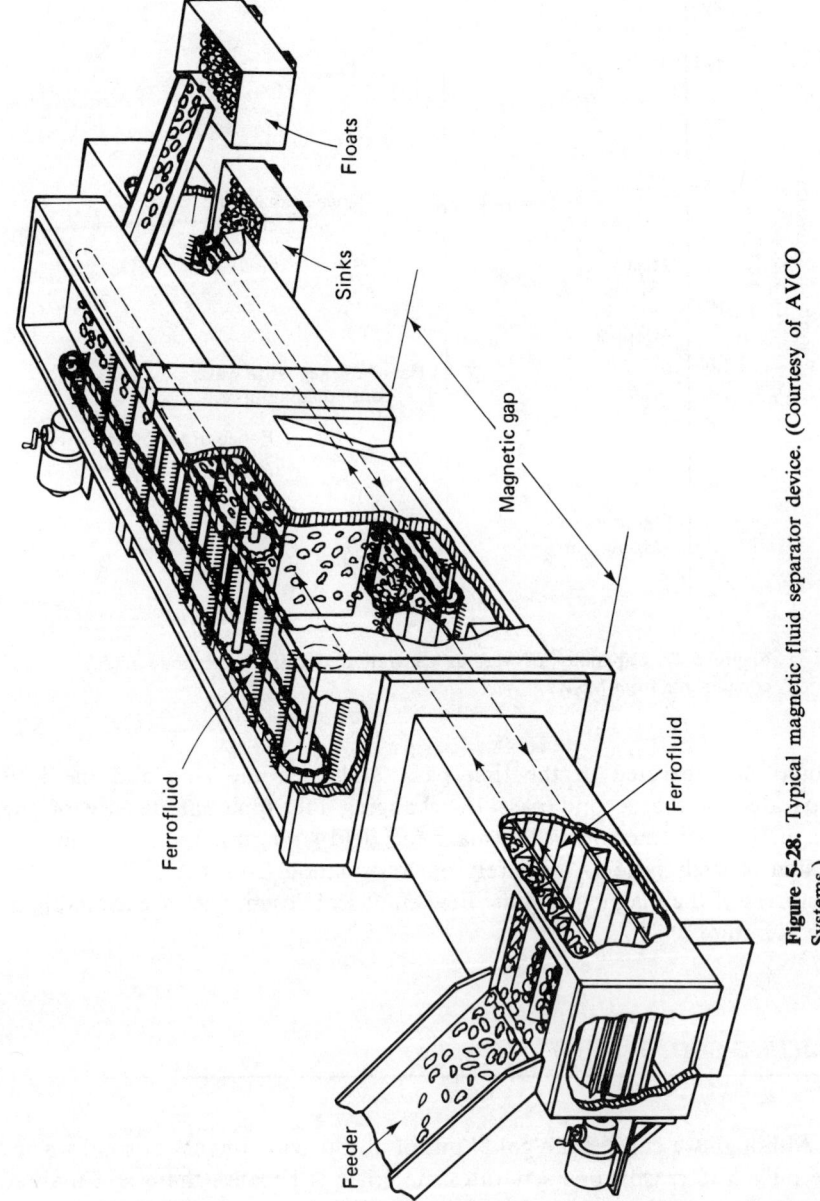

Figure 5-28. Typical magnetic fluid separator device. (Courtesy of AVCO Systems.)

PROBLEMS

5-1. In the expression for purity [see Ch. 4] it is assumed that the product extracted by a magnet is "pure"; that is, we define a magnetic particle as one picked up by a magnet. How then can the extract from a magnetic separation not be always 100% pure?

5-2. Assuming that the material from the heavy fraction of air-classified MSW results in the following characteristics, compute the purity and recovery.

$$\begin{aligned} \text{Feed:} \quad & \text{Mags} = 2000 \text{ kg/h} \\ & \text{Tails} = 5000 \text{ kg/h} \\ \text{Extract:} \quad & \text{Mags} = 1800 \text{ kg/h} \\ & \text{Tails} = 300 \text{ kg/h} \\ \text{Reject:} \quad & \text{Mags} = 200 \text{ kg/h} \\ & \text{Tails} = 4700 \text{ kg/h} \end{aligned}$$

5-3. (a) Develop equations for a polynary separation process that characterizes the purity of the tails, mags, and middlings.
(b) What would similar equations be for recovery?

5-4. Speculate on the effect the thickness of MSW on the belt-holding-type separator would have on recovery and purity.

5-5. Explain why zinc had such a high weight ratio (in percent) to aluminum as opposed to all other materials separated in the LIM described in Table 5-3. See Table 5-1 for electrical conductivities.

5-6. Based on Fig. 5-10, can a mathematical relationship be developed between the al-mag feed rate, the belt speed, and the loose organics in the al-mag product?

5-7. What methods might a community use to control the moisture in their solid waste to permit the use of the permanent magnet system for separation of aluminum?

5-8. (a) How would the magnetic fluid device need to be operated to remove aluminum, zinc, and silver?
(b) Would it be useful for the removal of either the glass or plastics noted in Problem 4-8?

5-9. Describe where in a resource recovery facility you would install an electrostatic separation unit. Why did you place it there?

REFERENCES

[1] GUNTHER, C. G., *Electro-magnetic Ore Separation*, McGraw-Hill Book Company, 1909.

[2] GAUDIN, A. M., *Principles of Mineral Dressing*, McGraw-Hill Book Company, New York, 1939.

[3] ISRAELSON, A. F., *Magnetic Fundamentals Simplified*, Eriez Magnets, Erie, Pa.

[4] HALACSY, A. A., "Electromagnets," Nevada University, Reno, 1970.

[5] TAGGART, A. F., *Elements of Ore Dressing*, John Wiley & Sons, Inc., New York, 1951.

[6] SPENCE, P., "Magnetic Separation—A Laboratory Study," Duke University, Durham, N.C., Apr. 1977.

[7] Automatic Self-Cleaning Overhead Magnetic Separator, Promotional literature for Eriez Magnetics, Erie, Pa.

[8] "Magnetic Separation of Nonferrous Metal," Annual Report, Department of Physics and Astronomy, Vanderbilt University, 1971.

[9] SOMMER, E. J., JR., and R. KENNY, "An Electromagnetic System for Dry Recovery of Nonferrous Metals from Shredded Municipal Solid Wastes," *Proc. 4th Miner. Waste Util. Symp.*, ITT Research Institute, Chicago, 1975, pp. 78–84.

[10] ABERT, J. G., "Aluminum Recovery—A Status Report," *NCRR Bull.*, 7(2), 10 (1977).

[11] ALTER, H., S. NATOF, and L. C. BLAYDEN, "Pilot Studies, Processing MSW and Recovery of Aluminum Using an Eddy Current Separator," *5th Miner. Waste Util. Symp.*, ITT Research Institute, Chicago, 1976, pp. 161–168.

[12] BLAYDEN, L. C., "Aluminum Recovery with Eddy Currents," Paper presented at Soc. Min. Eng. AIME, Las Vegas, 1976.

[13] SPENCER, D. B., and E. SCHLOEMANN, "Recovery of Non-ferrous Metals by Means of Permanent Magnets," *Resour. Recovery Conserv.*, 1, 151–165 (1975).

[14] SCHLOEMANN, E., "Self-cleaning Non-ferrous Metal Separator," *IEE Trans. Magnetics*, **MAG-13** (5), 1496–1498 (Sept. 1977).

[15] SCHLOEMANN, E., "Separation of Nonmagnetic Metals from Solid Waste by Permanent Magnets I Theory," *J. Appl. Phys.*, **46**, 5012–5021 (Nov. 1975).

[16] SCHLOEMANN, E., "Separation of Nonmagnetic Metals from Solid Waste by Permanent Magnets, II Experiments on Circular Disks," *J. Appl. Phys.*, **46**, 5022–5029 (Nov. 1975).

[17] SCHLOEMANN, E., "Separation of Nonferrous Metal in Automobile Scrap by Means of Permanent Magnets," *Proc. 6th Miner. Waste Util. Symp.*, ITT Research Institute, Chicago, 1978.

[18] Nonferrous Metal Separator, Promotional literature for Iowa Manufacturing Co., Cedar Rapids, Iowa.

[19] SULLIVAN, P. M., M. H. STANCZYK, and M. J. SPENDLOVE, *Rep. Invest. RI 7760* (description of feed materials), Bureau of Mines, U.S. Department of the Interior, 1972.

[20] GRUBBS, M. R., and K. H. IVEY, *Recovering Plastics from Urban Refuse by Electrodynamic Techniques*, Tech. Prog. Rep. 63, Bureau of Mines Solid Waste Research Program, Department of the Interior, 1972.

[21] KNOLL, F. S., "Recovery of Aluminum by High Tension Separation," AIME Preprint 74-B-28, AIME Ann. Meet., Dallas (1974).

[22] ROSENSWEIG, R. E., "Material Separation Using Ferromagnetic Liquid Techniques," *J. AIAA*, 3 (483), 969 (Dec. 1969).

[23] ROSENSWEIG, R. E., R. KAISER, and G. MISKOLZXY, "Viscosity of Magnetic Fluids in a Magnetic Field," *J. Colloid Interface Sc.*, 29 (680) (1969).

[24] KAISER, R., and G. MISKOLCZY, "Magnetic Properties of Stable Dispersions of Subdomain Magnetic Particles," *J. Appl. Phys.*, 41 (1064) (1970).

[25] KHALAFALLA, S. E., "Separating Nonferrous Metals in Incinerator Residue Using Magnetic Fluids," *Sep. Sci.*, 8, 161–178 (1973).

[26] KHALAFALLA, E., "Magnetic Fluids," *Chemtech.*, 540–546 (1975).

[27] ROBERTS, K., Personal communication, AVCO Systems Division, Lowell, Mass., Aug. 1975.

[28] Midwest Research Institute, "Study of Processing Equipment for Resource Recovery Systems," EPA Contract No. 68-03-2387, 1979.

[29] MOREY, B. and B. RUDY, "Aluminum Recovery From Municipal Trash by Linear Induction Motors," Presented at 78th National Meeting of the AICHE, Salt Lake City, Utah, August 18–21, 1974.

[30] SHENK, C., discussion on visit to El Cagon research facility, *Occident. Res.*, Dec. 1977.

[31] MAKAR, H. V., and R. S. DE CESARE, "Unit Operations for Nonferrous Metals Recovery," *Resource Recovery and Utilization*, ASTM STP 592, 1975, pp. 71–88.

[32] SCHLOEMANN, E. Personal communication, Oct. 1979.

6 BIOCHEMICAL PROCESSES

Municipal refuse contains about 75% organic material. This material can be converted to useful products by simple combustion, as discussed in Chapters 7, 8, 9, or biochemical processes, which is the topic of this chapter.

The three components of MSW which are of most interest in the bioconversion processes are garbage (food waste), paper products, and yard wastes. The last two are especially valuable in biochemical processes as a source of cellulose.

The garbage fraction of refuse varies with geographical location and season. Dietary habits, of course, affect its composition and quantity, as does the standard of living. Kitchen garbage grinders in more-affluent communities transfer much of the putrescible waste from the refuse stream to the sewerage system and the reduction of the garbage fraction is a continuing trend in the United States and in many other countries.

The garbage fraction also has by far the highest moisture content of any constituent in MSW, but the moisture is rapidly transferred to absorbent materials such as newspapers as soon as contact is made. Garbage also tends to be well mixed in refuse and moisture transfer can be so efficient that it is often difficult to find identifiable bits of garbage in mixed refuse, other than the obvious large pieces, such as orange peels or apple cores. Garbage is even better distributed in refuse if the waste is shredded, a fact that allows shredded refuse to be disposed of on land without the necessity of earth cover, since vermin cannot find sufficient food for survival.

TABLE 6-1
Organic Analysis of Refuse

		Percent
Organics		54.4
cellulose, sugar, starch	46.6	
lipids (fats, oils, waxes)	4.5	
protein	2.1	
other organics (e.g. plastics)	1.2	
Inorganics		24.9
Moisture		20.7
		100.0

Source: Ref. 1.

The fraction of paper in refuse tends to remain fairly stable throughout the year, while yard wastes in most locations are obviously seasonal. The latter also vary from week to week, often reflecting the weather, and thus yard wastes are seldom a stable source of organics for resource recovery operations.

On the average, the organic components of refuse can be described by an analysis as shown in Table 6-1. The largest single constituent is, of course, cellulose.

The five methods of biochemical energy conversion discussed herein all utilize the organic fraction of refuse. The first three methods (two means of anaerobic digestion and composting) are broad-spectrum processes, where the specific organisms responsible for the bioconversion are not identified or isolated, and the processes are described by empirical data. The last two techniques, enzyme and acid hydrolysis, on the other hand, involve the use of known chemicals and specific chemical pathways to produce a known pure product (glucose). Because composting and anaerobic digestion do not begin with raw material comprised of only one chemical, the specific biochemical reactions involved in these processes are numerous. Accordingly, it is not possible to approach them as one would describe the hydrolysis of cellulose. The first two sections thus lack in scientific rigor from the biochemical viewpoint. Research has not provided the necessary information, and the variable nature of the raw material makes it unlikely that definitive analyses will ever be possible, or necessary.

METHANE GENERATION BY ANAEROBIC DIGESTION

When organic matter decays under anaerobic (without free oxygen) conditions, the end products include such gases as methane (CH_4), carbon dioxide (CO_2), small amounts of hydrogen sulfide (H_2S), ammonia (NH_3), and a few others. The recognition that methane is an excellent fuel long ago prompted wastewater treatment plant design engineers to digest (decompose) waste solids and capture this gas for use in heating buildings and running machinery in the treatment plant. The total quantity of methane generated in a wastewater treatment plant, however, is not sufficient to consider the production of pipeline gas. On the other hand, the ever-increasing shortage of natural gas has prompted the active study of the potential for producing methane by decomposing refuse.

The two ways of generating methane are to capture the gases produced in landfills, or to pretreat the refuse and digest it in tanks similar to those used in wastewater treatment plants. Methane generation in landfills is discussed in the following section, and methane generation in anaerobic digesters is discussed in this section. Much of the following anaerobic decomposition theory applies to both processes, however.

Theory of Anaerobic Decomposition in Mixed Digesters

The two basic metabolic pathways for the decomposition or degradation of wastes are *aerobic* (with oxygen) and *anaerobic* (in the absence of oxygen). While aerobic system might be generally represented as

$$\text{complex organics} + O_2 \xrightarrow{\text{microorganisms}} CO_2 + H_2O + NO_3^- + SO_4^{2-}$$
$$+ \text{other stable products}$$

the anaerobic decomposition of organics can be described as

$$\text{complex organics} \xrightarrow{\text{microorganisms}} CO_2 + CH_4 + H_2S + NH_4^+$$
$$+ \text{other partially stable products}$$

The end products in aerobic decomposition are all stable, possessing no additional energy to be used by decomposing organisms (they are at their highest oxidation state). The products of anaerobic decomposition, on the other hand, still contain energy. Ammonia and hydrogen sulfide could be still further oxidized, and methane contains considerable energy.

The microorganisms responsible for anaerobic decomposition can be divided into two rather broad categories. The first of these groups hydrolyzes and ferments the complex organic compounds to more simple organic forms such as acetic and propionic acids. These hardy organisms, which are facultative and anaerobic, are commonly known as *acid formers*, since their end products are mostly organic acids.

The second major group of organisms are known as *methane formers*, and as the name implies, their function is to convert the organic acids to methane. These organisms are strict anaerobes and have very slow growth rates—two characteristics that cause considerable problems in anaerobic processes in wastewater treatment, and will similarly plague anaerobic decomposition of refuse.

These methane-formers are very sensitive to various environmental factors. It was already noted that they are strict anaerobes; oxygen is, in fact, toxic to them. They are also quite sensitive to temperature. Two different groups of methane formers seem to exist, with one group (mesophilic) operating best around 30 to 38°C (85 to 100°F) and a second group (thermophilic) operating best around 50 to 58°C (120 to 135°F).

The methane formers also require stable and neutral pH. Sufficient alkalinity (resistance to pH drop) should be present to prevent the pH from falling below 6.2. Finally, methane formers are extremely sensitive to the presence of toxic materials, such as heavy metals and pesticides.

During the acid-forming stage, the first step in the process involves extracellular enzymes which break down the large complex organic molecules. For example, the enzymes cellobiase and cellulase break down cellulose to glucose, and lipase breaks fat to shorter-chained fatty acids. This process is energy-consuming.

Following this, the bacteria metabolize the glucose and other products into organic acids, mostly acetic and propionic. This methane formation is performed by a number of organisms which have specific substrates and roles in the overall reaction. It is recognized that the two reactions

$$CH_3COOH \rightarrow CH_4 + CO_2$$
$$4CH_3CH_2COOH + 2H_2O \rightarrow 7CH_4 + 5CO_2$$

for acetic and propionic acids respectively are actually the net results of a large number of steps. The prevailing theory states that the organics are first, in fact, fully oxidized to CO_2, and that this is followed by reduction reactions which convert some of the CO_2 to CH_4.

The resulting gas from this process varies in composition but averages about 60% methane, with a heating value between 4700 kJ/m^3 and 6500 kJ/m^3 (500 and 700 Btu/ft^3) [2].

The total amount of gas theoretically available from the anaerobic digestion of MSW is considerably more than has been captured to date in pilot-plant facilities. About 54% of the volatile solids have been found to escape the digester [3] and have not been converted to CO_2 and CH_4.

The digestion of refuse involves hardware and a flow diagram not unlike the anaerobic digestion used in wastewater treatment. In the case of MSW, the organics are first separated from the refuse and are slurried with sewage sludge or some other suitable liquid. The resulting mixture is digested in a heated and enclosed tank. The gas is captured either under a floating cover or in a separate tank. The residual of the digestion process is a dark odiferous slurry which must be disposed of.

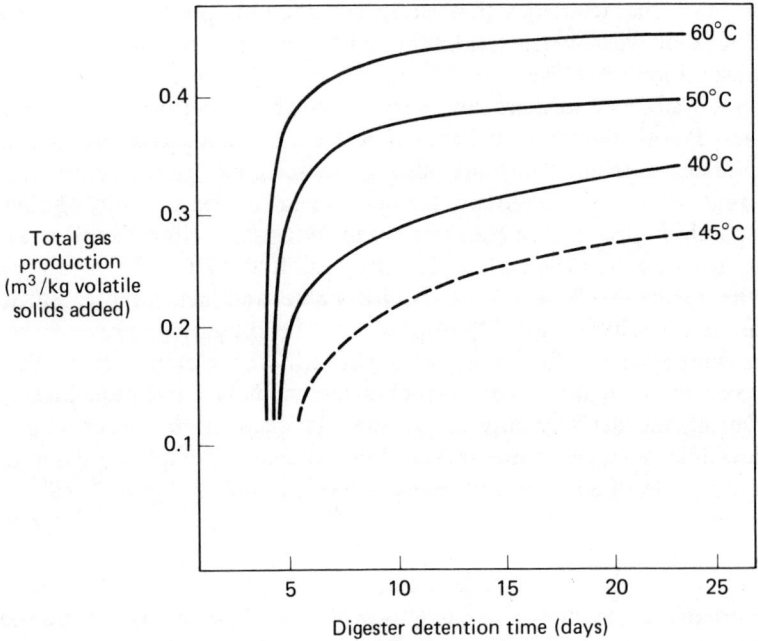

Figure 6-1. Gas production from digestion of MSW (gas at 0°C, 1 atm). (From Ref. 3.)

Pilot-plant studies have shown that the total gas produced is strongly influenced by detention time in the digester and the digester temperature as shown in Fig. 6-1. Note that the 45°C result is lower than 40°C, suggesting the existence of both a mesophilic (40°C) and a thermophilic (60°C) operating range.

In addition to temperature and residence time, other variables are important in this process, such as the maintenance of total anaerobiasis. It is also necessary to maintain a neutral pH level; never below 6.2 at which point methane production ceases. Since this is a biological process, the provision of adequate nutrients, such as nitrogen, is required. Usually, the C/N ratio of shredded waste is not sufficient for full decomposition, and another source of N is needed, such as sewage sludge rich in nitrogen. The C/N ratio for typical MSW has been reported as 24:1, with some values [4] as high as 40:1, while a ratio of 20:1 as a minimum is required for active anaerobic decomposition. Raw primary sludge [5] has a C/N ratio of about 16:1.

Finally, the entire system must be free of toxic materials. This is best achieved by removing potential toxins before they get to the digester, or if once there, precipitating them out. The latter method has been used in the removal of metals by precipitation with sulfide in wastewater treatment plants [6].

A number of rather sophisticated models have been developed for the anaerobic digestion process as used in wastewater engineering. Most of these models are bases on enzyme kinetics and a mathematical description of cellular growth rates [7, 8]. Although these models have value, they do not address the major objective of anaerobic digestion as applied to MSW —the conversion of volatile (organic) solids to gas. Neither the correctness nor the rigor of these models is in question—just their practical applicability to the aerobic digestion process for organic refuse.

For MSW digestion, it might be more useful to describe the process with the following model.

The decay or reduction in volatile (organic) matter in a digester might be reasonably expressed as

$$\frac{dS}{dt} = -K_d S$$

where S = concentration of the biodegradable material (measured as volatile suspended solids, or a specific material if the system feed is controlled), mg/liter at time t
K_d = decay constant, days^{-1}

Obviously, this is simply a first-order decay equation, stating that the rate of decay is proportional to the organics remaining, which is a reasonable assumption if the process rate is not time-dependent. After integration,

$$\frac{S}{S_o} = e^{-K_d t}$$

where S_0 is the original organic solids concentration ($t = 0$), mg/liter, the materials balance within a completely mixed continuous digester would be

input − output − change that occurs in digester = net change

If the digester is operating at steady state, the net change is zero, and

$$\frac{QS_0}{V} - \frac{QS}{V} - K_d S = 0$$

where Q = flow rate through digester, m^3/day
V = volume of digester, m^3

The hydraulic residence time $\bar{t} = Q/V$, or

$$\bar{t} = \frac{S_o - S}{K_d S}$$

Hence, if K_d is known, the required residence time for any reduction in solids can be calculated. Batch laboratory experiments can be used to obtain values of K_d by plotting the values of $\log S/S_o$ versus time and measuring the slope as $(K_d/2.303)$. Values of K_d for refuse slurries have not been reported.

The process kinetics may also be described in terms of the gas produced instead of the volatile matter destroyed. Using a similar mass balance, Pfeffer [10] found that it was possible to describe the reactor performance by the model

$$\frac{G_0 - G}{G} = K_g \bar{t}$$

where G_0 = maximum gas production attainable, estimated at 0.547 liters gas per gram volatile solids in the reactor [10].
G = daily gas production, liters/g volatile solids
$\bar{t}$ = hydraulic residence time, days
K_g = rate constant, days^{-1}

Interestingly, K_g was found to have two distinct values. The initial rate was rapid and lasted between 5 and 10 days, followed by a significantly lower rate. Table 6-2 is a listing of the K_g values found. It should also be noted that at 45°C there is a substantial drop in K_g from 40°C, indicating again the existence of mesophilic and thermophilic regimes in anaerobic digestion.

The quantity of gas generated can be estimated by entering a plot such as Fig. 6-1 at the calculated $\bar{t}$ and reading off the gas production. Because of the heterogeneous nature of the waste and the fact that not all the organics decompose, any theoretical calculations would probably be fruitless.

TABLE 6-2

Rate Constants, K_g, for Gas Production in Anaerobic Digesters

Temperature °C	Rate Constant (day^{-1})	
	Initial	Final
35	0.055	0.003
40	0.084	0.043
45	0.052	0.007
50	0.117	0.030
55	0.623	0.042
60	0.990	0.040

Source: Ref. [10].

Potential for Application

Since the anaerobic digestion of refuse process has never been successfully operated on a prototype scale, design guidelines are not available. It is reasonable, however, to try to assess the future of this process.

The shortage of natural gas makes production of methane extremely attractive. Bench-scale data indicate that one 900-tonnes/day (1000-tons/day) plant could produce about 16,000 m^3 (3.6 million ft^3) of methane per day [9]. There are about 65 metropolitan areas in the United States that could support such a plant, and thus the total potential production is 1,000,000 m^3/day [8]. Since the total use of natural gas in the United States is about 100×10^9 m^3/day, the impact of methane from waste would be a substitution of only 0.001% of the national need—not a very significant fraction.

In addition to the miniscule impact, the process is plagued by potential problems. There is no way to ensure the removal of toxic materials before the waste goes into the digesters, and "sour" digesters, such as those encountered in wastewater treatment plants, are a definite possibility.

The problem of mixing a paper slurry has not yet been solved. Even pilot-plant scale mixing with fairly dilute slurries has been found to be a problem. The desired solids concentration for these digesters is 10%, which is a highly viscous and thixotropic slurry [11]. In wastewater treatment practice, where solids concentrations normally range from 3 to 5%, mixing has always been a problem. Recent tracer studies have shown that typical primary digesters seem to have only 25% of their volume mixed; the remaining being dead space [12]. Such problems will surely plague refuse digestion facilities as well.

Large land areas are required by the digesters, a minimum 5 ha (12 acres) for a 900-tonne/day (1000 ton/day) plant [9]. This can be a problem where transport costs prohibit long-range refuse movement and the treatment facility must be located on expensive urban land.

Finally, the problem of what to do with the residue has not been solved. The sludge does seem to dewater readily [3], as it should, with all the fiber in it, but its ultimate disposal is an additional problem in the application of this process.

METHANE GENERATION FROM LANDFILLS

Methane generation from landfills differs from the previously discussed methane generation in digesters in one important aspect—there is almost no operational control of the process when gas is captured from landfills.

The only control available is moisture concentration, and very little information is available on this aspect.

The fact that landfills produce methane gas has been known for a long time, and many accidental explosions have occurred when the gas has seeped into basements and other enclosed areas where it could form explosive mixtures with oxygen. Only recently, however, has the capture and production of pipeline-quality gas been practiced.

The first system was placed in operation in California after a number of gas-extraction wells were driven around a deep landfill to prevent lateral migration of gas. These wells burned off about 44 m^3 of gas/min (1000 ft^3/min) without the need for auxiliary fuel [13]. Based on this experience, it was thought reasonable to capture and use this gas instead of just wastefully burning it off. The indications were that the following assumptions would hold:

1. The in-place refuse contains about 50% decomposable material.
2. Fifty percent of this material would be actually decomposed to produce gas.
3. Fifty percent of this gas is captured.
4. 0.37 cubic meter of gas is generated per kilogram of refuse decomposed (6 ft^3/lb).*
5. The heating value of the gas is 530 kJ/m^3 (500 Btu/ft^3).

Using these assumptions, it is obvious that huge landfills (such as Palos Verdes, California, 15 million tons in place) could produce substantial amounts of usable gas. In the case of the Palos Verdes landfill, the heating value of the gas would be 10 trillion Btu.

The gas generated from landfills contains only about 50% methane and cannot be fed directly into a pipeline because of the low heating value. Upgrading the gas can be accomplished by molecular sieves, and the product has a heating value of 950 kJ/m^3 (1000 Btu/ft^3).

Theory of Methane Generation in Landfills

The microbial activity relative to methane formation within a landfill occurs in four stages [15].

1. *Aerobic stage*, which may last from a few days to several months, depending on the rate of decomposition. If the waste has sufficient moisture content, the aerobic decomposition will quickly strip the deeper part of the landfill of oxygen, and the

---

*This is an unusually high value. Other investigators suggest 0.19 m^3/kg. (3 ft^3/lb) or even 0.06 m^3/kg (1 ft^3/lb) as the maximum [14].

decomposition will enter stage 2. The major end product during stage 1 is CO_2.

2. *Anaerobic nonmethanogenic stage*, during which the microorganisms use nitrates and sulfates as sources of oxygen, producing sulfides, nitrogen gas (denitrification), and CO_2. There is a rapid buildup of CO_2 during this stage, but almost no methane is produced since the reduction phase has not yet begun. Again, the time for this phase varies with environmental conditions, with warm moist landfills being the quickest to go through this stage into stage 3.

3. *Anaerobic methane production buildup stage*, where the percentage of CH_4 progressively increases as the methane formers begin to take hold. The temperature of the landfill, incidentally, also increases to about 55°C (135°F). At the conclusion of stage 3, the steady-state condition is reached.

4. *Anaerobic steady state stage*, where the production of CO_2 and CH_4 has stabilized.

These four stages are pictorially represented in Fig. 6-2.

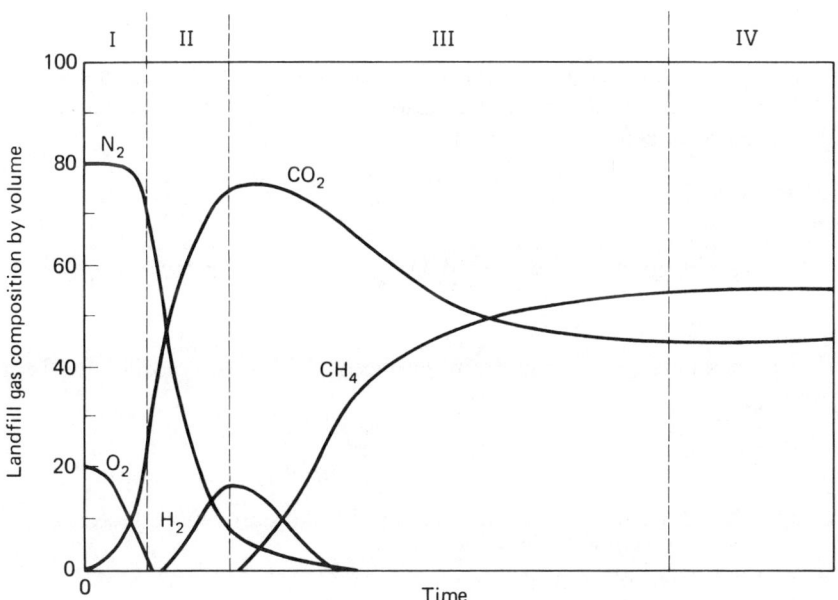

Figure 6-2. Landfill gas production patterns. (From Ref. 15.)

The production of methane and carbon dioxide can be assumed to follow the empirical equation [15].

$$CH_aO_bN_c + \tfrac{1}{4}(4 - a - 2b + 3c)H_2O \rightarrow \tfrac{1}{8}(4 - a + 2b + 3c)CO_2$$
$$+ \tfrac{1}{8}(4 + a - 2b - 3c)CH_4$$

Using this equation, it is possible to calculate the amount of gas produced per unit weight of raw material, as illustrated in Example 6-1.

Example 6-1

Estimate the production of CO_2 and CH_4 during the anaerobic decomposition of glucose.

The general formula for glucose is $C_6H_{12}O_6$, hence by the equation above, $a = 2$, $b = 1$, and $c = 0$.

$$CH_2O + \tfrac{1}{4}(4 - 2 - 2)H_2O \rightarrow \tfrac{1}{8}(4 - 2 + 2)CO_2 + \tfrac{1}{8}(4 + 2 - 2)CH_4$$
$$CH_2O \rightarrow \tfrac{1}{2}CO_2 + \tfrac{1}{2}CH_4$$

Note that the equation balances. The molecular weights are $30 \rightarrow 22 + 8$, hence 1 kg of glucose produces 0.73 kg of CO_2 and 0.27 kg of CH_4. Recalling that 1 gram molecular weight of a gas at standard temperature and pressure occupies 22.4 liters, the production of CO_2 and CH_4 from 1 kg of glucose is 746 liters each, or about 12 ft³/lb of glucose.

If sulfur is included in the analysis, and ammonia and hydrogen sulfide included as end products, the following equation for anaerobic biodegradation has been suggested [16]:

$$C_vH_wO_xN_yS_z + \left(v - \frac{w}{4} - \frac{x}{2} + \frac{3y}{4} + \frac{z}{2}\right)H_2O \rightarrow$$
$$\times \left(\frac{v}{2} + \frac{w}{8} - \frac{x}{4} - \frac{3y}{8} - \frac{z}{4}\right)CH_4 + \left(\frac{v}{2} - \frac{w}{8} + \frac{x}{4} + \frac{3y}{8} + \frac{z}{4}\right)CO_2$$
$$+ (y)NH_4 + (z)H_2S$$

The methane production from southern California landfills has been estimated as [17]:

$$Q = 18.77 \times 10^6 \left(\frac{Ah}{R^2}\right)$$

where Q = production rate of methane in standard cubic feet per day ($T = 520°R$, $P = 14.7$ lb/in.²)
A = area of landfill, acres
h = depth of landfill, ft
R = radius of influence of the wells, ft

This empirical equation was developed for a 14-m (40-ft)-deep 12-ha (30-acre) landfill, where the radius of influence for the 20 wells was 40 m (130 ft). The calculated production rate for such a landfill is 33×10^3 standard m^3 (1.17×10^6 scf) per day, which is optimistic (see below).

The rate of decomposition within a landfill can be controlled somewhat by manipulating the moisture content, the size of the refuse particles, and the density of the compacted refuse. Present indications are that if methanol production is to be enhanced, the landfill should have

    High moisture content.

    Small particle size.

    Low density.

Results of a landfill decomposition study using small enclosed cells is shown in Fig. 6-3 [16].

While it is possible to calculate the total amount of gas generated, the rate of gas formation is quite difficult to predict. The main problem is that the rate of gas production is dependent on environmental factors which are often uncontrollable. One limited study concluded that the half-life of a sanitary landfill gas production capability is about 20 years [18], while another study estimated that most of the useful gas production occurs within the first 6 years [19].

It is reasonable to assume that the production of gas can be defined by an exponential relationship

$$\frac{dG}{dt} = -K'(L - R)$$

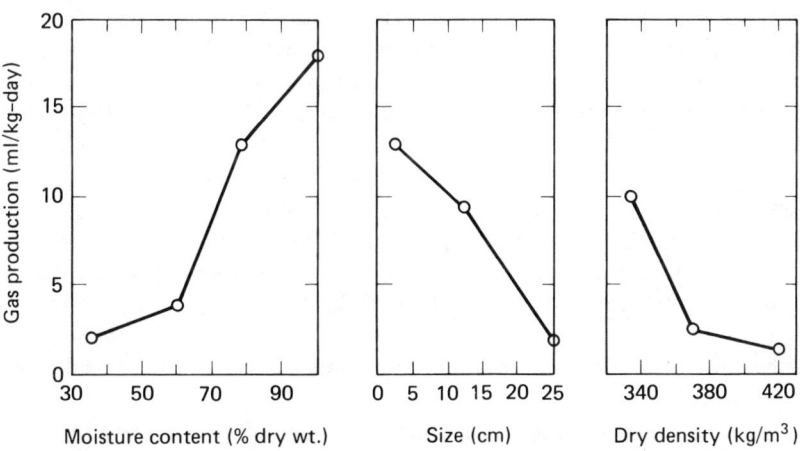

Figure 6-3. Effect of moisture, particle size, and density on gas production in a model landfill. (From Ref. 16.)

where G = total gas produced at time t, m³/kg
 K' = rate constant, y^{-1}
 L = maximum possible gas production m³/kg

or in integrated form,

$$G = L(1 - 10^{-Kt})$$

where $K = K'/2.303$. If the half-life of a landfill is estimated as 20 years, then $G = 0.5L$ and the constant $K = 0.015$ y^{-1}.

Using this analysis, it is possible to estimate the rate of generation by the procedure illustrated in Example 6-2.

Example 6-2

Assume that each kilogram of solid waste contains 50% by weight of carbohydrates, and that only these are decomposing to produce CO_2 and CH_4. According to Example 6-1, 1 kg of a carbohydrate produces 0.76 m³ of CH_4. Find the rate of this production.

The total amount of gas produced by 1 kg of solid waste during the first year is

$$G = L(1 - 10^{-Kt}) = 0.76(1 - 10^{-0.015(1)}) = 0.0258 \text{ m}^3/\text{kg}$$

If we assume that the average production rate is one-half of this value (not an unreasonable assumption considering other more serious inaccuracies),

$$Q = \frac{0.0258}{2(365)} \frac{(\text{m}^3/\text{kg})/\text{yr}}{\text{days/yr}} = 3.5 \times 10^{-5} (\text{m}^3/\text{kg})/\text{day}$$

These calculations give some feeling for the rate of gas production in landfills. It is thus unreasonable to expect a landfill to produce much more than this, even with an optimum number of wells. If higher pumping rates are attempted, air will seep into the landfill and the oxygen will kill the strictly anaerobic methane formers. A maximum 1.4 m³/min (50 ft³/min) has been reported for one 9- to 12-m (30- to 40-ft)-deep landfill, with a radius of influence of 40 m (130 ft) [14]. As could be predicted, higher pumping rates reduced the methane production due to air seeping into the landfill. Although the spacing of wells and the resulting areas of influence are also important, such information is not available. Theoretically, if a tight cover is placed on a landfill (e.g., clay) the spacing of wells should not be critical, since the limiting aspect is the production of methane, not its withdrawal. Practically, the spacing of wells is no doubt of substantial importance.

An alternative analysis of gas formation, proposed in part by Alpern [20], and based on work with anaerobic digesters [21] is that gas formation is actually a two-stage process, with the first part characterized by an increased rate of gas production, followed by a decreased production rate. Both of these rates can be described by exponential functions.

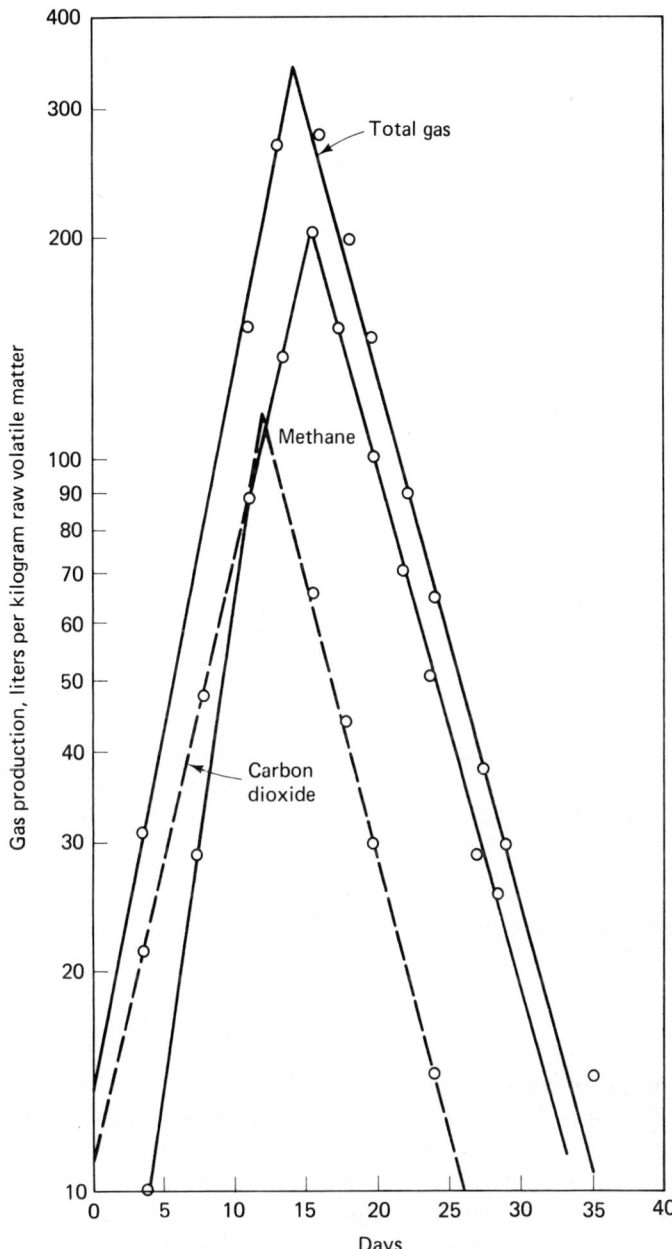

Figure 6-4. Gas production in a digester. (From Ref. 20; after Ref. 21.)

The first stage of growth can be expressed as

$$\frac{dG}{dt} = K_1 G$$

where G = amount of gas produced in time t, m³/kg
       K_1 = constant, time^{-1}

The second stage is described as

$$\frac{dR}{dt} = -K_2 R$$

where R = amount of gas *remaining* to be produced, after digestion has proceeded to time t
       K_2 = constant, time^{-1}

After integration, the two equations yield

$$\log G = K_1' t + \log C_1$$

and

$$\log R = -K_2' t + \log C_2$$

where C_1 and C_2 are constants of integration.

Plotting these two equations on semilog paper, as in Fig. 6-4, it is noted that both equations can be assumed to be discontinuous at a point where gas production has reached its ultimate value, and $G = R = \frac{1}{2}L$, or half of the gas has been produced. This can then be assumed to be the half-life of the landfill. (Note that the data in Fig. 6-4 are for sludge digestion, and the half-life is in terms of days. A similar plot for landfills would of course have a half-life in terms of years.)

Design of Methane Extraction from Landfills

The design of methane extraction from landfills involves the proper spacing of wells, the type of wells, and the gas cleaning or processing facility.

The spacing of wells, as previously mentioned, is still an art. For shallow landfills (10 m), a 1.4-m³/min (50 ft³/min) pumping rate, coupled with a production rate of 3.5×10^{-5} (m³/kg)/day, as previously calculated, translates into a spacing of 110 m apart, which agrees with the field experience of about 40 m (130 ft) radius of influence. For deep landfills, of about 38 m (125 ft), a pumping rate of 9 m³/min (320 ft³/min) has been found acceptable [13].

The wells used for the Palos Verdes landfills are described in Fig. 6-5. The wells could be installed as the landfill is developed, although it is probably less trouble to drill after completion. A certain amount of care is

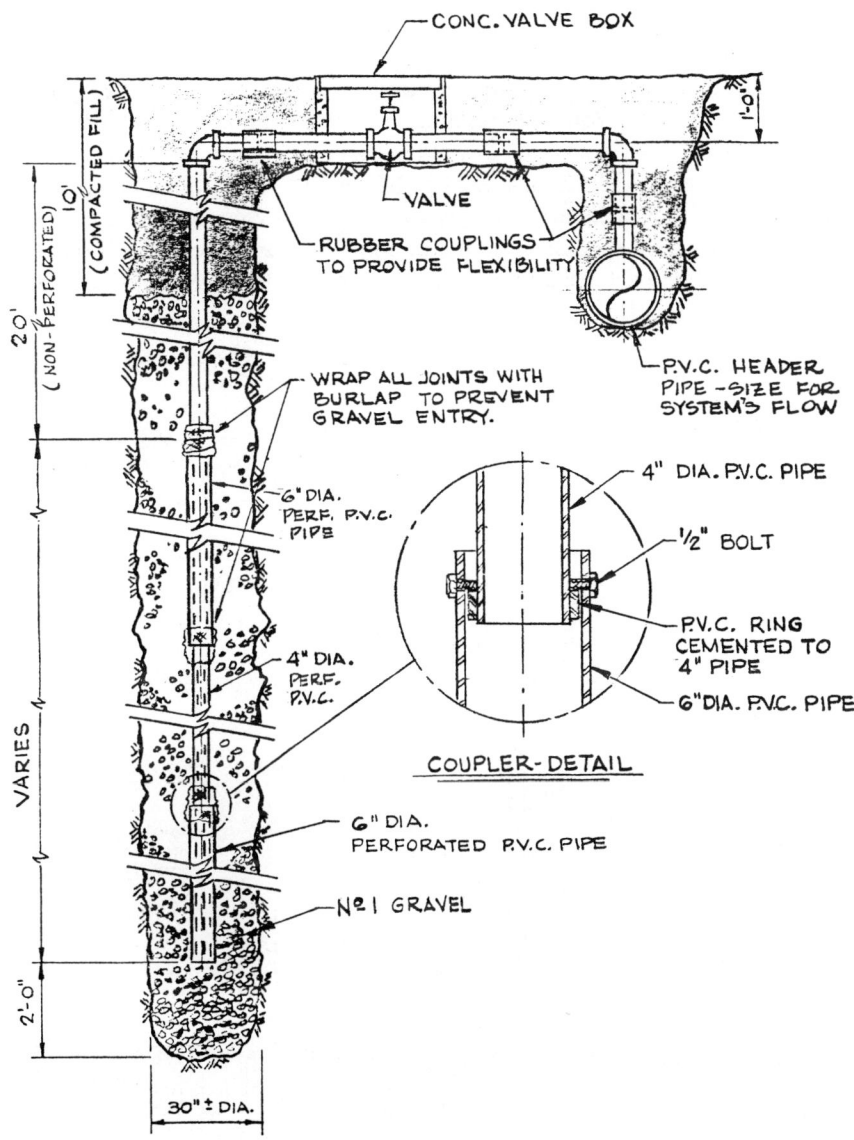

Figure 6-5. Well and pipe construction at the Palos Verdes landfill. (From Ref. 13.)

required during the drilling operation, since methane gas is escaping and mixing with oxygen.

The gas cleaning process using molecular sieves is shown in Fig. 6-6. The gas is pumped from the wells so as to maintain a slight negative pressure, and the water removed. Following preheating, the gas enters the molecular sieves, which in this case are hydrated metal aluminum silicates characterized by a structure such that it allows only certain molecules to enter, depending on their size, shape, and polarity, and excludes all others. The gas is circulated through a vessel that contains the absorbent until the

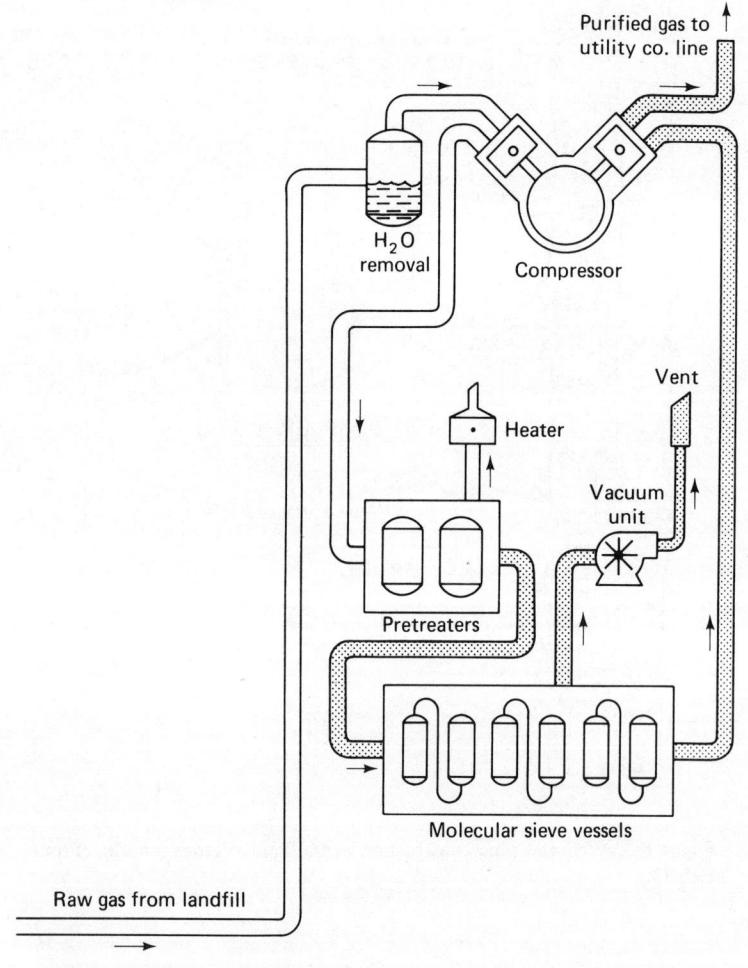

Figure 6-6. Molecular sieve system for cleaning landfill gas. (Courtesy of NRG Nu Fuel.)

capacity of the silicate to absorb the impurities, in this case CO_2, is reached. As the gas flow is switched to a second vessel, the first one is regenerated by depressurizing. Eventually, simple depressurization no longer completely cleans the screen, and the silicate has to be heated in order to be regenerated. The cleaning is cyclical but wholly automated, so that a continuous flow of gas is produced.

Potential for Application

Interestingly, little consideration has thus far been given to the use of landfill gas as a fuel without prior cleaning. Estimates show that it is unlikely that landfill gas will ever contribute more than about 0.5% of our national gas use [19] so it is not a major source of pipeline gas. It is, however, a useful fuel as it is extracted from the landfill, and the direct use of this 4.4×10^6 cal/m^3 (500 Btu/ft^3) gas should increase in the future.

COMPOSTING

Composting differs from the previous two processes primarily in that it is an aerobic process, and the end product of interest is the partially decomposed organic fraction. The process holds a great deal of attraction for the public, because it is often promoted as a "natural" process of solid waste treatment. One reason for this reputation is that compost piles can be readily constructed in the backyard, and the product is a useful soil conditioner. It is little wonder, therefore, that municipal engineers and city councils are besieged by citizens groups urging that composting be initiated in their community in place of alternative solid waste disposal schemes such as landfilling and incineration, which many people view (perhaps correctly) as wastes of money and natural resources.

Composting on a municipal scale is indeed an uncomplicated process. At its simplest, screened refuse is placed in long parallel piles, called windrows, and the moisture content is maintained near 50%. The piles are periodically aerated by fluffing and moving the material around. After several weeks, such accelerated aerobic decomposition results in a dark brown earthy-smelling material which has low nutrient value but is an excellent soil conditioner. The ranges of nitrogen, phosphorus, and potassium in finished compost from MSW are 0.4 to 1.6, 0.1 to 0.4, and 0.2 to 0.6, as N, P and K respectively [22].

Composting can also be complicated. Many composting plants are highly mechanized, with shredding, screening, and aerated digesters. Over

30 mechanized composting systems are known today which are identifiable by inventors or trade names [23]. The product in most of these systems is identical to the windrow method.

A typical mechanized composting system is shown in Fig. 6-7. The initial composting is accomplished in a mechanically aerated aerobic digester with a retention time of about 5 days. This is followed by windrow curing for several months. The best documented example of the windrow system in the United States is the now-closed plant in Johnson City, Tennessee, owned and operated by the EPA. This plant, shown in Fig. 6-8, produced an excellent product but was closed because of the lack of a market for the finished compost and the lack of local support for the project.

In the United States and in other Western countries, the major problem of composting operations is, in fact, always the same—a lack of a market for the finished product. Many large municipal composting plants have come on stream, and operated successfully, but closed down because of the inability to dispose of the product (even giving it away, free of charge).

On the other hand, successful composting plants have operated for many years in Europe and India, but always because a market for the finished product exists. In India, where food production is critical and soil conditioners are in great demand, over 2000 plants are reported to be in operation producing over 4 million tons of compost annually [23].

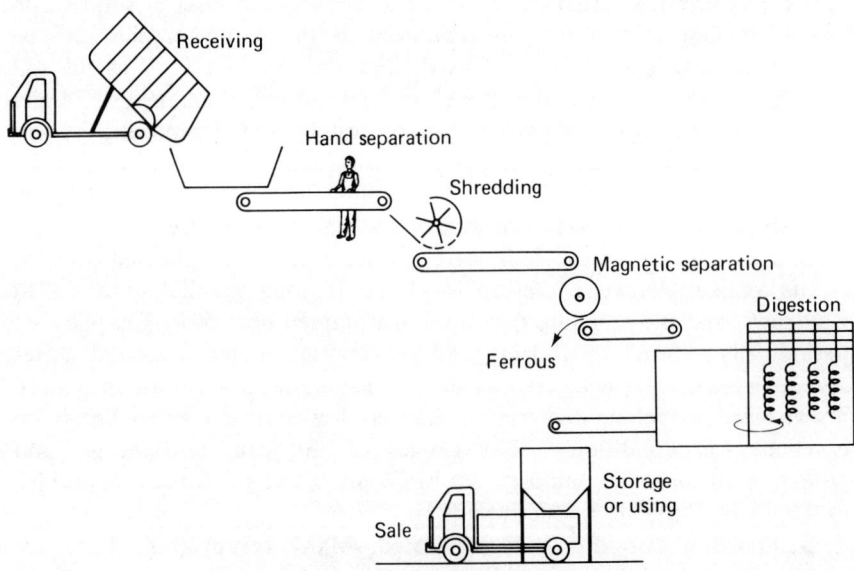

Figure 6-7. Typical mechanized composting system.

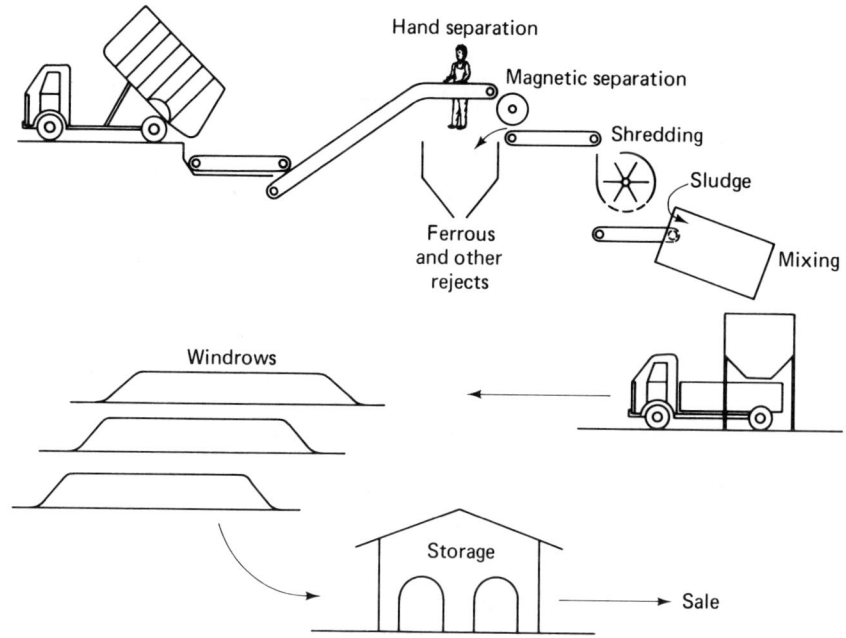

Figure 6-8. Johnson City, Tennessee, windrow composting system.

Fundamentals of Composting

Aerobic microorganisms extract energy from the organic matter and through a series of exothermic reactions break the material down to simpler materials. The basic aerobic decay equation holds:

$$(\text{complex organics}) + O_2 \xrightarrow[\text{microorganisms}]{\text{aerobic}}$$

$$CO_2 + H_2O + NO_3^- + SO_4^{2-} + \text{other less complex organics} + \text{heat}$$

During this decomposition the temperature increases to about 70°C (160°F) in most well-operated composting operations. The rise in temperature within the windrows at the Johnson City compost plant is shown in Fig. 6-9. As the reaction develops, the early decomposers are mesophilic bacteria followed in about a week by thermophilic bacteria, actinomycetes, and thermophilic fungi [25]. Above 70°C, spore-forming bacteria predominate. As the decomposition slows, the temperature drops and mesophilic bacteria and fungi reappear. Protozoans, nematodes, millipedes, and worms are also present during the later stages. The concentration of dead and living organisms in compost can be as high as 25% [26].

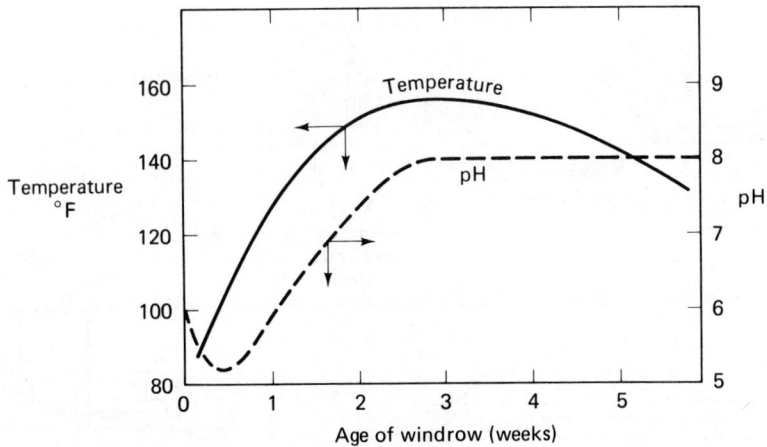

Figure 6-9. Variation in the windrow temperature at the Johnson City, Tennessee, compost plant. (From Ref. 24.)

TABLE 6-3

Destruction of Some Common Pathogens and Parasites

Organism	Observations
Salmonelia typhosa	No growth beyond 46°C; death within 30 min. at 55–60°C and within 20 min. at 60°C; destroyed in a short time in compost environment
Salmonelia sp.	Death within 1 h at 55°C and within 15–20 min. at 60°C
Shigella sp.	Death within 1 h at 55°C
Escherichia coil	Most die within 1 h at 55°C and within 15–20 min. at 60°C
Entamoeba histolytica cysts	Death within a few minutes at 45°C and within a few seconds at 55°C
Taenia saginata	Death within a few minutes at 55°C
Trichinella spiralis larvae	Quickly killed at 55°C; instantly killed at 60°C
Brucella abortus or *Br. suis*	Death within 3 min. at 62°–63°C and within 1 h at 55°C
Micrococcus pyogenes var. *aureus*	Death within 10 min. at 50°C
Streptococcus pyogenes	Death within 10 min. at 54°C
Mycobacterium tuberculosis var. *hominis*	Death within 15–20 min. at 66°C or after momentary heating at 67°C
Corynebacterium diphtheriae	Death within 45 min. at 55°C
Necator americanus	Death within 50 min. at 45°C
Ascaris lumbricoides eggs	Death in less than 1 h at temperatures over 50°C

Source: Ref. 27.

The elevated temperatures destroy most of the pathogenic bacteria, eggs, and cysts. Some of the more common pathogens and their survival at elevated temperatures are shown in Table 6-3. The product of thermophilic composting is essentially free of pathogens. Several extensive studies have been conducted on the fate of pathogens in composting [28, 29]. In all cases, the authors concluded that all potential pathogens, including resistant parasites such as *Ascaris* eggs and cysts of *Endamaeba histolytica*, are destroyed.

Clarence Golueke of the University of California (Berkeley) has pointed out that all biochemical conversion processes such as composting are in essence two-step operations [30]. The first step is the decomposition of complex molecules of waste materials into simpler entities. If there is no nitrogen available, this is the full extent of the process. If nitrogen is available, however, the second step is the synthesis of the breakdown products into new cells. These new microorganisms contribute to the process and the system operates in balance.

Because of the high rate of microbial activity, a large supply of nitrogen is required by the bacteria. If the reaction were slower, the nitrogen could be recycled, but since many reactions are occurring concurrently, a sufficient nitrogen supply is necessary. The requirement for nitrogen can, as before, be expressed as the C/N ratio.

Most experts agree that a C/N of 20:1 is the ratio at which nitrogen is clearly not limiting the rate of decomposition. Above a C/N of 80:1, thermophilic composting cannot occur because the nitrogen severely limits the rate of decomposition. Most systems operate between these extremes.

At C/N ratios less than 30:1, nitrogen is lost to the atmosphere as ammonia [27]. Some of the carbon, however, is tied up in very slowly biodegradable cellulose and lignin, while almost all of the nitrogen is available to the organisms. Hence the practical C/N ratio that prevents nitrogen loss [31] is probably closer to 35 to 40:1. The C/N ratios for various materials used in composting are shown in Table 6-4.

The pH also affects nitrogen loss, because ammonia escapes as ammonium hydroxide above a pH of 7.0. Thus efficient compost operations, which operate around a pH of 8.0, cannot retain nitrogen at a greater concentration than C/N of about 35:1.

The pH of the compost pile varies with time, showing an initial drop and then a leveling off at about 8.0, as shown in Fig. 6-9. If the compost heap becomes anaerobic, the pH continues to drop. There is sufficient buffering within the compost to allow the pH to stabilize at an alkaline level provided that the reaction stays aerobic. For educational purposes, the progression of pH and temperature in a compost pile can be readily demonstrated by laboratory-scale apparatus [32].

TABLE 6-4

Carbon/Nitrogen Ratio for Various
Materials Used in Composting

Material	C/N
Garbage	
Raleigh, N.C.	15.4
Louisville, Ky.	14.9
MSW (including garbage)	
Berkeley, Calif.	33.8
Savannah, Ga.	38.5
Johnson City, Tenn.	80
Raleigh, N.C.	57.5
Chandler, Ariz.	65.8
Sewage sludge	
Waste-activated	6.3
Mixed digested	15.7
Wood (pine)	723
Paper	173
Grass	20
Leaves	40–80
Sawdust	511

Source: Compiled in Ref. 22.

The time required for a compost pile to mature depends on such factors as the putrescibility of the feed, the insulation and aeration provided, the C/N ratio, the particle size, and other conditions. Usually, 2 weeks is considered the minimum time for the adequate composting of shredded municipal refuse in windrows [31]. Mechanical composting plants, using inoculation of previously composted materials, can accomplish decomposition in 2 or 3 days. This material is still quite active, however, and usually requires further stabilization in windrows [27].

The completion of composting is judged primarily on the basis of a slight drop in temperature and a dark brown color. A more accurate measure is the determination of starch concentration in the compost. Starch is readily decomposable, and thus its disappearance is a good indicator of mature compost. A simple laboratory method for measuring starch in compost is available, although the technique yields only qualitative information [33, 34]. This technique can also be applied to a composting demonstration project for the classroom [32]. A more rigorous measure of the end point is the drop in the C/N ratio to perhaps 12:1. Higher C/N ratios will result in continued decomposition of the compost after it is applied, and the subsequent robbing of nitrogen from the ground.

Potential for Composting Refuse

Composting is an old process and quite well understood. The practical application of this process to MSW, however, is limited by three serious problems:
1. Lack of markets for the finished product.
2. Small reduction in the total refuse volume requiring disposal.
3. Environmental factors of composting plants, specifically odor.

The first is the most serious of the three problems. The majority of the municipal composting plants in the United States closed because they could not sell (or even give away) the product. When composting plants are designed, however, the economic analysis invariably includes a profit from the sale of compost.

The second problem is in the main amenable to technical solution, since the process can be changed to actually capture such items as glass, metals, and so on. In fact, composting may be considered as simply another process within a complete resource recovery plant. One pilot-scale process has been run to produce a putrescible material from the screening (7-mesh) of the light fraction of air-classified refuse [35]. This material consisted of 6% of the raw refuse, and contained 74% volatile solids. Although in this study the material was digested for methane production, it could also be readily composted.

The fraction of putrescibles in refuse has been steadily declining over the years, and will no doubt continue to do so. As the organic fraction is reduced, the compostable fraction similarly drops, and it is unlikely that more than 20% of refuse could ever be recovered as compost. The 80% must be disposed of by normal means, and this cost must be added to the total cost of the composting facility.

The last problem is mostly one of odor production. Although some people insist that the odor from compost piles is pleasant, these people are in the distinct minority. Compost plants do smell, there is no doubt of that, and thus the plants must be located fairly far from residential areas. This requirement adds another cost—transportation.

In conclusion, it seems unlikely that municipal refuse will be composted on a large scale, at least in developed countries. Whenever composting of refuse on a municipal scale is suggested, the decision makers should always bear in mind the dismal record of past composting efforts. All of these plants were built under an aura of optimism and enthusiasm, and all of the economic analyses promised success. The engineer and decision maker should ask what is so different in the proposed new compost plant that would allow it to succeed while all the others have failed.

Composting Wastes Other than Refuse

This discussion has thus far been mainly concerned with the composting of municipal refuse. It should be strongly pointed out that many other wastes can be composted at a much lower cost, and will produce excellent composts that might have significant value.

Leaves, for example, are composted in many communities. Air pollution control regulations have eliminated the practice of burning leaves in most communities, and these leaves are now picked up by specially constructed trucks. The quantities of leaves can be staggering. In one town in North Carolina, with a population of 27,000, three leaf vacuum trucks work every day for 3 months out of the year. In a New Jersey community, 25,000 shade trees provide enough leaves to produce about 1500 to 2300 m^3 (2000 to 3000 yd^3) of compost annually [36].

A typical leaf-composting operation consists of the following. Leaves are gathered and wetted thoroughly, then gently placed in windrows about 2.5 m (8 ft) wide. Periodic aeration is required, and since leaves are low in nitrogen, a nutrient such as sodium nitrate, at a rate of 0.6 kg/m^3 (1 lb/yd^3) of stacked leaves, is added [36]. After about 6 months, the leaves can be ground in a simple compost grinder and sold to the public. The entire operation can be inexpensive and produces a high-quality soil conditioner.

Various types of industrial and agricultural wastes can also be readily composted. A combination such as sawdust and chicken manure, for example, produces a superior compost, high in nitrogen.

Sludges from sewage treatment plants have for many years been used as sources of nutrients in composting. A new process developed by the U.S. Department of Agriculture uses the sludge as the primary source of organic material as well as nutrients [37]. The problem with composting sludge has always been that the solids compact too tightly, leaving no spaces for air to enter the pile. The USDA has solved this problem by mixing the sludge with wood chips and composting the mixture. The wood chips are composed of poorly decomposable cellulose and lignin and can be readily removed from the stabilized material by simple screening and is then reused. Raw sludges can be composted to a passable soil conditioner or a high-grade top soil in about 2 weeks by this method.

GLUCOSE PRODUCTION BY ACID AND ENZYMATIC HYDROLYSIS

Cellulose is abundant in nature, comprising at least one-third of all the material in the plant world. The approximate empirical formula of cellulose is $C_6H_{10}O_5$, with a molecular weight of about 500,000. Pure cellulose has a low solubility in water, and the molecules are tightly bound together by hydrogen bridges to form chains.

It is possible to break down cellulose to form glucose by either acid or enzymatic hydrolysis. One kilogram of cellulose is capable of being converted to 0.5 kg of glucose. This process is, in fact, used by the ruminantics (cows, deer, sheep, etc.) for the production of glucose, whereas other animals cannot digest cellulose and must ingest sugars directly.

Cellulose is a renewable resource, its energy coming from the sun. This does not, however, imply that the source of cellulose is unlimited. One constraint to the production of cellulose is available land, since land that could be used for cellulose production may instead have to be used for other purposes, such as food production. On the other hand, the cellulose found in waste is a "free" source, as no additional land is needed for its production and no additional resources must be allocated. The reclamation of cellulose from wastes thus makes a great deal of sense.

The quantity of waste cellulose produced annually is impressive—about 100 billion tonnes per year, or about 70 kg of cellulose per person per day [38]. Not all of this waste cellulose is in the form of MSW, however.

Foremost among cellulosic wastes other than MSW are agricultural wastes, which are usually free of contaminants and often occur in large quantities. The ideal use of agricultural wastes occurs when the normal procedure is to collect the residues as part of the harvesting or processing operation. One example of this is bagasse (sugar cane after the sugar has been squeezed out). This is a clean, shredded, homogeneous material which still contains considerable energy, as illustrated by the fact that it is used as a primary electric fuel in power production in several sugar plantations in Hawaii. It could just as readily be used as a source of cellulose for the production of glucose.

Wastes from the wood-processing industries also yield clean products that might be used as raw materials for cellulose production. At the present time, only about 20% of forest wastes are being put to some use [39].

In MSW, cellulose is found mostly in paper and paper products, which constitute a major fraction of refuse. Wood and cotton are also sources of cellulose, but their fraction within the waste stream is small. In fact, wood is actually mostly lignaceous in composition, and, relative to paper, is not a good source of cellulose.

The process of converting cellulose to glucose and then to various other useful products is shown in Fig. 6-10. The various ways by which glucose is further processed, such as fermentation and microbial and chemical conversion of glucose, are not covered here, and may be found in standard biochemistry or modern chemical engineering texts. The process of interest here is the conversion of cellulose to glucose.

Cellulose belongs to a family of organic compounds called polysaccharides, which are polymers of simpler compounds called mono-

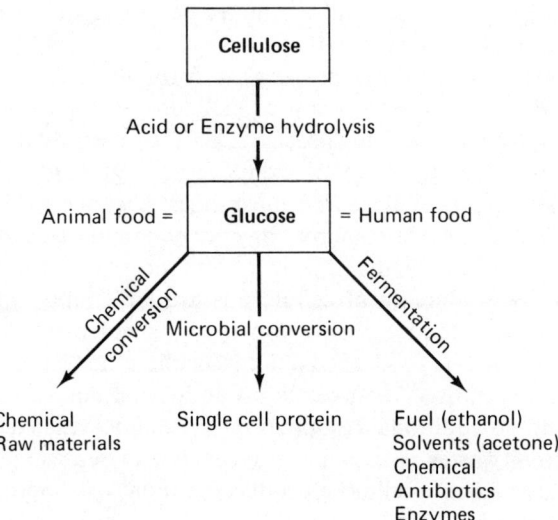

Figure 6-10. Cellulose to glucose to further products. (From Ref. 38.)

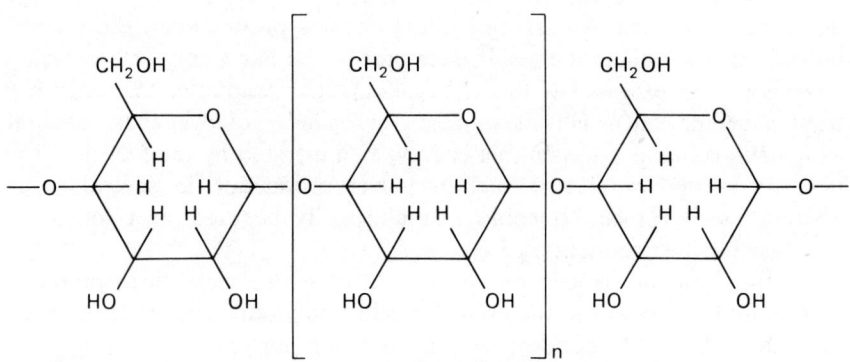

Figure 6-11. Simplified structure of cellulose.

saccharides, such as glucose. The structure of cellulose, which has an approximate empirical formula of $C_6H_{10}O_5$, is shown in Fig. 6-11. The number of monomers making up the chain is very large, in the range 2000 to 3000. The cellulose molecules are bound together tightly by a large number of hydrogen bridges along the chain.

The conversion of cellulose to glucose is through a process known as *hydrolysis*, which in organic chemistry denotes any reaction in which water is involved. The empirical equation for this process is

$$C_6H_{10}O_5 + H_2O \rightarrow C_6H_{12}O_6$$

where the product is glucose. In structural terms, hydrolysis involves the severing of the bonds ($\beta - 1:4$ glucosidic) between the cellulose monomers, and the reaction of these with water.

This reaction can be achieved by several processes, particularly by the use of either acids or enzymes. Under carefully controlled conditions, mild aqueous acids will break down the cellulose into glucose. In practice, acid hydrolysis requires corrosion-proof equipment, and the high temperature and the low pH can cause decomposition of the resulting sugars. The process must therefore be so balanced as to produce more sugar than that which is destroyed during the process [38]. Glucose yield of about 50% per weight of cellulose have been obtained, but this invariably contains unwanted by products [40].

Enzymatic hydrolysis of cellulose, on the other hand, seems to perform well with high glucose yields obtained without unwanted by products [38]. The process is based on the use of cellulose enzymes derived from mutant strains of the fungus *Trichodema viride* or other organisms which are capable of hydrolyzing insoluble cellulose.

The first step in the process, as illustrated in Fig. 6-12, is growing the fungus in a culture medium containing pure cellulose and various nutrients. The fungus is then filtered out and the enzymes are captured in the filtrate. The waste-containing cellulose is then mixed with the enzyme broth in a reactor at 50°C and at pH adjusted to 4.8. The glucose is produced in the reactor and filtered out.

The enzymes that hydrolyze glucose are called cellulases. In strict chemical terms, cellulase is known as $\beta - 1:4$ glucanase, since these enzymes hydrolyze the $\beta - 1:4$ glucosidic linkages of cellulose. The two main cellulases are called C_1 and C_x. The C_x component attacks amorphous cellulose, whereas C_1 together with C_x is required for hydrolyzing insoluble crystalline cellulose.

The action of the cellulase enzyme complex (C_1, C_x) on cellulose can be approximated as

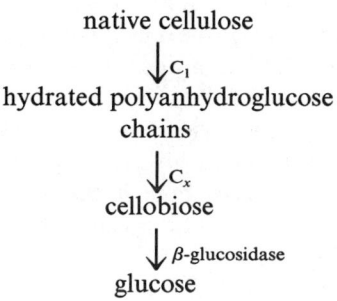

The actual reactions are quite complex, with multicomponent enzymes and many reaction pathways [41].

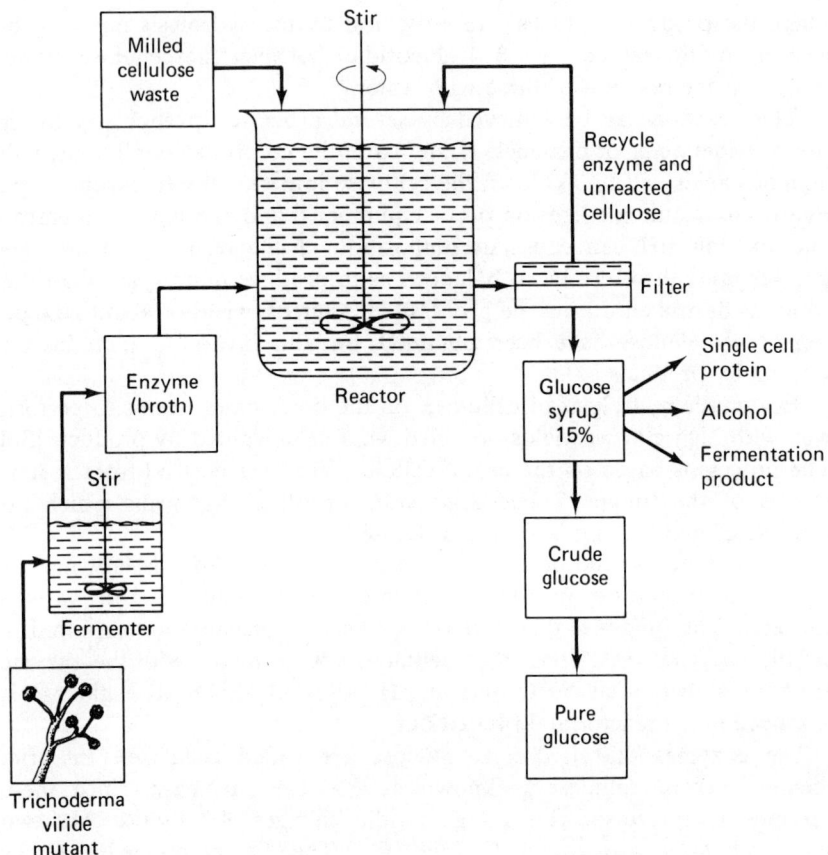

Figure 6-12. Process for the conversion of cellulose to glucose. (From Ref. 38.)

The susceptibility of native cellulose to enzymatic hydrolysis is affected by a number of variables [42], the most important being the degree of crystallinity, the presence of lignin and its association with cellulose, and the degree of polymerization of the native cellulose [43].

X-ray diffraction analysis has shown the cellulose has four stable crystal lattices. Any of these crystalline highly ordered forms of cellulose are more resistant to enzymatic hydrolysis than is the amorphous or less-ordered variety. The crystalline lattice structures can be broken down physically by crushing or grinding the cellulose.

The presence of lignin can interfere with the enzymatic hydrolysis of cellulose. The exact nature of the association between lignin and cellulose is still unknown, and the possibilities of a covalent chemical bond, hydrogen bonding, and incrustation of the cellulose in a lignin network, thus preventing the access of the enzyme molecules, all have some scientific evidence [43]. Whatever the association, however, it is possible to break down the complex by various chemical means as well as by mechanical grinding.

The third important variable affecting the rate of enzymatic hydrolysis is the degree of polymerization. Again, the degree of polymerization can be significantly reduced by mechanical milling of the cellulose.

OTHER BIOCHEMICAL PROCESSES

A number of other biochemical processes have been used for converting cellulose to useful products.

A process developed at Louisiana State University produces single-cell protein by the bacterial fermentation* of bagasse [44]. The Environmental Protection Agency has experimented with the use of fungus to convert cellulose to protein [45]. This process can be speeded up by the treatment of cellulose with inorganic salts and ultraviolet light, but the long required irradiation time makes the practical application of the process questionable [46].

The production of ethanol by the fermentation of glucose has also received considerable interest [47]. Ethanol can be utilized as a substrate for the growth of single-cell protein, with up to 93% yield [48].

PARTING SHOTS

The wide variety of bioconversion processes prompts a valid skepticism about the applicability of all of these to solid waste. What types of systems therefore seem to have the greatest potential?

Recognizing the difficulties involved in the processing of a heterogeneous material such as solid waste, Golueke [30] has suggested that the viability of a process be judged on the basis of some characteristics of the organisms employed. The organisms used in the process should be:

Not fastidious—they will work under adverse conditions (e.g., wide temperature range) and be tolerant of environmental change.

Ubiquitous—they should exist in nature, since pure stock cultures degenerate with time and rarely stay pure.

---

*The term "fermentation" is used in biochemical engineering to mean any process where CO_2 is produced.

Persistent—they must grow in the environment without special assistance.

Not picky—they should be able to use a broad spectrum of substrates.

If these criteria are used to judge the applicability of the several bioconversion processes discussed above, only a few would be considered applicable. Although this evaluation ignores some other aspects important to resource recovery (e.g., markets for the finished product), it points out a very important aspect of resource recovery engineering: The process used must be able to handle all kinds of materials and be as insensitive as possible to changes in flow rates and characteristics. As is the case with the treatment of all waste streams (e.g., sewage) the variability of the feed is not always known and often even the mean values of the characteristics are difficult to establish. It is the job of the engineer to design a system that will accept literally everything, and still operate and produce a desired end product. This provides a true challenge to resource recovery engineers.

REFERENCES

[1] BELL, J. M., "Characteristics of Municipal Refuse," *Proc. Natl. Conf. Solid Waste Res., A.M. Public Works Assoc.*, Feb. 1964.

[2] PFEFFER, J. T., *Reclamation of Energy from Organic Refuse*, U.S. EPA 670/2-74-016, Cincinnati, Ohio, 1974.

[3] PFEFFER, J. T., and J. C. LIEBMAN, "Energy from Refuse by Bioconversion, Fermentation and Residue Disposal Processes," *Resour. Recovery Conserv.* **1**, 295 (1976).

[4] HAGERTY, D. J., J. L. PAVONI, and J. E. HEER, *Solid Waste Management*, Van Nostrand Reinhold Company, New York, 1973.

[5] VESILIND, P. A., *Treatment and Disposal of Wastewater Sludge*, Ann Arbor Science Publishers, Ann Arbor, Mich., 1979.

[6] LAWRENCE, A. W., and P. L. MCCARTY, "The Role of Sulfide in Preventing Heavy Metal Toxicity in Anaerobic Treatment," *J. Water Pollut. Control Fed.*, **37**(113) (1965).

[7] ANDREWS, J. F., "Dynamic Model of the Anaerobic Digestion Process," *J. Sanit. Eng. Div. ASCE*, **95**(SA1), 1969.

[8] LAWRENCE, A. W., and P. L. MCCARTY, "Kinetics of Methane Fermentation in Anaerobic Treatment," *J. Water Pollut. Control Fed.*, **41**(2) (1969).

[9] HILLE, S. J., *Anaerobic Digestion of Solid Waste and Sewage Sludge to Methane*, U.S. EPA OSWMP SW-15g, Washington, D.C., 1975.

[10] PFEFFER, J. T. "Temperature Effects on Anaerobic Fermentation of Domestic Refuse," *Biotech. and Bioeng.*, 16, p. 77, 1974.

[11] GEISSER, R. "The Pompano Project," presented at Eng. Found. Conf., Henniker, N.H., July 1979.

[12] MONTEITH, H. D. and J. P. STEPHENSON, "Mixing Efficiencies in Full-Scale Anaerobic Digesters by Tracer Methods," Symp. on Sludge Treatment, Wastewater Tech. Centre, Burlington, Ontario, Canada, 1977-78.

[13] DAIR, F. R., and R. E. SCHWEGLER, "Energy Recovery from Landfills," *Waste Age*, Mar.-Apr. 1974, p. 6.

[14] JAMES, S. "Methane Generation from Landfills" presented at Eng. Found. Conf. at Henniker, N.H., July 1979.

[15] CHEREMISINOFF, P. N., and A. C. MORRESU, *Energy from Solid Waste*, Marcel Dekker, Inc., New York, 1976.

[16] CHIAN, E. S. K., F. B. DEWALLE, and E. HAMMERBERG, "Effect of Moisture Regime and Other Factors on Municipal Solid Waste Stabilization," in *Management of Gas and Leachate in Landfills*, S. K. Banerji (Ed.), U.S. EPA 600/9-77-026, Washington, D.C., 1977.

[17] CARLSON, J. A., City of Mountain View, Calif., as reported in M. J. Blanchet, *Treatment and Utilization of Landfill Gas*, U.S. EPA OSWMP SW-583, Washington, D.C., 1977.

[18] KUFEL, J. M., "Generation of Methane in Sanitary Landfills," M.S. thesis, New Jersey Institute of Technology, Newark, 1975.

[19] BOWERMAN, F. "Methane Generation from Deep Landfills" presented at Eng. Found. Conf., Henniker, N.H., July 1979.

[20] ALPERN, R., Personal communication.

[21] FAIR, G. M., and E. W. MOORE, "Heat and Energy Relations in the Digestion of Sewage Solids," *Sewage Works J.*, 1932.

[22] POINCELOT, R. P., "The Biochemistry and Methodology of Composting," *Conn. Agr. Exp. Sta., New Haven, Bull.* 754, 1975.

[23] MACARTHUR, R. S., "Municipal Composting: Techniques, Policy and Politics," *Soil and Health*, Journal of the Soil Association of New Zealand, undated.

[24] BREIDENBACH, A. W., *Composting of Municipal Solid Waste in the United States*, U.S. EPA SW-47r, Washington, D.C., 1971.

[25] CHANG, Y., and H. J. HUDSON, "The Fungi of Wheat Straw Compost," *Trans. Br. Mycol. Soc.*, **50**(4), 649 (1967).

[26] SATRIANA, N. J., *Large Scale Composting*, Noyes Data Corp., 1974.

[27] GOLUEKE, C. G., *Composting*, Rodale Press, Inc., Emmaus, Pa., 1972.

[28] WILEY, J. S., "Pathogen Survival in Composting Municipal Wastes," *J. Water Pollut. Control Fed.*, 34, 80 (1962).

[29] SCOTT, J. C., *Health Aspects of Composting with Night Soil*, WHO Exp. Comm. Environ. Sanit., Geneva, Switzerland, 1953.

[30] GOLUEKE, C., "State of the Art of Bioconversion Processes," Presented at the Engineering Foundation Conference in Rindge, N.H., July 1978.

[31] *Reclamation of Municipal Refuse by Composting*, Univ. Calif. Berkeley, Tech. Bull. 9, 1953.

[32] VESILIND, P. A., "A Laboratory Exercise in Composting," *Compost Sci.*, Sept.–Oct. 1973.

[33] LOSSIN, R. D., "Compost Studies," *Compost Sci.*, Mar.–Apr. 1971.

[34] VESILIND, P. A., *Solid Waste Engineering Laboratory Manual*, Department of Civil Engineering, Duke University, Durham, N.C., 1973.

[35] DAIZ, L. F., F. KURZ, and G. J. TREZEK, "Methane Gas Production as Part of a Refuse Recycling System," *Compost Sci.*, 1976.

[36] WALTER, R., "How to Compost Leaves," *Am. City*, June 1971, p. 116.

[37] WILLSON, G. B., and J. M. WALKER, "Composting Sewage Sludge: How?" *Compost Sci.*, Sept.–Oct. 1973.

[38] SPANO, L. A., J. MEDEIROS, and M. MANDELS, "Enzymatic Hydrolysis of Cellulosic Wastes to Glucose," *Resour. Recovery Conserv.*, 1, 279 (1976).

[39] TREZEK, G. J., and C. G. GOLUEKE, "Availability of Cellulosic Wastes for Chemical or Biochemical Processing," *Energy, Renewable Resour. New Foods, AIChE Symp. Ser.*, 72 (158), 52.

[40] GOLDSTEIRN, I. S., *The Potential for Converting Wood into Plastics and Polymers or into Chemicals for the Production of These Materials*, School of Forest Resources, North Carolina State University, Raleigh, N.C., 1974.

[41] HAJNY, G. J., and E. T. REESE (Eds.), *Cellulases and their Applications, Adv. Chem. Ser.* 95, American Chemical Society, Washington, D. C., 1969.

[42] CROWLING, E. B., "Physical and Chemical Constraints in the Hydrolysis of Cellulose and Lignocellulosic Materials," *Biotech. Bioeng. Symp.* 5, 1974, p. 163 (as reported in ref. 38).

[43] SPANO, L. A., et al., *Pretreatment and Substrate Evaluation for Enzymatic Hydrolysis of Cellulosic Wastes*, U.S. EPA 600-7-77-038, Cincinnati, Ohio, 1977.

[44] CALLIHAN, C., and C. E. DUNLAP, *Construction of a Chemical-Microbial Pilot Plant for Production of Single Cell Protein from Cellulosic Wastes*, U.S. EPA, SW-24C, Washington, D.C., 1974

[45] ROGERS, C. J., E. COLEMAN, D. F. SPINO, T. C. PURCELL, and P. V. SCORPIOS, "Production of Fungal Proteins from Cellulose and Waste Cellulosics," *Environ. Sci. Technol.*, **8**, 715 (1972).

[46] COOKSON, A., and G. FROHNSDORFF, *The Nitrite Accelerated Photochemical Degradation of Cellulose as a Pre-treatment for Microbiological Conversion to Protein*, U.S. EPA. 670-2-73-052, Washington, D.C., 1973.

[47] PORTEUS, A., "WP Disposal Process Turns Cellulose Material into Alcohol," *Paper Trade J.*, Feb. 7, 1972.

[48] MCABEE, M. K., "Japan Pushes Alcohol for Protein Process," *Chem. Eng. News*, Dec. 9, 1974, p. 11.

PROBLEMS

6-1. Assume that refuse has a C/N ratio of 24:1. If raw sludge, with a C/N of 16:1 is to be added to refuse in order to reach a required C/N of 20:1, how much of each, refuse and sludge, is needed? (*Note*: use a weight basis.)

6-2. Assume that a city of 50,000 people uses, on the average, 20 million m^3 of natural gas per year. What fraction of this demand could be met by digesting the refuse from this community?

6-3. Using the chemical analyses of refuse described in Chapter 1, develop an empirical formula for the organic fraction of refuse, and estimate the theoretical production of methane from this hypothetical compound.

6-4. Estimate the methane gas production from a local landfill. Show all assumptions made.

7 COMBUSTION

HISTORICAL BACKGROUND

The process of combustion dates back to the first fire that was knowingly created by the caveman. Although we do not know whether that fire was lit to cook food, to keep man warm (or even to burn his solid waste!), it is from that first effort that we began to develop the art and science of burning various fuels for a number of different reasons.

In solid waste engineering, combustion was originally used to reduce the volume and putrescibility of solid waste. In fact, the first municipal incinerators were originally known as "reducers." Before incinerators, burning in open dumps had been practiced by many communities, and backyard trash burning was a time-honored tradition in many places.

The push by the states and the federal government to close burning dumps and eliminate private trash burning provided a strong impetus for the construction of more incinerators. Concurrently, however, the strict air pollution control requirements imposed on incinerators required intricate and expensive control measures, and drove up the cost of municipal refuse incineration.

More recently, the energy shortages have rekindled interest in combustion as a means of processing solid waste since under controlled conditions refuse can be a valuable source of energy. Modifications of classical incineration have further encouraged the use of combustion in the processing of solid waste.

Principles of Combustion

In this chapter the basic concepts of combustion are introduced first, followed by descriptions of thermal and materials balances, and a description of the major process variables. These basic concepts are then applied to the process of incineration in Chapter 8 and to pyrolysis in Chapter 9.

PRINCIPLES OF COMBUSTION

Simply stated, complete combustion is defined as the rapid oxidation of organic materials as follows:

$$(HC)_n + O_2 \rightarrow CO_2 + H_2O + \text{heat}$$

Actually, however, the process is more complex, even for pure hydrocarbons, and is quite complex if contaminants such as sulfurous and nitrogenous compounds are included. In addition, some components of air other than oxygen enter into the reaction, such as nitrogen:

$$N_2 + O_2 \xrightarrow{\text{heat}} 2NO$$

This reaction is temperature dependent, with little oxidation occurring below 1500°C (2700°F). The nitric oxide thus formed oxidizes to nitrogen dioxide, which is one of the major compounds associated with the formation of photochemical smog.

Another element commonly found in fuels is sulfur. Although it is of minor significance as a source of heat, it may cause corrosion problems in combustion equipment and represents a major air pollution concern. Sulfuric compounds are oxidized to sulfur dioxide and trioxide, both of which are readily soluble in water (also a product of combustion). In simple chemical terms,

$$SO_3 + H_2O \rightarrow H_2SO_4$$

The resulting sulfuric acid is corrosive and a definite hazard to health and vegetation.

Ideally, all organic material oxidizes during combustion, and the amount of oxygen necessary to allow the combustion to proceed can be calculated. This theoretical oxygen is known as "stoichiometric oxygen." For example, if pure carbon is combusted,

$$C + O_2 \rightarrow CO_2 + \text{heat}$$

One mole of oxygen is required for every mole of carbon.

If inorganic materials are in the fuel and these are oxidized as well, the oxygen required must be added to the total oxygen requirement.

Since air is about 21% oxygen by volume (and 23.15% by weight), the stoichiometric air required for combustion is calculated by dividing the oxygen requirement by 0.21 (or 0.2315 if it is in terms of weight).

Example 7-1

Calculate the required stoichiometric air for the combustion of 1 kg of methane gas.

$$CH_4 + 2O_2 \rightarrow CO_2 + 2H_2O$$

One mole of methane requires 2 moles of oxygen. The molecular weight of methane is 16, and for O_2 the gram molecular weight is $2 \times 16 = 32$. Hence, for 16 g of CH_4, 32 g of O_2 is required. By proportions, for 1 kg of CH_4,

$$\frac{16 \text{ g}}{32 \text{ g}} = \frac{1000 \text{ g}}{x \text{ g}} \qquad x = 2000 \text{ g of oxygen is required}$$

This is equal to

$$\frac{2000 \text{ g}}{0.2315} = 8639 \text{ g of air.}$$

Thus far, the discussion has centered on the complete combustion of organics. Under certain conditions, it is possible to achieve partial combustion with the end products still retaining some energy. For example, if insufficient air is provided in the combustion of carbon, carbon monoxide would be the main end product. If a hydrocarbon was combusted in the partial or complete absence of oxygen, the end product might be methane, CH_4, which can combust further and yield substantial heat. Under different conditions (temperature and time of combustion) the same hydrocarbon feed may produce carbon, C, and hydrogen gas, H_2. The partial combustion of organic materials is commonly called pyrolysis, although some forms of this process are more accurately known as gasification or starved air combustion. These processes are discussed in Chapter 9.

While incomplete combustion is a unit operation used in many experimental and a few prototype resource recovery facilities, the following discussion of process fundamentals is applied most easily to classical complete combustion such as incineration.

The amount of heat generated by the combustion of an organic fuel is independent of how the combustion occurs, provided that the process goes to completion. For example, the combustion of carbon,

$$C + O_2 \rightarrow CO_2 + \text{heat}$$

produces 32,800 J/g (14,100 Btu/lb) of heat. Should this reaction go in two steps, however, the heat of combustion would be

$$C + O \rightarrow CO + 10,100 \text{ J/g (4345 Btu/lb)}$$

and

$$CO + O \rightarrow CO_2 + 22,700 \text{ J/g (9755 Btu/lb)}$$

The sum of the two heats released is still, however, 32,800 J/g (14,100 Btu/lb). [1]

TABLE 7-1
Typical Heats of Combustion for Various Fuels

Fuel	Heat of Combustion	
	J/g	Btu/lb
Carbon (to CO)	9,300	4,000
Carbon (to CO_2)	32,800	14,100
Hydrogen	142,000	61,100
Sulfur (to SO_2)	9,300	3,980
Methane	55,500	23,875
Ethylene	50,300	21,635
Carbon monoxide	10,100	4,345
Municipal solid waste	9,300–18,600	4,000–8,000

Source: Ref. 2, 7.

TABLE 7-2
Heat of Combustion of Homogeneous Samples of Dry Shredded Refuse[a]

	Heat of Combustion	
	J/g	Btu/lb
A	18,500	7954
B	19,850	8535
C	20,040	8616
D	20,050	8621

[a]The refuse was essentially free of glass and metals.
Source: Ref. [3]

The heat of combustion is commonly determined in a bomb calorimeter. In this device, the fuel is placed in a metal container (bomb) immersed in water, and ignited. The heat absorbed by the water is a measure of the heat of combustion. Heats of combustion for common fuels are shown in Table 7-1. As suggested by the last line in this table, the heat of solid waste combustion is more difficult to evaluate. Table 7-2 is a listing of some heats of combustion of homogenized samples of a dry, shredded solid waste, essentially free of glass and metals. The energy equivalent of the samples tested range from approximately 18,500 J/g (7950 Btu/lb) to about 20,000 J/g (8600 Btu/lb) on a moisture and ash free basis.

For approximate heat of combustion values, the Dulong formula has long been found acceptable. For coals, it yields answers to within 2 or 3% of the actual heat value [2].

In English units, the Dulong formula reads

$$\text{Btu/lb} = 14{,}544\text{C} + 62{,}028\left(\text{H}_2 - \frac{\text{O}_2}{8}\right) + 4050\text{S}$$

where the symbols C, H_2, O_2, and S represent fractions by weight of the constituents in the fuel. To convert to metric, multiply Btu/lb by 2.32 to obtain J/g.

One of the end products of the combustion of fuels containing hydrogen is water. The water produced is in vapor form, but may condense to a liquid if allowed to do so. If this water condenses, it releases heat, which is what occurs inside a bomb calorimeter. Accordingly, any heat of combustion measurements conducted with a bomb calorimeter are gross or high heat values of the fuel, with the heat of vaporization included in the value. While the high-heat-value determination is a standard ASTM procedure, measurement of the net or low heat value is difficult [4]. Hence all heat-value data are assumed to be high heat values unless specified otherwise. In Europe, however, the convention is to use low heat values.

The calculation from high to low heat values can be approximated by [2]

$$Q_L = Q_H - 1040 W$$

where Q_L = low heat value, Btu/lb
Q_H = high heat value, Btu/lb
W = lb water formed/lb fuel

Another important thermodynamic consideration in combustion calculations is that of the *enthalpy* (often called the *sensible heat*) of combustion gases. The amount of work that can be done, or the amount, temperature, and pressure of steam that can be produced, is a function of the change in enthalpy of the combustion gases. The first law of thermodynamics (conservation of energy) states that any change in the internal energy (U) of a system must be accounted for by changes in heat (q) or work (w) (the only two ways in which energy can enter or leave a system).

$$dU = \delta q - \delta w \qquad (7\text{-}1)$$

If the only work is $P\,dV$ work (expanding or compressing a gas),

$$dU = \delta q - P\,dV \qquad (7\text{-}2)$$

where P is pressure and V is volume, and where a positive sign implies energy entering the system and a negative sign implies energy leaving the system. [The sign convention of Eqs. (7-1) and (7-2) dates from the days of the steam engine when *heat* was *added to* the system (the engine) and *work* was *produced* by the system.] Equation (7-2) can be rewritten as

$$dU + P\,dV = \delta q$$

which suggests that the left-hand side can be expressed as a differential of a new quantity—called *enthalpy* (H)—for processes taking place at *constant pressure*.

$$H = U + PV$$
$$dH = dU + P\,dV + V\,dP$$

At constant pressure,

$$dH = dU + P\,dV = \delta q$$

so the change in enthalpy is a measure of the heat that enters or leaves a system at constant pressure. When $dH < 0$, heat leaves the system; when $dH > 0$, heat enters the system.

The *specific heat capacity* (often called just *specific heat*) of a substance is another thermodynamic property of importance in combustion processes. The specific heat provides a measure of the amount of heat required to raise the temperature of a unit mass of material 1°C.

$$C = \frac{1}{m}\frac{dq}{dt}$$

In English units (Btu/lb/°F), the specific heat of water is unity (because the British thermal unit is defined in terms of water's specific heat), but in SI units, this is no longer the case. Nevertheless, common practice is still to represent specific heats of other substances in terms of the specific heat of water (Table 7-3), and this convention is usually followed in incinerator-design calculations. The specific heat of a given gas depends on whether it is measured at constant volume, C_v, or at constant pressure, C_p.

TABLE 7-3
Specific Heats of Various Materials

Material	Mean specific heat (Btu/lb/°F)
Water	1.000
Brass, yellow	0.089
Glass, crown	0.161
Glass, flint	0.120
Brick	0.20
Clay, sand, stone	0.20
Wood	0.60
Aluminum	0.225
Copper	0.100
Iron, cast	0.119
Iron, hard-drawn	0.134
Lead	0.034
Zinc	0.104

Source: Adapted from standard tables [17].

At constant pressure,

$$C_p = \frac{1}{m}\left(\frac{dH}{dT}\right)_p$$

The importance of the concept of specific heat can be illustrated by comparing the use of water and air for gas cooling. Water can carry away almost four times as many J/g than can air (up to 100°C), since the specific heat of air is approximately 25% that of water. In addition, as the water vaporizes to steam at atmospheric pressure, it absorbs even more heat, because of its enthalpy of vaporization and therefore is an efficient medium for gas cooling. The enthalpy of vaporization is also called the *latent heat of vaporization*. Latents heats usually exist for any change of phase: sublimation, fusion, or melting.

Ignition of a material occurs when the temperature in the combustion process rises to the point where more heat is generated by the process than is lost to the surrounding combustion chamber. Thus the *ignition temperature* has been reached and the oxidation process becomes self-sustaining. Table 7-4 lists some typical ignition temperatures. Ignition temperatures are influenced by many factors, including the moisture content, the shape and size of the combustion chamber, and the pressure in the chamber.

TABLE 7-4
Ignition Temperatures for Some Common Fuels

Fuel	Temperature (°C)	Temperature (°F)
Fixed carbon (as in bituminous coal)	410	770
Hydrogen	575–590	1065–1095
Sulfur	240	465
Methane	630–750	1170–1380
Ethylene	480–550	900–1020
Carbon monoxide	610–660	1130–1220
Municipal solid waste	260–370	500–700

Source: Ref. 2.

One final consideration in combustion calculations is the *adiabatic flame temperature*. This term may be defined as the temperature reached by the products of oxidation of a fuel/air combination when there are no heat losses. "Adiabatic" means perfectly insulated, or so rapid that no heat transfer takes place. While the heat of combustion of the fuel is a predominant factor in determining the temperature of the flame, if the temperature of the combustion air is increased by preheating, the flame temperature will also be increased. The adiabatic flame temperature would

Thermal Balance in Combustion 287

obviously be higher than that experienced under actual field conditions for several reasons. Because it takes a certain amount of heat to begin the process of oxidation and a portion of this heat is lost to the surrounding combustion chamber, this heat loss lowers the flame temperature, but the adiabatic flame temperature provides a theoretical maximum.

THERMAL BALANCE IN COMBUSTION

The law of conservation of energy states that in a closed system, the energy input must equal the energy output. We can, therefore, develop an energy balance for a combustion operation by including the various sources of energy. Items in the heat balance include [5, 6, 7]

Heat input	Heat output
Heat of combustion	Heat due to vaporization of water (latent heat from moisture and water from the combustion of cellulose.)
	Heat losses due to radiation
	Heat lost in ashes (unburned fuel)
	Heat lost in stack gases (sensible heat)

Example 7-2

For a MSW containing 28% moisture and 55.9% organics, calculate the sensible (useful) heat produced at a feed rate of 10,000 kg/h. Assume a heat of combustion of 19,193 J/g for the dry organic fraction.

Gross heat of combustion:

$$10,000 \text{ kg/h} \times 0.559 \times 19,193 \text{ kJ/kg} = 107.3 \times 10^6 \text{ kJ/h}$$
$$\text{(organics)} \quad \text{(moisture)}$$

Latent heat loss from hydrogen (assuming organics are cellulose and 55.6% of cellulose becomes water)

$$10,000 \text{ Kg/h} \times 0.559 \times 0.556 \times 2420 = 7.5 \times 10^6 \text{ Kg/h}$$

Latent heat losses due to vaporization of water

$$(10,000 \text{ kg/h} \times 0.28) \times 2575 = 7.2 \times 10^6 \text{ kJ/h}$$

Note: 2420 kJ/kg is the approximate enthalpy (or latent heat) due to vaporization.

Heat losses due to radiation: assume that this is 5% of heat input, or

$$0.05 \times 107.3 \times 10^6 \text{ kJ/h} = 5.3 \times 10^6 \text{ kJ/h}$$

Heat lost in ashes (unburned fuel): assume that the residual contains 10% organics, so that the heat loss is

$$0.1 \times 107.3 \times 10^6 \text{ kJ/h} = 10.7 \times 10^6 \text{ kJ/h}$$

Heat lost in stack gases (sensible heat): this is the unknown, and is calculated from

$$\text{heat input} = \text{heat output}$$
$$107.3 \times 10^6 = (7.5 + 6.8 + 5.3 + 10.7) \times 10^6 + \text{sensible heat}$$
$$\text{sensible heat} = 82.3 \times 10^6 \text{ kJ/h}$$

Many assumptions have been made in the calculations above, such as no sensible heat in the ashes, and so on. The final answer will, however, be useful for rough preliminary design purposes.

When steam is produced as a useful end product, the energy input must include the heat in the boiler feed water, while the energy output must include the heat of the steam. The balance is then:

Heat input	Heat output
Heat of combustion of the feed	Latent heat due to evaporation
Heat in boiler feed water	Heat losses due to radiation
	Heat in ashes (unburned fuel)
	Heat in stack gas (sensible heat)
	Heat in steam

Example 7-3

Using the same data as in Example 7-2, if the feed water temperature is 100°C, and the desired steam temperature and pressure are 300°C and 4×10^6 N/m² (or 4×10^3 kPa), respectively. Calculate the heat in the flue gas exit temperature if 20,000 kg/h of steam is required.

Heat of combustion (from Example 7-2) = 107.3×10^6 kJ/h
Heat in boiler feed water:

$$20{,}000 \text{ kg/h water} \times (273 + 100)°\text{K} \times 1 \times 0.00418 \text{ kJ/kg°K} = 31.183 \text{ kJ/h}$$

Note: The temperature of the water is 100°C, or 373°K. The specific heat of water is 0.00418 kJ/kg/°K (or °C).

Heat in the steam: at 4×10^6 N/m², for saturated and superheated steam, from steam tables (see any thermodynamics textbook), the heat in the steam is 2975 kJ/kg.

$$20{,}000 \text{ kg/h} \times 2975 \text{ kJ/kg} = 59.5 \times 10^6 \text{ kJ/h}$$

Heat lost due to vaporization, radiation, and ashes from Example 7-2 is 30.3×10^6 kJ/h

The heat balance then yields $107.3 \times 10^6 + 0.031 \times 10^6 = 59.5 \times 10^6 + 30.3 \times 10^6 +$ heat in flue gas.

The flue gas heat = 17.5×10^6 kJ/h.

Process Variables

MATERIALS BALANCE IN COMBUSTION

With homogenous fuels such as coal, the products of combustion are fairly well known. Even for these fuels, however, the products are dependent on the method of operation. Predicting the materials balances for refuse combustion is therefore difficult [8].

A materials balance requires that the material fed to the combustion process plus the air must equal the sum of the fly ash, bottom ash, and stack gas produced. In incineration, the amount of each of these end products is highly dependent on the operating conditions.

PROCESS VARIABLES

An examination of general characteristics of the combustion process required to yield satisfactory ignition and combustion of the fuel reveals several important considerations. It is helpful if the following conditions are met in order to ensure adequate combustion [9-14].

1. The thickness and density of the material burned in the combustion chamber is most important when related to the length of time the material is retained in the active combustion zone.
2. The refuse should be fed to the fuel bed in such a way as to ensure that raw fuel is not dumped on top of the flame.
3. The air supplied to the combustion process must be properly regulated. Ideally, the flow of the air should be balanced between the flow from the raw fuel into the ignited fuel or vice versa.
4. The temperature of the furnace must be maintained above the ignition temperatures of the material to be combusted.
5. The gases in the incinerator must be properly mixed. Thus the overfire air must be thoroughly mixed with the hot gases flowing from the fuel bed (underfire air).
6. Excess air may be used to lower the exit gas temperatures and must be properly mixed in the combustion chamber so as to attain complete combustion.

To ensure complete combustion, adoption of the three T's principle is important. The three T's are time, temperature, and turbulence. All three of these components (time, temperature and turbulence) are interdependent, but each may be considered independently to evaluate their impact on combustion.

Time

If one assumes that dry refuse particles are introduced into a combustion chamber, the initial reaction requires only the time necessary for the heating of the particles [12]. The time required for the oxidation reaction is that necessary to burn the solid pieces. Generally, the larger the particles that must be oxidized, the longer the required residence time in the combustion chamber. It is important, therefore, to consider particle size when selecting the residence time in the combustion chamber.

Turbulence

Mixing in the combustion chamber can be enhanced by the injection of underfire and overfire air and by the action of the grate systems. The interaction of the gaseous products that result from the burning of the solid fuel in the bed also aid in mixing the bed [15]. The injection of overfire air has the primary objective of assisting in the oxidation of the gaseous products which evolve during the combusion of the solid fuel.

Temperature

Time and turbulence both play a very important part in the maintenance of adequate temperatures within the combustion zone. Temperature is dependent on the characteristics of a fuel, such as its ignition temperature, its moisture content, the configuration of the furnace, and the combustion and excess air distribution [16].

As temperatures are increased in the combustion chamber, the time required for the combustion of a waste is generally reduced, but problems with refractories and boiler tubes are increased. It has been suggested that if the chamber is hot enough, the burning rate is mass-transfer-controlled [12], and the time required for combustion is no longer as important. Conversely, if the temperatures are lower in the combustion chamber, the rates may be controlled by chemical reactions and combustion times may become very important. Air flow, since it can be regulated in the incinerator, is the primary means of controlling the temperature in refractory lined furnaces.

PARTING SHOTS

Based on the concepts of energy and materials balances and of equilibrium presented above, what approach must one take to the computations that will properly characterize the combustion process? Figure 7-1 illustrates one iterative procedure that might be adopted. It is important to note, however, that when dealing with the calculations that relate solid wastes to the various equations outlined, a high degree of variability in facts (as well as assumptions) is to be expected. Numerous approaches to calculating the same values is recommended to help "zero in" on the appropriate answer.

There is no "right" method for the computation of combustion values for MSW—all are approximations. In Chapter 8, the process of incineration is outlined and in Chapter 9, pyrolysis is discussed. These chapters should help to put some practical perspectives on the more theoretical aspects of combustion.

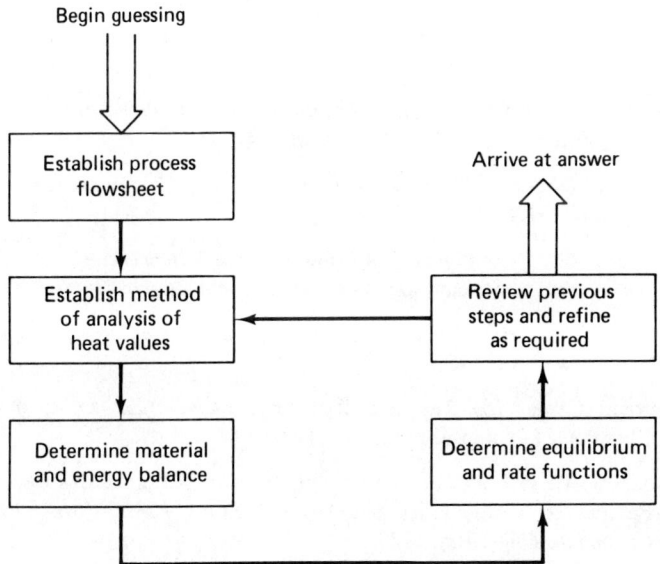

Figure 7-1. Proposed methodology for the computation of combustion calculations.

REFERENCES

[1] MONROE, E. S., "Combustion Fundamentals—An Engineering Approach to the Design of Industrial Incinerators," ASME Winter Annu. Meet., New York, Nov., 1972.

[2] *Steam—Its Generation and Use*, Babcock & Wilcox, Inc., New York, 1976.

[3] JANSEN, T. E., D. J. EATOUGH, and L. D. HANSEN, "Discarded Energy: The Heat of Combustion of Garbage," *J. Chem. Ed.*, Nov., 1977.

[4] HASLAM, R. T., and R. P. RUSSELL, *Fuels and Combustion*, McGraw Hill Book Company, New York, 1926.

[5] NISSEN, R., *Combustion and Incineration Processes*, Marcel Dekker, Inc., New York, 1978.

[6] COHAN, L. J., and J. H. FERNANDES, "Potential Energy Conversion Aspects of Refuse," ASME Winter Meet., Pittsburgh, Pa., 1967.

[7] WILSON, D. L., "Prediction of Heat of Combustion of Solid Wastes from Ultimate Analyses," *Environ. Sci. Technol.*, 13, (June 1972).

[8] ESSENHIGH, R. H., "Incineration: A Practical and Scientific Approach," *Environ. Sci. Technol.*, 2, 526 (1968).

[9] NICHOLLS, P., "Underfeed Combustion, Effect of Preheat, and Distribution of Ash in Fuel Beds," *U.S. Bur. Mines Bull. 378* (1934).

[10] MANTEL, C. L., *Solid Wastes—Origin, Collection, Processing, and Disposal*, John Wiley & Sons, Inc., New York, 1975.

[11] SKITT, J., *Disposal of Refuse and Other Waste*, Charles Knight Co. Ltd., London, 1972.

[12] HOWARD, J.B., "Combustion of Solid Residue," *Incinerator and Solid Waste Technology*, American Society of Mechanical Engineers, New York, 1976.

[13] CAREY, R. C., *Principles and Practices of Incineration*, John Wiley & Sons, Inc., New York, 1969.

[14] National Center for Resource Recovery, *Incineration*, D. C. Heath and Company, Lexington, Mass., 1974.

[15] RAO, S. T. T., T. J. KUO, and R. H. ESSENHIGH, "Characterization of Stirring Factors by Cold-Model Simulation," *Proc. ASME Nat. Incin. Conf., Cincinnati, Ohio, May 1970.*

[16] STERNNITZKE, R. F., and M. DVIRKA, "Temperatures and Air Distributions in Large Rectangular Incinerator Furnaces," *Proc. ASME Nat. Incin. Conf.*, New York, May 1968.

[17] KENT, R. T., *Mechanical Engineers Handbook*, John Wiley & Sons, Inc., New York, 1979.

PROBLEMS

7-1. Discuss the significance of increased combustion temperatures and the relative production of SO_x and NO_x.

7-2. Calculate the low heat value if the higher heat value of the fuel is 16000 J/g and 0.2 kg of water is formed per kilogram of waste fuel.

7-3. Based on the specific heats of the materials, explain why common brick materials are not as satisfactory as refractory bricks for lining a combustion chamber.

7-4. In Example 7-2 if the MSW were predried to a moisture concentration of 20%, what would be the sensible heat and the heat lost due to vaporization?

7-5. Develop a materials balance for the typical combustion of MSW.

8 INCINERATION AND ENERGY RECOVERY

INTRODUCTION

In Chapter 7, it was mentioned that controlled combustion dates back to the days of the caveman. We have come a long way since that time and in many applications have developed the "art" of combustion into a science. In this chapter the process of incineration and related areas such as waste heat recovery and prepared fuels are discussed.

Combustion has most often been defined as any chemical process which is accompanied by the evolution of light and heat, commonly the union of various substances with oxygen. The principles of combustion are commonly applied in a specialized process called incineration, which involves the burning of solid, liquid, or gaseous organic materials to form gases and a residue which should contain little additional combustible material. In recent years, interest has centered on the means by which this process of combustion can be used to gain benefits from the energy sources, and not just to destroy the caloric value of the fuels.

This chapter is organized into three sections: incineration of raw refuse, waste heat recovery from incineration, and the preparation and use of refuse and prepared fuels.

A number of excellent textbooks have been written on the process of incineration and incinerator design. The intent of this chapter is not to reproduce the work in such volumes, but rather to generally describe the important process considerations of incinerator facilities, especially as they relate to the recovery of energy from MSW. Study of the references, coupled with the general approaches noted herein, should assist all but the most detailed designer with an understanding of the fundamental design considerations of the incineration process.

INCINERATOR DESIGN CONCEPTS

Background

The first large municipal incinerator (or destructor, in the British terminology) was constructed in Nottingham, England, in 1874. Although many large incinerators were constructed in England during subsequent years, it took nearly 50 years before a substantial number of incinerators were constructed on the continent and in America. A thorough description of the early phases of the development of incineration can be found in a work by two American engineers, Hering and Greeley, which traces the history of municipal incineration essentially from its inception [1].

The first incinerator on record in the United States was built in 1885 on Governor's Island in New York Harbor [2], and also in 1885 the first municipal incinerator was built in Allegheny City, Pennsylvania. By the early 1900s over 200 municipal incinerators had been built in the United States, but within a decade, over half of those units had been abandoned [3], primarily because they did not efficiently reduce the waste volumes.

The most popular unit constructed during this period—called a batch feed incinerator—had a hearth onto which the municipal solid waste was charged in batches and over which the hot gases passed. Most of these incinerators employed an auxiliary fuel source, since only garbage was being burned and this could not sustain combustion. When rubbish was mixed with the garbage beginning in the late 1940s and early 1950s, the requirement for auxiliary fuel was eliminated in many incinerators.

One of the earliest continuous feed furnaces built in the United States was in Atlanta, Georgia, in 1951. New York City pioneered in this design during the 1950s. Almost all incinerators now constructed in the United States employ the continuous-feed principle.

Modifications to the design of incinerators have evolved over the years from singular rectangular boxes which enclose the grates to designs which now incorporate variously shaped combustion chambers. The newer chambers are generally smaller, which reflect heat back into the furnace, and increase the exposure of the solid fuel to the hot furnace environment. New grate modifications conserve the amount of air required for combustion. More recently, the concepts employed in the design of incinerators have been altered significantly in the United States and Europe by the change in the composition of an average municipal waste. In the past, this waste was predominantly garbage, but in Chapter 1 it is noted that the percentage of garbage has steadily declined in municipal solid waste (MSW), resulting in a more highly combustible material.

The goal of incineration is to combust solid waste so as to reduce its volume and to produce nonoffensive gases and non-combustible ash residues. Typical volume reductions amount to anywhere from 80 to 95% and change its composition from a public health menace to a basically sterile, inert ash [4, 5]. Weight reductions may range from 70 to 80% depending on the quality of the burnout and the amount of non-combustibles in the charge. Since volume reductions of 10 to 1 are common, significant reductions in required landfill capacities result.

Incineration has certain drawbacks, the most notable of which are the high capital and operating costs. Operating problems may also result from the variability of the wastes that are burned. The public also perceives incinerators as a nuisance because of the environmental problems they sometimes cause, particularly those related to air pollution.

Incineration Processes

A typical incinerator system generally contains certain basic elements, including a feed system; a combustion chamber(s), an exhaust gas system, and a residue disposal system. Ancillary equipment may include shredders and materials sorters at the front end, air pollution control devices, and a heat recovery system at the back end of the incinerator. Modern incinerators in the United States use continuous-feed systems and moving grates in a primary combustion chamber which is lined with refractory materials (heat-resistant silica-based materials). Secondary combustion chambers are used to burn the gases and solid particles not burned in the primary combustion chamber prior to discharge to the atmosphere or to air pollution control devices.

Until the 1950s, conventional incinerators were often plagued by inefficient burning rates and high operating costs. One of the major problems associated with incineration today is the lack of a sound theory about the combustion process with respect to MSW. Because the design is based on generalized data on MSW and lack of complete engineering information, the approach to the design should encompass both theoretical as well as practical considerations.

The Incinerator Institute of America (now disbanded) has established seven incinerator classifications, noted in Table 8-1. These units are classified by the Institute as noted in Table 8-2.

TABLE 8-1
Incineration Classification

Class I—portable, packaged, or completely assembled, direct-fed incinerators, having not over 0.14 m³ (5 ft³) storage capacity, or 11.33 kg (25 lb) per hour burning rate. These incinerators are suitable for the combustion of Type 2 wastes.

Class IA—portable, packaged or job-assembled, direct-fed incinerators having 0.14 to 42 m³ (5 to 15 ft³) primary chamber volumes; or a burning rate of 11.33 kg (25 lb) per hour up to but not including 45.3 kg (100 lb) per hour of Type 0, 1, or 2 wastes. Type 3 wastes may be burned at a rate of 11.33 kg (25 lb) per hour up to, but not including, 34 kg (75 lb) per hour.

Class II—fuel-fed, single-chamber incinerator with more than 0.19 m² (2 ft²) of burning area suitable for Type 2 wastes. (This type of incinerator has normally been installed in apartment houses.)

Class IIA—chute-fed multiple-chamber incinerators for apartment buildings with more than 0.19 m² (2 ft²) of burning area, which is suitable for Type 1 or 2 wastes.

Class III—direct-fed incinerators with a burning rate of 45.3 kg (100 lb) per hour and over. These incinerators are suitable for Type 0, 1, or 2 wastes.

Class IV—direct-fed incinerators with a burning rate of 34 kg (75 lb) per hour or over. These incinerators are suitable for the combustion of Type 3 wastes.

Class V—municipal incinerators which are continuously fed and are rated in tonnes per hour or tonnes per 24 h. These incinerators are suitable for the combustion of Type 0, 1, 2, or 3 wastes, or a combination of all four.

Class VI—these incinerators are for crematory pathological purposes and suitable for Type 4 wastes.

Class VII— these incinerators are designed for specific by-product wastes and generally burn Type 5 or Type 6 wastes.

Source: Ref. 7.

Incinerators may be classified as either batch-fed or continuously fed. Continuously fed incinerators have many advantages over batch-fed units, not the least of which is the large furnace capacities. The process can also be more easily controlled and more uniform temperatures can be maintained in the combustion zone. All large modern incinerators are continuous; thus the discussion in this chapter is limited to such continuous systems (except for modular incineration, page 324).

Two major classes of continuously fed furnaces are employed in the United States today: refractory-lined and waterwall furnaces. The waterwall furnace recovers waste heat as well as reduces waste volumes, while the typical refractory furnaces are usually designed only for volume reduction. In addition to these two major types of processes, there are

TABLE 8-2
Solid Waste Classifications To Be Incinerated

Classification of Wastes	Principal Components	Approximate Composition (% by Weight)	Moisture Content (%)	Incombustible Solids (%)	J/g of refuse as fired[b]	kcal of Auxiliary Fuel per Jg of Waste to be Included in Combustion Calculations[b]	Recommended Min. J/h Burner Input per Waste
Trash,[a] Type 0	Highly combustible waste, paper, wood, cardboard, cartons, including up to 10% treated papers, plastic, or rubber scraps; commercial and industrial sources	Trash, 100	10	5	19,650 (8500)	0	0
Rubbish,[a] Type 1	Combustible waste, paper, cartons, rags, wood scraps, combustible floor sweepings; domestic, commercial, and industrial sources	Rubbish, 80 Garbage, 20	25	10	15,050 (6500)	0	0
Refuse,[a] Type 2	Rubbish and garbage; residential sources	Rubbish, 50 Garbage, 50	50	7	10,000 (4300)	0	3300 (1500)

Classification	Description	Principal components	% Moisture	% Incombustible solids			
Garbage,[a] Type 3	Animal and vegetable wastes, restaurants, hotels, markets, institutional, commercial, and club sources	Garbage, 65 Rubbish, 35	70	5	(5850) (2500)	3300 (1500)	7100 (3000)
Animal solids and organic wastes, Type 4	Carcasses, organs, solid organic wastes, hospital, laboratory, abattoirs, animal pounds, and similar sources	Animal and human tissue,	85	5	2300 (1000)	7000 (3000)	18,800 (8000) 11,700 primary (5000) 7100 secondary (3000)
Gaseous, liquid, or semiliquid wastes, Type 5	Industrial process wastes	Variable	Dependent on predominant components	Variable according to wastes survey	Variable according to wastes survey	Variable according to wastes survey	Variable according to wastes survey
Semisolid and solid wastes, Type 6	Combustibles requiring hearth, retort, or grate burning equipment	Variable	Dependent on predominant components	Variable according to wastes survey	Variable according to wastes survey	Variable according to wastes survey	Variable according to wastes survey

[a]The figures for moisture content (ash), incombustible solids, and kcal (Btu) as fired have been determined by analysis of many samples. They are recommended for use in computing heat release, burning rate, velocity, and other details of incinerator designs. Any design based on these calculations can accommodate minor variations.
[b]Values in parentheses are Btu and pound equivalents.
Source: Ref. 7.

many other types of incinerators and waste heat recovery units available:

Modular incinerators.

Multiple-hearth incinerators.

Wet oxidation.

Fluidized bed incinerators.

Slagging incineration.

Open-pit incinerators.

All of these processes have factors in common and some concepts of incinerator design apply to all the units. In order to establish these concepts, however, the refractory and waterwall incinerators are used to exemplify typical incinerators with and without waste heat recovery. The specific details of each of the specialized incinerators are then covered.

The refractory incinerator is a proven incineration system. A typical refractory-type incinerator is illustrated in Fig. 8-1. The name is derived from the use of refractory materials which line the combustion chamber, flues, and stacks. The incinerators rely on air for cooling, which must be supplied in large quantities (far in excess of that chemically required simply for combustion) to prevent overheating of the refractory lining and

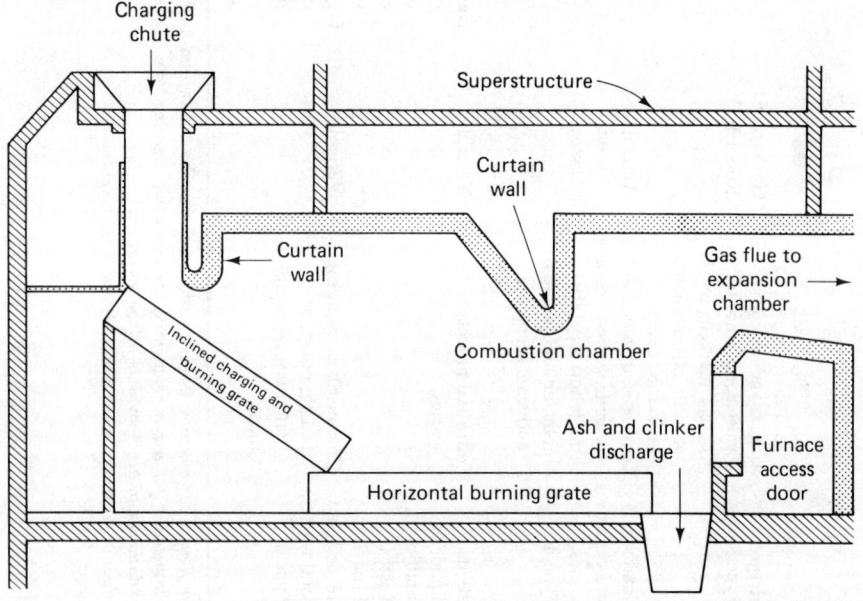

Figure 8-1. Typical municipal incinerator. (From Ref. 5)

to aid in the prevention of excessive slagging of the refuse on the walls of the combustion chamber. Since this excess air becomes part of the off-gases, it must be passed through the air pollution control equipment just as the combustion gases are, and these larger volumes increase the size of air pollution control equipment. In short, refractory-lined furnaces, while popular, have certain associated design and operating problems which waterwall furnaces have mostly overcome.

Waterwall furnaces employ water-filled tubes instead of refractory material to line the combustion chamber. As the burning refuse transfers the heat through the wall to the water in the tubes, they form a cool wall which is in contact with the flame and hot gases. These cooler walls prevent the accumulation of slag on the side of the combustion chamber and produce steam.

While an understanding of the combustion of municipal solid wastes has not developed to the advanced state that it has with other fuels, the principles of solid fuel combustion can generally be stated as follows [8]:

Air and fuel must be mixed in proper proportions for combustion.

Air and fuel, especially combustible gases, must be thoroughly mixed.

Temperatures in the incinerator must be sufficient for the ignition of both the solid and gaseous components of the fuel.

The volume of the combustion chamber must be large enough to provide for the detention time needed for complete combustion.

With the general design concepts in mind, the examination of the following incinerator components is made: waste storage, feed arrangements, combustion chambers, incinerator grates, combustion air, refractories, stacks, residue, instrumentation, general design considerations, and finally, an example of the design of an incinerator.

Waste Storage

When the MSW is delivered to an incinerator, provision must be made for its storage prior to combustion in the incinerator. The storage provided is dependent on the variations in the rate of delivery of refuse to the facility, and the planned burning schedule. Large storage areas are normally required for MSW since it is quite bulky, with a bulk density of approximately 150 to 210 kg/m^3 (250 to 350 lb/yd^3). Refuse also tends to flow very poorly and can maintain an angle of repose of greater than 90°. The storage and conveyance of municipal solid waste is more thoroughly discussed in Chapter 3. Storage of the MSW can create problems at the front end of incineration facilities and should be carefully considered in the design.

The installation of storage facilites (or pits) at an incinerator facility is necessary to allow required flexibility in plant operation when considering the burning capacity of the incinerator and the rate of the incoming wastes. Storage permits the retention of refuse during peak loads and thus allows the combustion chambers to be sized for a lower average capacity. While there are no hard-and-fast rules for determining the required storage capacity, provision is often made for the storage of as much as a week's refuse at smaller incinerators to allow for downtime and other operating problems. Provision for the storage of from 2 to 3 days' refuse is more common in larger installation (> 500 tons/day). The mass diagram technique can be used to estimate storage requirements if the collection data are available.

Feed Arrangements

For continuously charged furnaces, the solid waste is normally fed to the furnace directly through a furnace that is kept filled at all times in order to maintain an air seal between the furnace and the outside atmosphere. While the furnace may be fed directly by pushing material into the furnace by a ram, more often the refuse is lifted from pits by an overhead crane and dropped into the chute. The basic requirement of the feed system is to supply a flow of refuse to the furnace and to protect against burnbacks of fire from the combustion chamber through the chute to the pit area.

The most commonly employed type of feeding system is a traveling bridge crane with a grapple which lifts and carries the refuse above the furnace and releases it into a hopper. The hopper leads to the chute through which the refuse slides by gravity into the furnace. The traveling bridge and grapple is also used to mix the MSW (Fig. 8-2). The mixing of the MSW facilitates control of the combustion processes, particularly if an unusual amount of one type of waste is discharged to one part of the storage area.

A clamshell, grapple, or orange-peel type of bucket is attached to the overhead trolley and normally has a capacity of 1.5 to 4.5 m^3 (2 to 6 yd^3). The unit is most often operated by a person in an enclosed cab mounted on the bridge or trolley, although remote operation from a control room is possible.

The charging hopper into which the grapple discharges leads to a chute which is smooth and lined with steel plate and is water cooled near the furnace. The chute, which extends several feet into the throat of the combustion chamber, terminates just above one end of the grate system. It is often flared at the bottom to prevent bridging of the waste in the chute. Such a configuration forms a column of refuse above the grate which

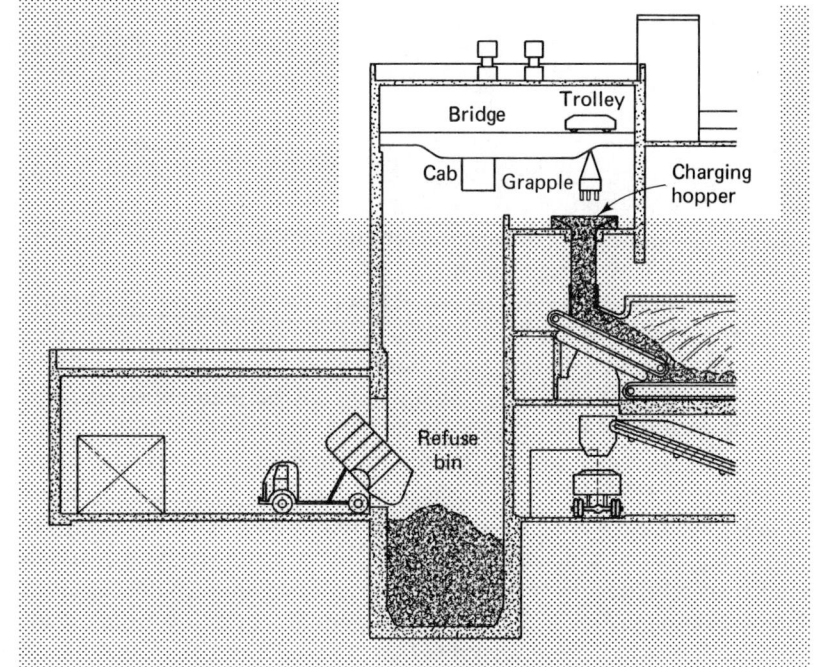

Figure 8-2. Traveling bridge crane. (From Ref. 9)

serves as an air seal. The lower end of the column of refuse is exposed to the heat of the furnace. As the grate system moves the refuse into the furnace, new refuse is continuously forced down the chute.

Combustion Chambers

The combustion zones in a refuse incinerator are commonly referred to as the *furnace*. Today the most common configurations include the rectangular furnace, the multicell furnace, the vertical-circular furnace, the combined rectangular furnace and rotary kiln. They are commonly constructed of refractory material or watertube walls and may employ any one of a number of different types of grate systems. Since the combustion process is believed to occur in two overlapping stages (zones) known as primary combustion and secondary combustion (burnout), design of a typical incinerator must accommodate these processes. In the primary combustion stage the processes of drying, volatilization of the waste, and finally ignition, occur. The design of the combustion chamber (with respect to its shape and size) is predicated on the assumption that it is desirable to be able to maintain temperatures in the ignition zone between approximately 700 and 1000°C (about 1300 and 1800°F), at the operator's discretion. It has also been suggested that for rectangular incinerators, the size of the combustion chamber should be between 0.4 and 0.6 m^3/tonne (15 and 23 cu.ft/ton) of waste processed per day [8]. Alternatively, it has been suggested that the heat release rate be between 4.6×10^5 and 9.2×10^5 kJ/m^3/hr (12,500 and 125,0000 Btu/ft^3/hr) [5].

In the secondary combustion stage, the remainder of the smaller unburned particulate matter which is released from the primary combustion stage, as well as the gases passing from that zone, are oxidized. The secondary combustion zone is either a separate chamber or additional volume in the primary combustion zone above the grates. The temperature in the secondary combustion chamber should be maintained at no more than 600 to 1000°C (about 1100 to 1800°F).

The number and size of the combustion chambers installed in an incinerator is an important design consideration. In the past, individual furnace capacities were limited to about 135 tonnes/day (150 tons/day) so that a 225 to 450-tonnes/day (about 250 to 500-tons/day) plant might require anywhere from two to four such furnace units. With the introduction of continuous feed, the size of furnaces has been increasing, and it is not uncommon to see units rated at 450 tonnes/day (500 tons/day), although units in the size 225 to 270 tonnes/day (250 to 300 tons/day) are more common [9]. If a furnace is too small, all the combustible matter may not be volatized, which can create air pollution and residual disposal problems. If it is too large, the waste may burn inefficiently.

Incinerator Grates

One of the more important components of an incinerator and of the combustion chamber is the *grate* (*stoker*). It serves a dual function: the grate must transport the solid waste and residue through the furnace and, at the same time, promote combustion by providing proper waste agitation and by permitting the passage of underfire air through the fuel bed. This agitation is further promoted by tumbling action which occurs when the burning solid waste drops from one level of the grate to another. Such tumbling may, if too violent, on the other hand, contribute to an excessive entrainment of particulate matter in the underfire air which is passing through the solid waste [10]. Grates must therefore be selected and designed to provide continuous but gentle agitation of the waste material.

Grate systems are generally classified by either their mechanical type or their function. Various mechanical classification schemes have been devised, but in general there are three types of grates commonly used in the United States: reciprocating grates, rocking grates, and traveling grates.

Incinerator Design Concepts 305

Other types of mechanical grates that are used include the rocking rotary kiln; vibrating, oscillating, and reverse reciprocating grates; and multiple rotating drums, rotating cones with arms, and variations or combinations of these types.

Grate systems may also be classified by their function: the drying grate, the ignition grate, the combination grate, and the burnout grate. The classification of the grate by function is less popular because most, if not all, of the mechanical grate systems defined can be utilized for any of the purposes described above.

Grate selection

Since there is a wide choice of grates available, what criteria are considered when choosing an appropriate grate, given certain waste characteristics? Generally, such a question is subjective, but Eberhart proposed 10 elements on which to base a judgment about the choice of a grate system [11]:

1. The adaptability of the combustion process to handle wide variations in the radiation effects.
2. The adaptability of the refractory to handle wide variations in the radiation effects.
3. Provision for the control of air quantity and temperature.
4. Provision for an adjustable retention time according to the other material being burned.
5. Adjustable height of waste layer to be burned.
6. A controllable, stabilizing heat supply (auxiliary fuel).
7. A controlled cooling of the residue (by quenching).
8. A controlled flue gas temperature prior to impinging on the radiation heating surface.
9. Ability to observe the fire layer and the fire gases.
10. Technical design, including:
    Prevention of reignition.
    Positive conveyance of the refuse mass.
    The serviceability and replaceability of worn-out parts.
    Proper measuring and control systems.

Grate systems

Older incinerators are commonly equipped with a *stationary grate*, which is simply a series of bars or castings forming the floor of the furnace and extending between the walls of the combustion chamber. Only manual stoking is possible with this system, and since they do not permit a sufficiently flexible control of the combustion process, they are seldom built and used.

A typical modern stoker is the *traveling grate*, which consists of continuous moving chain belts, similar to a conveyor belt. The grates, which are carried on sprockets and covered with separated but relatively small metal pieces called keys (Fig. 8-3), convey the waste material through the furnace. Most often two (or more) grates are provided at different elevations because as the material drops from one grate to another the agitation enhances combustion.

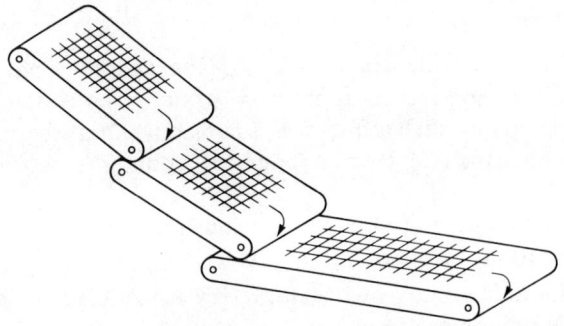

Figure 8-3. Traveling grate.

Another common grate, the *reciprocating grate*, is illustrated in Fig. 8-4. This grate system has been characterized as stacked overlapping roof shingles [5]. The grate consists of a bed of bars, or more commonly plates, which are arranged in alternate layers and reciprocate in a horizontal fashion, sliding over one another. The reciprocating action pushes the refuse along the grate surface. They are normally driven hydraulically.

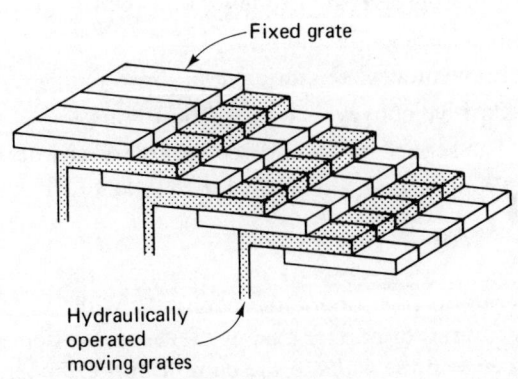

Figure 8-4. Reciprocating grate.

Rocking grates, illustrated in Fig. 8-5, are pie-shaped grates arranged in rows across the width of a furnace and at right angles to the solid waste flow. By rocking the axles of alternate rows of the grates in an upward fashion, a forward motion is imparted to the refuse, moving it through the incinerator. The relative drop between grate sections is commonly 5 to 10 cm (2 to 4 in.).

The *circular grate*, illustrated in Fig. 8-6, is most often used in combination with a central rotating rabble arm. Extended rabble arms are provided

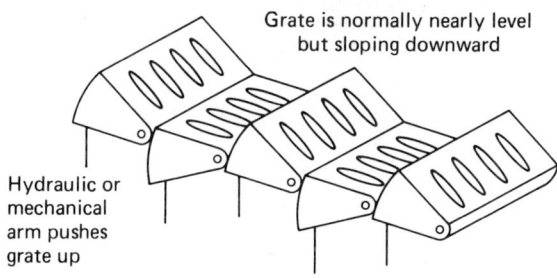

Figure 8-5. Rocking grate.

A Rotating cone
B Extended staking arm (rabble arm)
C Stationary circular grate
D Peripheral dumping grate

Figure 8-6. Circular grate. (From Ref. 5)

to agitate the fuel bed to promote combustion. These units are employed in circular batch-fed furnaces.

The *rotary kiln*, widely utilized in Europe, is a combination of a furnace and a grate system, which causes the solid waste to move in a slowly cascading and forward pattern through the unit (Fig. 8-7). Commonly, a reciprocating grate is installed to feed the rotary kiln because it can evenly distribute the waste material to the kiln and serve as both drying and ignition grates. By the time material gets to the rotary kiln, most of the moisture and volatile constituents have been driven off and the residue is fed into the kiln for final burnout.

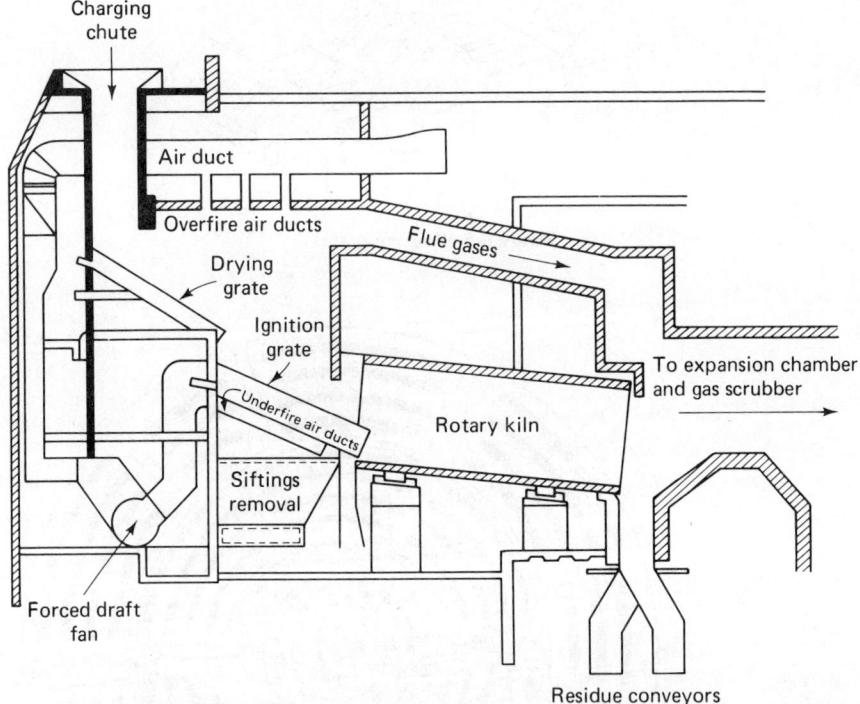

Figure 8-7. Rotary kiln. (Courtesy of Ametek, Inc., Ref. 12)

Many other types of grates have been designed and utilized in Europe. Three of the most popular include the Dusseldorf drum grate system, the Martin reverse action grate, and the Von Roll grate [13]. The latter two grate systems have been installed in North American incinerators.

The Dusseldorf *drum grate system* is comprised of a series of drums placed at a slope of approximately 30% (Fig. 8-8). The 1.5-m (about

Incinerator Design Concepts

Figure 8-8. Typical drum grate. (Courtesy of Grumman, Inc.)

5-ft)-diameter drums are placed on approximately 1.75 m (about 7 ft) centers. Each drum is built of bars (cast iron) in the form of arched segments which are keyed to a central element below. Each drum, in turn, rests over a separate chamber beneath it for the control of underfire air. The rolls rotate in the direction of discharge at an adjustable peripheral speed which may be varied according to the constituents of the wastes being burned. The drums turn independently of one another with the drying and ignition grates at the front end of the incinerator generally rotating at speeds of up to 15 m/h (50 ft/h). The burnout grates generally rotate at much slower speeds, normally 5 m/h (about 16 ft/h)^2, since they have little waste material to move.

The *Von Roll grate* consists of alternate fixed and moving beams set side by side across the width of the furnace (Fig. 8-9). These moving beams are coupled together by cross members and are actuated by synchronized rams. Like other grates, the Von Roll system has the capacity to vary the speed of the grates over a wide range of speeds and to provide good air distribution.

The Martin grate, a reciprocating grate, is comprised of sections which alternatively push the refuse uphill against the inflow of refuse. This action

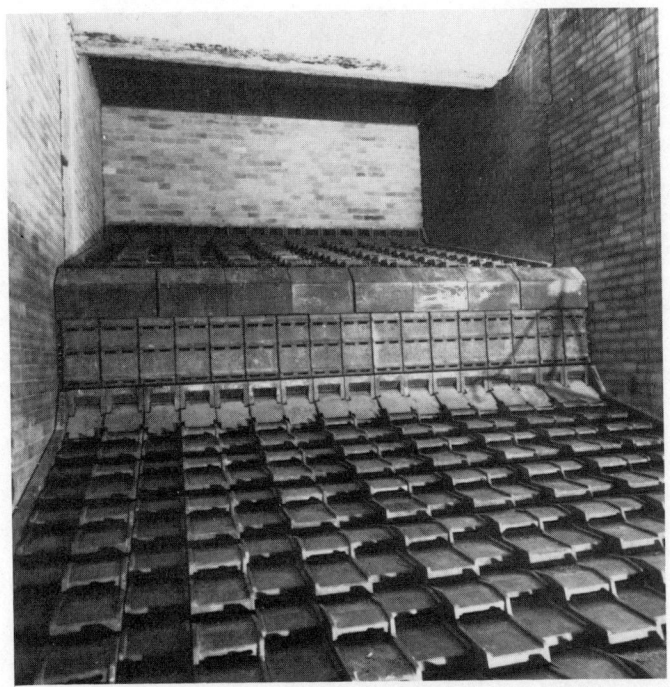

Figure 8-9. Von Roll grate system. (Courtesy of Wheelebrator-Frye, Inc.)

enhances turbulence in the fuel bed and is claimed to accelerate the combustion process.

Grate design

The sizing of a specific grate system depends on the type and weight of material to be burned and the capacity of that grate. The required grate area for an incinerator may be approximated by the following:

$$\text{grate area} = \frac{\text{kg/h solid waste burned}}{\text{kg/m}^2/\text{h grate capacity}} \quad (8\text{-}1)$$

Under ordinary conditions, the design values for this loading factor vary from approximately 240 to 340 kg/m^2/h of MSW (about 50 to 70 lb/ft^2/h). This appears to be the most widely adopted range of grate loadings and has become a commonly accepted design factor [2, 5].

Another common method of expressing grate loading is to describe it in terms of kJ/m^2/h (Btu/ft^2/h). Common loadings range from 2.8 × 10^6 to 3.4 × 10^6 kJ/m^2/h (about 250,000 to 300,000 Btu/ft^2 of grate/h) [5, 14]. Since grate systems may be designed on the basis of the heat release value

Incinerator Design Concepts

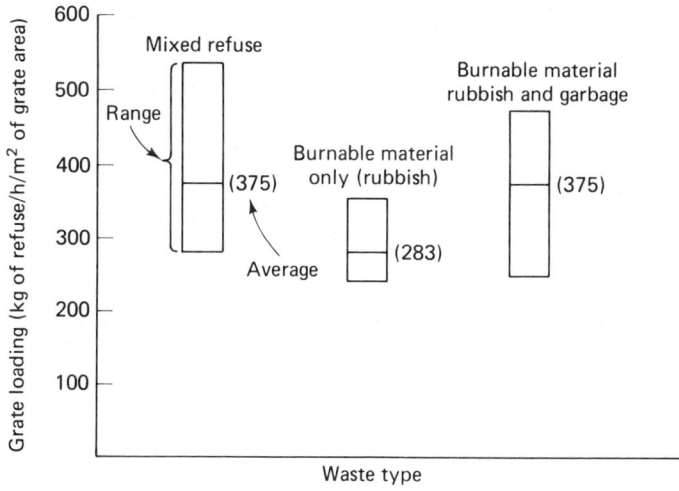

Figure 8-10. Grate loading for incinerators (Range and Average Values). (Adapted from Ref. 17)

of the refuse, Fig. 8-10 can be of assistance in the design process. The data illustrated were adopted from work by the American Public Works Association and denote the design loadings for grates for three grades of refuse.

The grates create agitation in the combustion chamber, which is also an important design consideration. Should the mixing be more rapid, or should more gentle stirring of the waste be provided? No definitive and empirical data have been collected in this area to answer this question satisfactorily, but agitation tends to promote better combustion of the MSW [14].

Another interesting aspect of grate design is the percentage of air openings to be provided in the grate. According to several investigations, and depending on the particular type of grate, air openings varied from 2% to over 30% of the grate area [14]. This is an apparent large variation in something that might be considered a fundamental design factor. Intuitively, one end of the range might seem to be more correct than another. Proponents of the larger air openings feel that the siftings (the ash from the fuel bed) should be permitted to fall below the grate as soon as possible and permit large amounts of air to pass through the bed to meet the combustion requirements of varying fuel characteristics. Those advocating the use of the smaller air openings cite advantages such as the small volume of siftings that must be dealt with, the relatively small amount of underfire air that is required, and the resulting shorter combustion flames, all of which assist in reducing particle entrainment in the escaping gases [14].

The grate system installed in modern incinerators generally must withstand very high temperatures, thermal shock, embedment, abrasion, slagging, and heavy loads. Because of such severe operating conditions, grates must be designed to avoid misalignment of the parts and to minimize the wear of moving parts such as bearings. Heavy castings that will not wrap or crack are used almost exclusively, and new metallurgies must be developed to withstand environmental factors.

Grate design must also be based on each manufacturer's stipulated design criteria, and about the only design-consistent criteria utilized by the various manufacturers is the specified kilograms (pounds) of waste that may be loaded per square meter (square feet) of grate area. Other than that, little similarity exists between the various design procedures adopted for each of the grate systems. More empirical data are necessary for proper design and a more rational approach developed to select the proper grate.

Combustion Air

The combustion process requires oxygen to complete the chemical reactions involved in the burning process. As outlined in Chapter 7, the air necessary to supply the exact amount of oxygen required for burning of the municipal waste is called the stoichiometric air requirement. Additional air supplied to the burning process is termed *excess air* and is generally expressed as a percentage of these stoichiometric requirements.

Excess air may be supplied to a typical incinerator as overfire air, underfire air, or secondary air. Overfire air is generally introduced to the furnace above the burning fuel bed. Its primary purpose is to provide turbulence and to complete the combustion of volatile gases driven off the solids.

Underfire air is admitted through the grates from underneath the combustion chamber. Its primary purpose is to control the combustion process and to keep the grates cool.

Secondary air, which is generally added for temperature control, is most frequently injected at the upper end of the primary combustion stage or at the transition between the primary and secondary stages. The point of injection of the air is largely dependent on the shape of the combustion chambers.

Each furnace configuration has its own air requirements. There seems to be a wide divergence of opinion as to what percentage of the total combustion air should be introduced as under fire air versus overfire air [14]. A range of 25 to 100% of the total combustion air is often provided as underfire air through the grates. If small quantities of combustion air are provided as underfire air, it is unlikely that there would be significant entrainment of particulate matter in the gases [15]. There appears to be little justification for putting 100% of the combustion air through the grates as underfire air, since a typical MSW requires some overfire air for process control.

Velzy has recommended that further testing and correlation of both old and new data on incinerator air supplies might be helpful in answering some of the design questions, because of inadequate data on these air relationships [14]. Such tests might correlate, for each grate type, the relationship among the quantity of underfire air, particulate matter entrained in the gases, and residue quality. Such studies might also determine the optimum range of air quantities and pressures to be provided and the more appropriate, or desirable, locations in which to inject the air.

The provision of excess air is important in incinerator design. The maximum temperature in a combustion process is obtained only when the theoretical air quantity is supplied. Any excess air reduces the furnace temperature in proportion to the amount of excess air. Thus, it is used to control the temperature of the gases exiting the refractory furnace. Too much excess air may actually lower the temperature to the point where the fire will not burn and the fire is "frozen" out.

The amount of excess air provided in an incinerator is a function of the type of incinerator. In the case of refractory furnaces, quantities of excess air of up to 200% of the theoretical requirements are often provided. In waterwall incinerators, the excess air may only amount to 50 to 100% of the theoretical value [14]. These values may be computed as described in Chapter 7.

The quantities of excess air that must be used to cool the gases emitted from the primary and secondary combustion chambers in a refractory furnace are illustrated in Fig. 8-11. The use of this figure will permit the determination of an average furnace exit temperature which can then be used for evaluating other design parameters, such as the size of the air pollution control equipment.

The combustion and gas cooling air is distributed under pressure to different parts of the furnace by fans. The influent air is usually at ambient temperature, normally assumed to be approximately 27°C (80°F). Once in the combustion chamber, the temperature rises very rapidly. It may get as high as 1650°C (2100 to 2500°F) in the immediate proximity of the flame. When the gas leaves the combustion chamber, the temperature should be reduced to between 760 and 1000°C (1400 to 1800°F). Gas temperatures entering a stack should not exceed 540°C (about 1000°F). If air pollution control devices are installed, induced draft fans must be used and the temperature should probably not exceed 260 to 370°C (500 to 700°F).

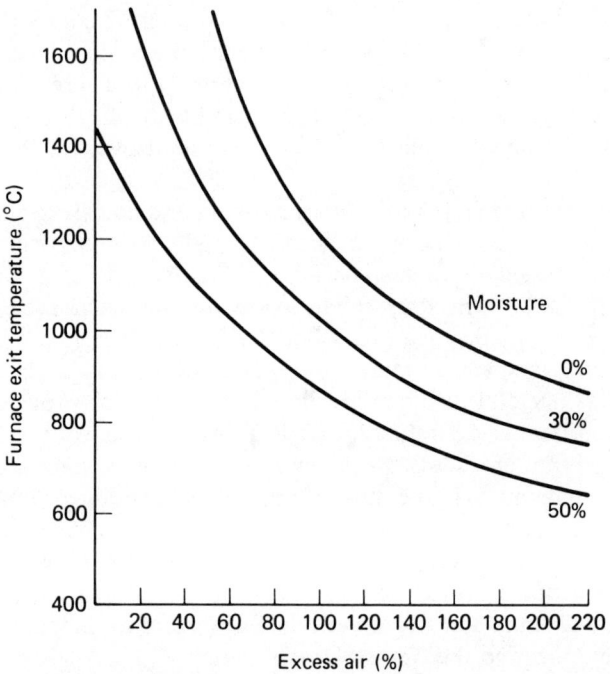

Figure 8-11. Exit gas temperature as a function of moisture content. (Adapted from Ref. 9).

Cooling of the hot combustion gases can be accomplished by either wet or dry methods. Water evaporation is a frequently employed method of gas cooling which is controlled by the quantity of water evaporated.

Wet or dry bottom methods may also be employed for gas cooling. In the wet bottom method, large quantities of water at low pressure are injected into the waste gases through coarse sprays. The water that is not evaporated falls to the bottom of the cooling area and is either recycled or wasted after treatment.

In the dry bottom method no excess water is used because all injected water is fully evaporated. In this system the water is finely atomized under high pressure. The advantage of such a system is that there is no attendant waste water treatment required, but the system must be operated carefully to ensure complete evaporation at all times.

Refractories

Refractories are materials used in various locations in the incinerator to efficiently confine the heat of combustion. Since many industries use refractories, refractory design has advanced significantly from the days when kaolin was first used. Table 8-3 lists typical refractories.

TABLE 8-3
Comparison of Incinerator Refractories

	Bulk Density (lb/ft^3)	Modulus of Rupture $(lb/in.^2)$
85% alumina phosphate bonded plastic (dried at 250°F)	172	1340
90% alumina phosphate bonded plastic (dried at 250°F)	178	1520
Superduty fireclay brick (burned)	148	1000
75% phosphate-bonded alumina brick (unburned)	175	1400
85% phosphate-bonded alumina brick (burned)	184	3500

Source: Ref. 18.

Refractories must have physical stability at high operating temperatures. When hot, they must withstand significant compressive forces as well as mechanical wear from the grates and moving refuse. Thermal stresses may also be placed on the refractories because of repeated exposure to short periods of high temperatures. The refractory may also be chemically attacked because of the liquids and gases formed in the combustion process.

Quite commonly, refractories are classified according to their physical and chemical properties. Characteristics of concern include their chemical resistivity, hardness, their physical strength, their heat conductivity, porosity, and the thermal expansion characteristics they exhibit. The refractory materials are normally cast in brick form in a wide variety of shapes. They are laid either with a refractory, air-setting, or thermal-setting mortar.

There are many conditions in the furnace that can destroy the refractories, including high temperatures, flame impingement, thermal shock, slagging, spalling, and abrasion. Slagging is caused by a buildup of a layer of ash (a deposit of flux) on the refractory surface [5]. It is normally caused by the overloading of the incinerator with waste or by excessive flame temperatures in some part of the furnace [9]. Refractory failure due to slagging is normally caused by the increased weight of the bonded material on the refractory.

Tests indicate that phosphate-bonded refractory brick and plastics appear to do well in resisting incinerator slag attack if wall temperatures are controlled [18, 19]. While 90% alumina refractories are resistant to slag attack, the 75% alumina-phosphate-bonded products appear to be about as resistant. Superduty fire clay brick also is a good choice for incinerator construction except in areas where heavy abrasion conditions are present.

Slagging can also be controlled by reducing the side-wall temperature at the lower part of the furnace to a point where the slag does not adhere to the wall. Wall cooling can be achieved by using air or steam [14].

Incinerator refractory materials are subject to *spalling*; the breaking away of the refractory, usually at the outer surface, because of mechanical and thermal stresses developed internally within the material and primarily because of differential expansion. Physically, it is similar to breaking and flaking of sidewalk concrete.

There are three types of spalling: thermal, mechanical, and structural. *Thermal spalling* results from the unequal thermal expansion and contraction of the refractory material. *Mechanical spalling* may result because provisions have not been made for thermal expansion of the material or because of too rapid drying of green (wet, uncured) firebrick. *Structural spalling* may result from the action of thermal stresses, slagging, or because of chemical reactions with the waste gases.

Abrasion of refractory materials most often occurs when the materials charged into the incinerator physically abuse the refractory material. The particulate matter in the waste gases, with sufficient velocity, may also cause abrasion of the refractory material.

Chemical reactions may result when the mineral constituents of the waste gases actually permeate the refractory material. These gases change the character of the refractory material and may ultimately permit the breakdown of the refractory by structural spalling.

The design of refractory incinerators has progressed significantly over the past several decades. Whereas older furnace walls were designed to conserve as much heat as possible and were quite thick, often 0.6 to 0.9 m (2 to 3 ft) at the base [8], emphasis is now placed on building walls with thinner cross-sections, 20 to 25 cm (9 in.), which will withstand temperatures in excess of 1100°C (about 2000°F) but which will not permit slagging and which will not spall easily.

In suspended wall and roof construction, the furnace and arch tiles are supported by metal clamps or hangers, which are, in turn, connected to a steel superstructure and to beams across the top of the arches. The suspension system has the advantage of permitting easy repair of portions of the furnace when problems occur.

There are two basic approaches which might be taken in the selection of refractories for an incinerator. If one wants to reduce capital costs, a high-heat-duty refractory in all but the combustion zone may be used, but maintenance costs may be high. If it is desirable to reduce maintenance costs, more expensive refractories should be used throughout the incinerator [14]. Consultation with a recognized manufacturer of refractories is encouraged when selecting a refractory for use [18].

Stacks

The heat of combustion of the refuse in an incinerator creates large volumes of gas which contain pollutants. All the equipment that must handle these hot exhaust gases, including the gas passages, air pollution control devices, and the stack, must be sized accordingly. Estimates of the gaseous products of combustion is developed in Chapter 7.

Creation of a negative pressure within the ignition zone of the combustion chamber is necessary to permit movements of gases through the furnace (draft) which result from the natural buoyance of the hot gases. Natural draft in a furnace is determined by the height and diameter of the stack, as well as the differential temperatures of the ambient and furnace air. An artificial draft may be induced by the installation of an induced draft fan. Stack design must therefore be based on whether one decides to use a tall stack (natural draft) or a short stack (with an induced draft fan).

The *natural draft* produced by a stack is directly related to the height of the stack, and can be calculated by using the following formulation [9, 20]:

$$D_t = 3.4 \times 10^{-2} P_b H_s \left(\frac{1}{T_0} - \frac{1}{T_s} \right) \tag{8-2}$$

where D_t = theoretical draft, atm
 P_b = barometric pressure, atm
 H_s = height of stack, m
 T_0 = ambient air temperature, °K
 T_s = average temperature entering stack, °K

For incinerators ranging in size from one to ten tons per hour, the draft would be about (0.8 to 1.0 cm) (0.3 to 0.4 inches) of water (7.4 to 9.8×10^{-4} atmospheres)[8].

The available draft created in a stack is affected by the velocity of the effluent gases and the cross-sectional area of the stack itself, as would be expected in any fluid-flow. Expansion losses are also a consideration but most often ignored. As the cross-sectional area decreases or as the velocity within the stack increases, frictional losses in the stack increase proportionally. The reduction of the available draft due to such losses can be calculated as follows [9, 20].

$$F_s = \frac{2.9 \times 10^{-6} H_s (v)^2}{dT_s} \tag{8-3}$$

where F_s = friction loss, atm
  H_s = stack height, m
  v = velocity, m/s
  d = average stack diameter, m
  T_s = stack temperature, °K

Natural draft is not appropriate for incinerators with air pollution control equipment, since the draft must be regulated more positively. A variable-speed-induced draft fan is a desirable control device, because it can better handle the continuously varying volumes of gases generated in the combustion process. Dampers may also be used for control in both natural draft stacks and in those stacks that use constant-speed induced-draft fans.

Stacks less than 30 to 40 m (100 to 140 ft) high are classed as short. Stacks above that height are considered tall and provide more positive natural draft and permit better diffusion of waste gases. Stacks may be constructed of unlined steel plate, sometimes using double-wall construction to resist corrosion failure due to condensation. Some stacks may be constructed of refractory-lined steel plate or be made completely of refractory and conventional masonary materials.

Residuals

Incineration is not an ultimate method of solid waste disposal! Even the most efficient incineration process leaves some residue, even if the feed is a liquid. The residue from incinerators is generally inert, relatively sterile, and makes good landfill material (although it does produce a leachate which may have significant concentrations of heavy metals). There are several different types of criteria used for characterizing incinerator residue, such as its potential impact on water quality, air quality, its factor as an environmental nuisance, or whether it will create a potential health hazard [21, 23]. Incinerator residue can also be classified according to the remaining combustible material. There are significant differences in the characteristics of residue from various incinerators, depending in part on the basic design [22]. The operation of an incinerator also has a marked effect on the overall character of the residue.

A measure of the efficiency of a furnace may be computed as a function of the combustible matter in the residue according to the following formula [22]:

$$E = 100\left(1 - \frac{V_r}{V_f}\right) \qquad (8\text{-}4)$$

where E = furnace efficiency, defined as the percent destruction of combustibles
V_f = weight of the combustibles in feed, mass/time
V_r = weight of the combustibles in residue, mass/time

Since it is difficult to measure the total mass per unit time of the combustibles in the feed or the residue, a useful equivalent formula, based on concentrations, is

$$E = \frac{1 - [V_r]}{[A_r] + [V_r]}$$

where $[A_r]$ is the concentration of ash in the residue, g ash/g residue, and $[V_r]$ is the concentration of combustibles in the residue.

Generally, the components of a typical residue include nonferrous and ferrous metals, glass, and other materials. The character of the residue depends, to a large part, on what kind of front-end processing is provided prior to the incineration process. Experience indicates that metals should be separated prior to incineration since the heat in the incinerator oxidizes or alters the character of the metals, thus reducing their scrap value. Residues generally have little economic value.

Residue generated in an incinerator must be handled and removed efficiently. With the advent of better combustion control, there should be little unburned combustible matter in the residue. In early incinerators, weight reductions of 2 to 1 were not uncommon. One estimate indicates that the residue may average approximately 0.50 m^3 (0.65 yd^3) or about 235 kg (520 lb) per long ton of refuse fired on a dry basis [9]. This is a weight reduction of approximately 4 to 1 and a volume reduction of nearly 10 to 1.

Systems for handling residues from continuous-feed furnaces are somewhat more complicated in that they generally include a continuous conveyor system. The hot residue from the last grate drops into a water-filled sump that has a conveyor at the bottom, as illustrated in Fig. 8-12. The conveyor moves the residue out of the trough and into a collection vehicle or to a container.

Instrumentation

Refuse incinerators require modern instrumentation techniques to ensure effective, efficient facility operation. The control problem is compounded by the fact that most municipal incinerators burn a fuel with widely varying characteristics. Adequate control systems are necessary to overcome many of the problems attendant with this combustion of MSW.

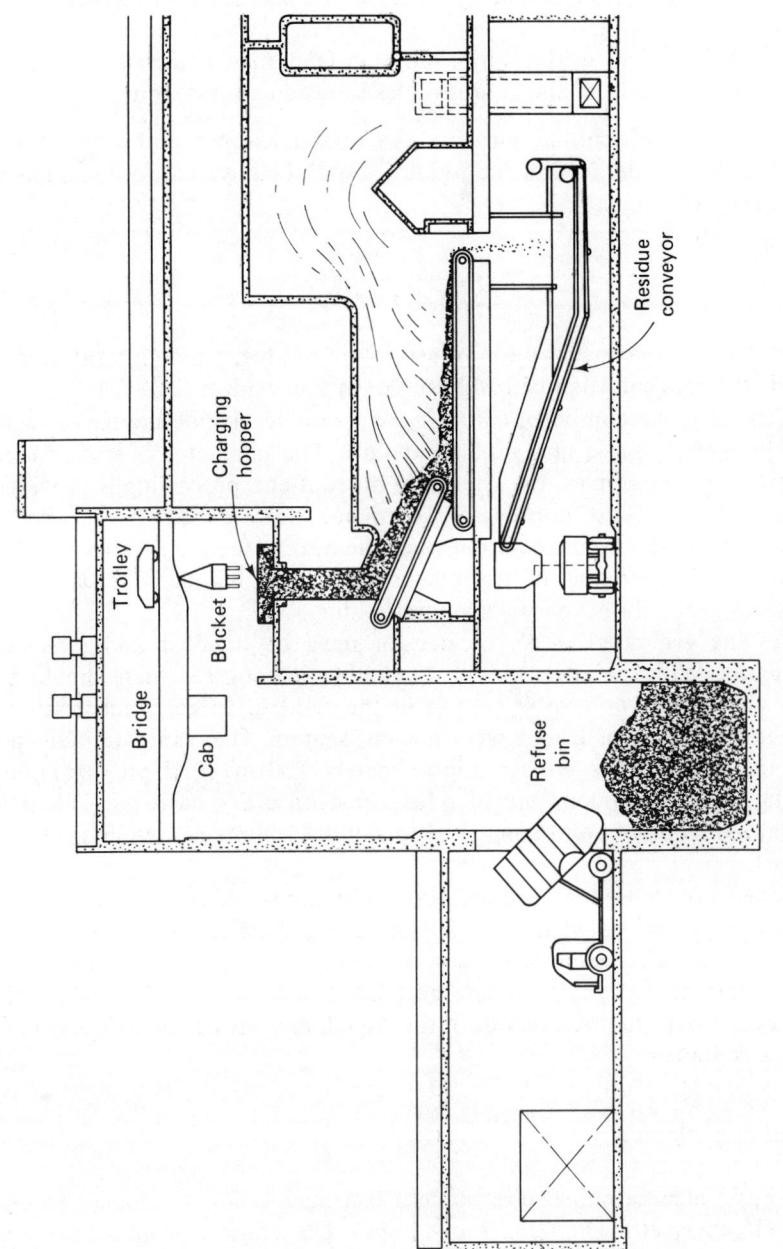

Figure 8-12. Residue disposal conveyor system. (From Ref. 9)

Four basic components are necessary for a control system and include:
1. The performance criteria required.
2. The sensor to determine the performance (an instrument).
3. The operation of the unit.
4. The control device necessary to make the unit comply with the performance criteria.

In control technology, this is a typical description of a feedback loop. Once performance criteria are established, the performance of the unit can be modified to accommodate that standard provided that the proper instrumentation and control is available.

The instrumentation-system configurations that should be employed include underfire air control, furnace temperature control, furnace pressure control, a cooling control system, a dust-collector capacity-control system, pressure indication, temperature indication, flow indication, smoke density, and alarm systems [25].

In a typical control system, there may be feedback loops for each subsystem, but often there are interrelationships between the variables being controlled. When, for example, smoke density increases, it is probable that sufficient overfire air is not being provided to develop complete combustion, or too much overfire air is added, thus prematurely quenching the fire. If an override control system is designed to permit the smoke density system to take over the control of the overfire air, the smoke density might be reduced by the provision of additional air, which is known as a cascade control system. One control variable depends on another and all are interrelated through the control system.

There should be a much greater emphasis placed on control systems in modern incinerators. The basic control technology for such systems is available. With the advent of more sophisticated analog-to-digital conversion equipment, and the decreasing costs associated with computerized systems, it is anticipated that control systems will become more sophisticated and yet more reliable. This should permit the development of better interrelationships among the operating variables to assure more efficient combustion [24, 26].

General Design Considerations

A number of other considerations that should be accounted for in the design of incinerators are important and not necessarily related to the actual combustion process. For example, the incinerator should be located within the service district, to the extent possible, so as to minimize the transportation of the largest quantities of waste.

Example 8-1

Given a refuse with a heat value at 11,500 kJ/kg (about 5000 Btu/lb), find the values for the size of the storage pit, the grate area, the volume of the combustion chamber, and the height of the stack. Assume an average feed rate of 450 tonnes/day (about 500 tons/day).

Assume a bulk density of 8000 kg/m^3. The volume to be stored per day is

$$\frac{450 \text{ tonnes/day} \times 1000 \text{ kg/tonne}}{8000 \text{ kg/m}^3} = 56 \text{ m}^3$$

Assume that 4 days' storage is required. Total volume = $4 \times 56 = 224$ m^3. The pit could therefore be about 10 m by 5 m wide by 6 m deep. The length of the storage pit will be based, in part, on the size of the combustion chamber and grates as determined below.

Using an average value of 300 kg/m^2/h for grate capacity, based on Eq. (8-1) we have

$$\text{grate area} = \frac{18,750 \text{ kg/h}}{300 \text{ kg/m}^2/\text{h}} = 62 \text{ m}^2$$

The heating value is 11,500 kJ/kg, and a net heat production is therefore 215×10^6 kJ/h. Checking the heat loading on the grate, we have

$$\frac{215 \times 10^6 \text{ kJ/h}}{62 \text{ m}^2} = 3.4 \times 10^6 \text{ kJ/m}^2/\text{h}$$

which is acceptable.

If a combustion volume of about 0.5 m^3/tonne of waste per day is needed, then the required volume of the combustion chamber is 0.5 m^3/tonne/day $\times$ 450 tonnes/day = 225 m^3. Using Fig. 8-12 and excess air of, say, 150%, and a moisture content of, say, 28%, the exit temperature of the gas would be about 850°C. This is too warm and will require that gas cooling be provided. Neither the stack nor the air pollution control equipment can withstand these temperatures.

If a natural draft is to be provided, the stack height is found using Eq. (8-2) and assuming a draft of 7×10^{-4} atm, an ambient temperature of 27°C (300°K) and a stack temperature of 300°C (573°K):

$$7 \times 10^{-4} \text{ atm} = 3.4 \times 10^{-2} \cdot 1 \text{ atm} \cdot H_s\left(\frac{1}{300} - \frac{1}{573}\right)$$

$$H_s = 12.6 \text{ m}$$

OTHER INCINERATION PROCESSES

In addition to the common refractory and waterwalled incinerators, there are other processes that use combustion to recover waste heat, or to reduce the volume of solid waste before disposal. A few of the more popular of these processes are outlined in this discussion.

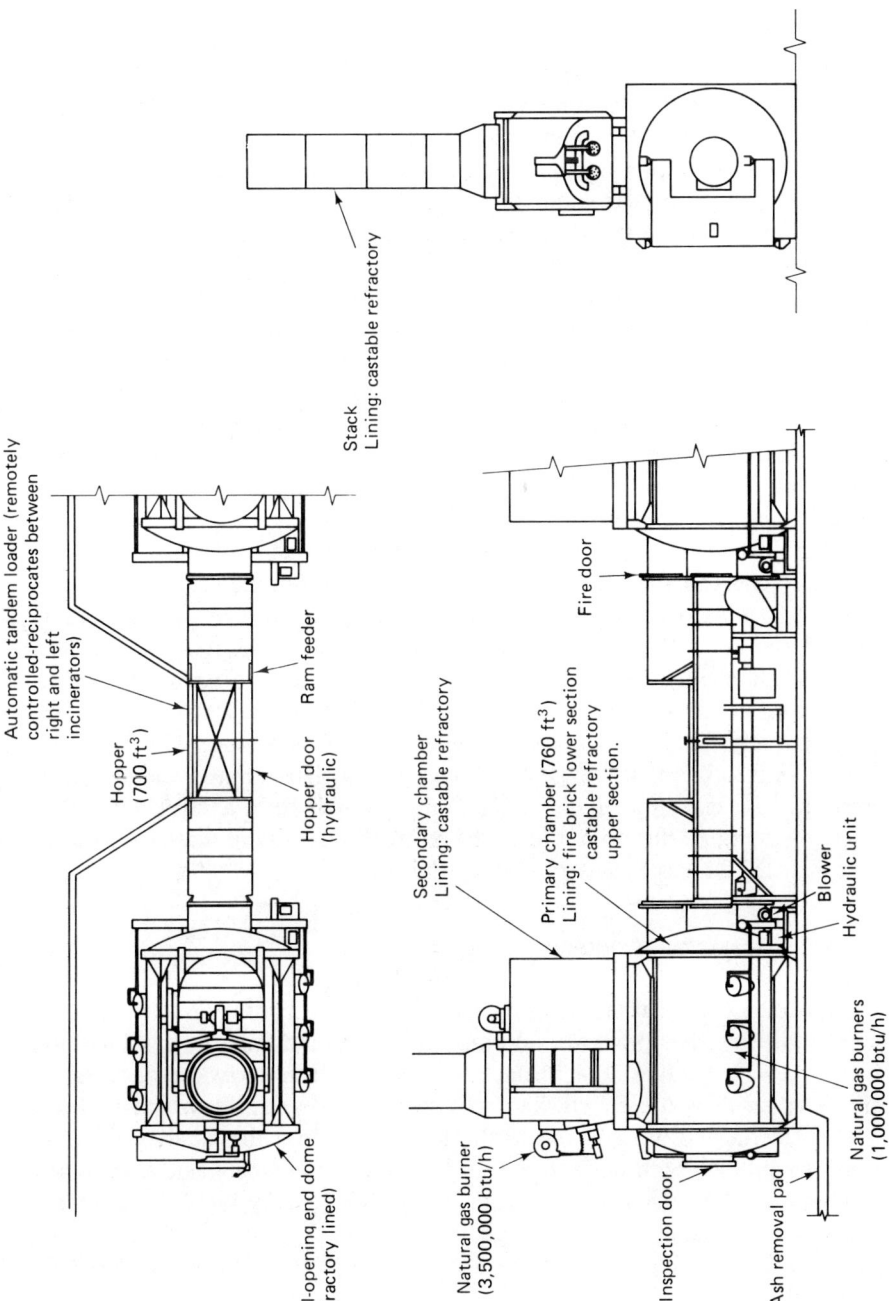

Figure 8-13. Typical modular incinerator. (From Ref. 27)

Modular Incineration

Modular incinerators are defined herein as those individual waste burning units with a capacity of less than 45 tonnes (50 tons) per day. They may be installed in municipal incinerator plants in identical modules of anywhere from two to eight units to achieve the desired plant capacity [27]. A typical modular incinerator, illustrated in Fig. 8-13, is used for burning municipal wastes that have not had any specific pretreatment such as shredding.

The modular incinerator uses a controlled-air principle where the material is burned as received in a primary combustion chamber in a reducing atmosphere (insufficient oxygen for combustion) which tends to minimize the particulate matter in the gas stream. The effluent gas from the primary combustion chamber is then burned in a secondary chamber, in an oxidizing atmosphere (excess air to burn the entrained particles and volatalize the materials in the gas). Auxiliary fuel such as gas or oil is required to assist in the combustion process.

These modular incinerators are fed on the batch-feed basis and are normally charged for 7 to 8 h. They commonly use auxiliary fuel during the first 3 hours of the burn cycle, are then allowed to cool overnight and the ash removed the following day before the start of the next 24-h cycle.

Modular incinerators can achieve weight reductions of about 68% and volume reduction of over 93% [27]. Such small units may also have waste heat recovery systems. A boiler efficiency of 72% was reported [27] but is more likely to be in the range of 60 to 65%. It is likely that these modular units will be used by smaller communities which have a market for the energy generated, but do not have the required quantity of waste necessary to construct the larger waterwall-type units.

Multiple-Hearth Incineration

Multiple-hearth furnaces may be used for the disposal of all forms of combustible waste materials, although they were originally designed to handle sewage sludges. The furnace is basically a refractory-lined, circular steel shell with vertically stacked refractory hearths, as illustrated in Fig. 8-14. Material is normally fed to the top hearth, and the rotating central shaft is equipped with horizontal ploughs which move the waste across the face of the hearth to a hole through which the partially combusted waste drops. As the waste passes through the furnace, dropping from hearth to hearth, it is burned and the residue ash falls to the furnace floor, where it is removed.

The furnace can be divided into three operating zones. The top part serves to dry the feed material to about 45 to 50% moisture at temperatures

Other Incineration Process

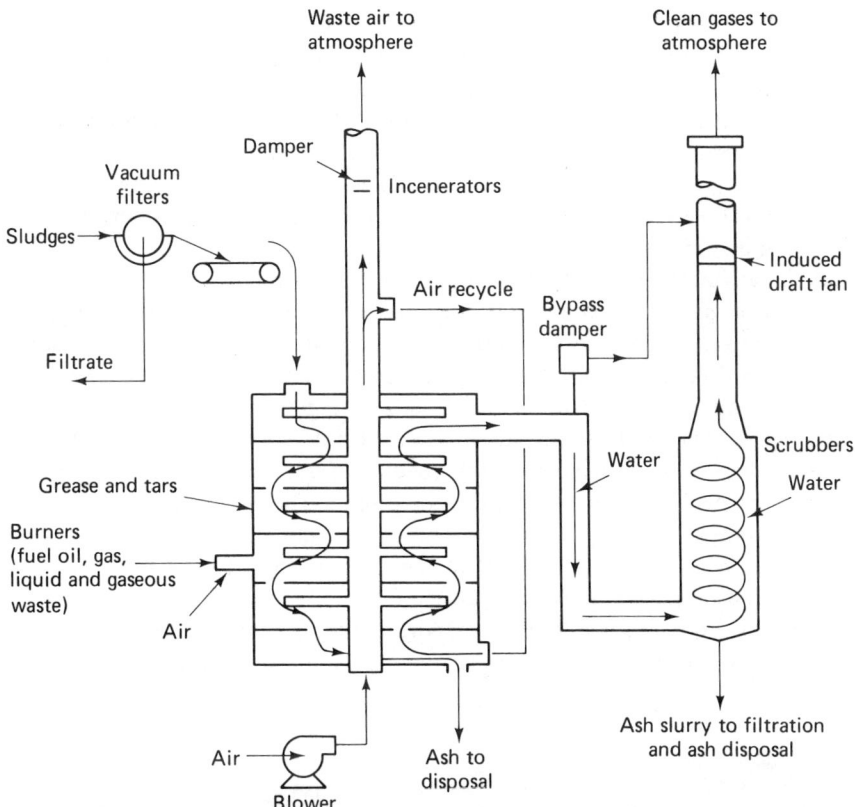

Figure 8-14. Typical multiple hearth furnace. (From Ref. 35)

that range from 310 to 540°C (600 to 1000°F). In the incineration zone (the second zone), which has a temperature range of from 760 to 980°C (1400 to 1800°F), the material is burned. The ash cooling zone, from which the ash is discharged at about 10% of the original feed volume, is generally much cooler. Gases normally exit the unit at about 260 to 540°C (500 to 1000°F).

Multiple-hearth units may find application in the co-combustion of MSW and sewage sludge.

Wet Oxidation

A process for oxidizing liquid or semi-solid organic materials (e.g., sludge) is known as wet oxidation and accomplishes the breakdown of organic materials by a flameless process at high pressures and moderate temperatures. In the process, the solids are solubilized and through the mediation of the hydrolysis reactions, the complex hydrocarbons are broken down. Once relatively simple hydrocarbons have been formed, the process further oxidizes them to alcohols, aldehydes, and ultimately to carbon dioxide. The reactions operate at temperatures of 150 to 350°C (300 to 675°F) and at pressures of 0.316 to 1.76×10^6 N/m² gauge (450 to 2500 psig). Residence time in the reactor normally amounts to between 10 and 30 min. [28].

A flow schematic of the wet oxidation processes is illustrated in Fig. 8-15. Waste materials enter the system under pressure and pass through a heat exchanger, where the temperature is brought up to that in the reactor. The waste flow from the heater to the reactor is supplemented by the air flow from an air compressor to serve as the oxygen source. The waste and compressed air mixture are then injected into the reactor. Once the material has been injected into the reactor vessel, the process is usually self-sustaining and no auxiliary heat is required. The waste passes from the heat exchanger to the separator, where the liquid and gas phases are separated. Waste heat is given off to the heat exchanger, and the material passes out as either a liquid or a gas.

The gas from the separator normally contains carbon dioxide, excess oxygen, and some nitrogen. The remaining materials include the salts of

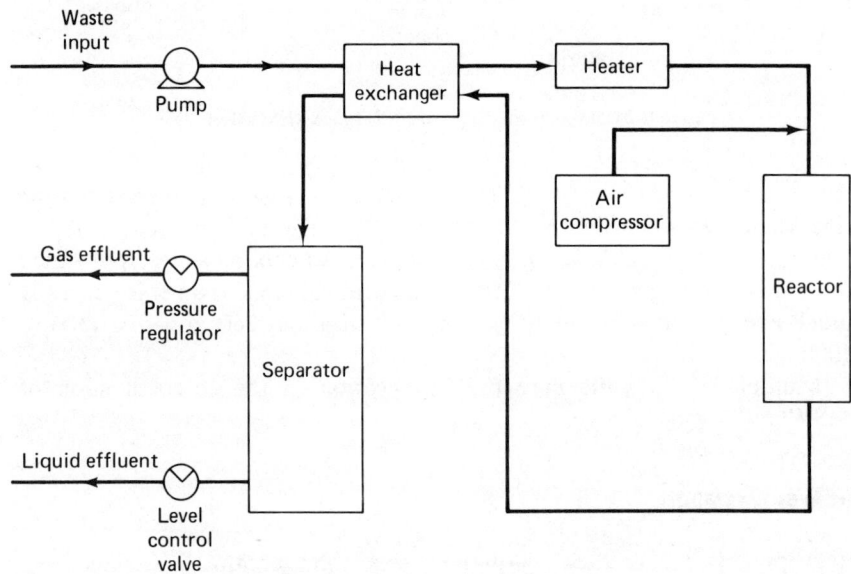

Figure 8-15. Wet oxidation system. (From Ref. 28)

Other Incineration Process

metals, which are unaffected by the reaction process, and other partially destructed organics, which normally exit in the liquid phase.

Since its introduction to the field of incineration in the late 1950s, the wet oxidation process has been used to stabilize sewage sludge, and has been applied to reclaiming potable water on NASA spacecraft [29]. It may be possible to use such a unit for the cocombustion of sludge and finely shredded MSW. Waste heat recovery is also possible with the unit.

Fluidized Bed Incineration

A fluidized bed reactor is essentially a cylindrical vessel with sand as the reactor bed inside the tube. Air, or some other gas, is injected at the bottom of the vessel and the sand becomes agitated, appears to become "fluidized," and behaves basically as a dense liquid medium. The wastes are then injected into the bed and rapidly combusted as a result of the rapid heat transfer within the bed, and the heat of combustion is subsequently absorbed by the bed. Figure 8-16 illustrates a typical schematic of a fluidized bed reactor.

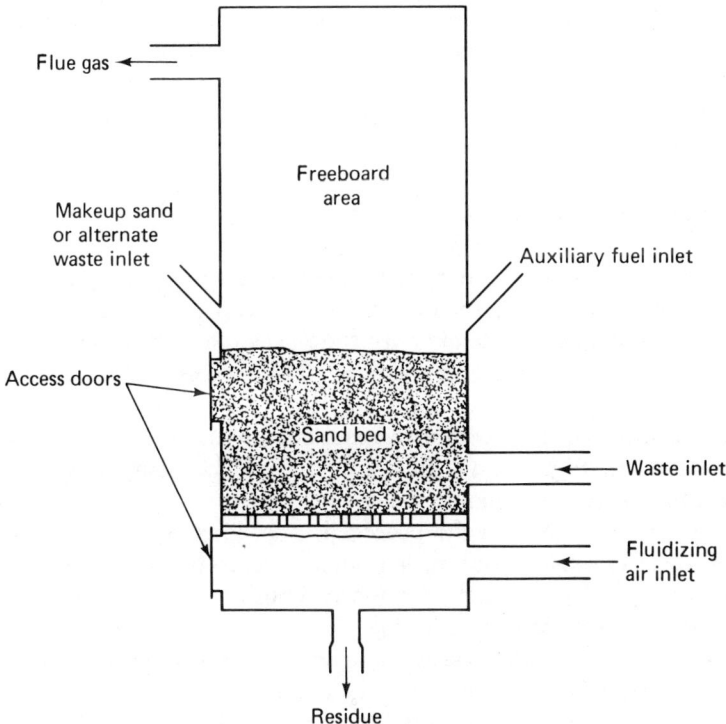

Figure 8-16. Fluidized bed incinerator. (From Ref. 28)

In the combustion process, the heat transfer between the bed materials and the injected waste materials occurs on a continuous basis and typical bed temperatures range from 760 to 870°C (1400 to 1600°F). It has been reported that the heat capacity of the bed material is approximately three orders of magnitude greater than the heat capacity of flue gases in a typical incinerator operating in the same temperature range [30]. Solid materials remain in the bed until they become small enough and light enough to be carried up through the fluidized bed and out through the stack with the flue gas as a particulate, or they may build up in the bed until the excess material must be drawn off to control the bed level.

Maintenance of gas velocities in a fluidized bed are also important and should be around 1.5 to 2.5 m/s (5 to 7.5 ft/s). Velocities higher than this will tend to carry too much unburned particulate matter with the effluent gas.

Reactor diameters range to approximately 15 m (approximately 50 ft) in size. Sand depths may range from approximately 0.5 to 3 m (1.5 to 10 ft).

Fluidized bed reactors have been used for the incineration of many types of liquid or gaseous organic wastes. Although it is reputed that fluidized beds will burn anything that can be fed into them, obviously there must be some limitations [30]. The combustion of municipal solid wastes in such units is limited by the necessity to shred the material to a very fine particle size and to thoroughly predry the material. Thus its application in resource recovery appears to be somewhat limited in the short term.

Slagging Incineration

Slagging incineration commonly refers to a process of combustion where temperatures of nearly 1500°C (about 3000°F) are achieved. These temperatures almost completely burn the waste and melt the entire residue to a slag which is drained and solidified. The process objectives include a maximum reduction of the solid wastes, complete combustion of all oxidizable material, and complete oxidation of the gaseous products of incineration [4]. Figure 8-17 illustrates one of the many types of total incineration processes available today.

Except for the fact that the process is dependent upon relatively high temperatures to achieve fusion, it is otherwise essentially a conventional incineration process. To expedite the combustion, a flux, such as limestone, may be necessary to aid in the fusion process. The process has been used for many years in industry and includes the electric furnace and oxygen-enrichment-type combustion units. Combustion of municipal solid wastes with the slagging incineration concept has not been successfully demonstrated, but waste heat recovery from such a unit should be possible, thus permitting a waste-to-energy facility.

Other Incineration Process

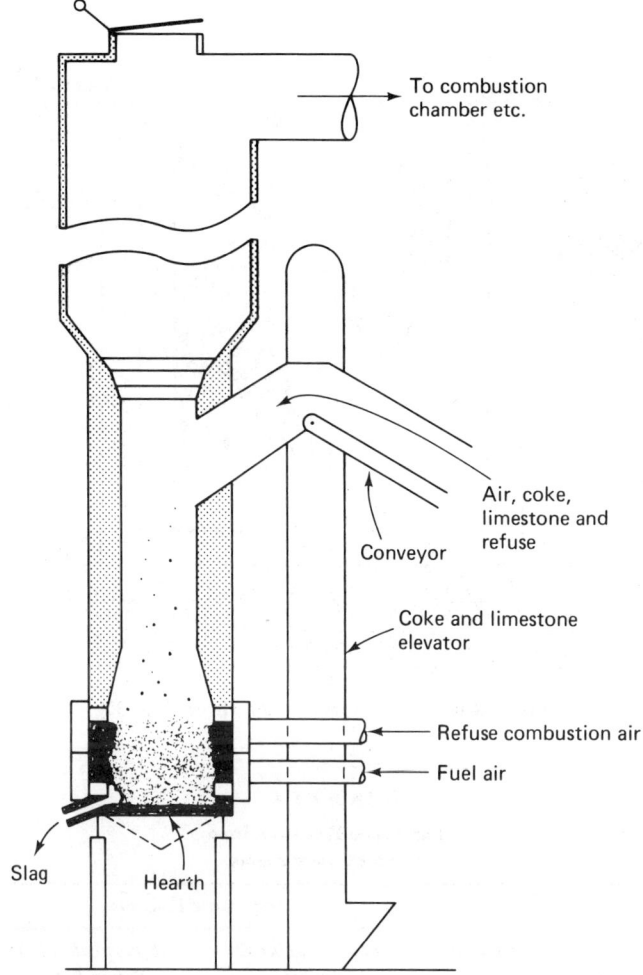

Figure 8-17. Typical slagging incinerator (after American Thermogen System, Ref. 31.)

Open-Pit Incineration

Open-pit incineration eliminates the need for an enclosed furnace structure for the incineration of wastes, as illustrated in Fig. 8-18, but does not appear to have any potential as a means of recovering waste heat. A refractory-lined open pit (or even an unlined pit) is used, and a manifold, with closely spaced nozzles, injects air into the pit, creating a rolling action of the air within the pit. This action promotes high burning rates, and relatively long residence times permit complete combustion of many types

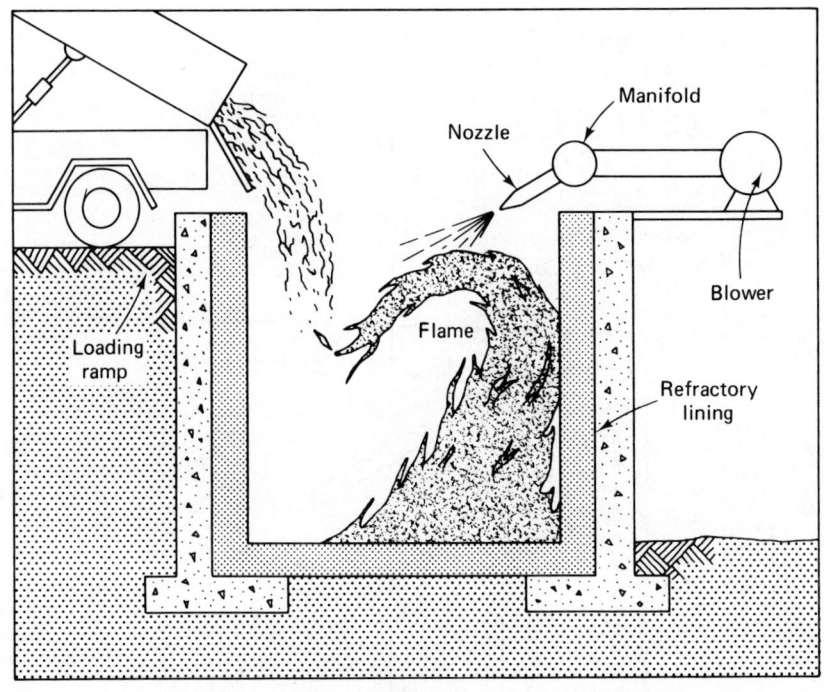

Figure 8-18. Open pit incinerator. (From Ref. 35)

TABLE 8-4

Particulate Emissions from Open-Pit Incinerators

Fuel	Air Rate (scf)	Particulate Emissions	
		g/scf at 12%CO_2 dry basis	lb Particulate/ton Refuse Burned
Cord wood	200	0.14	4.6
	420	0.53	12.7
	600	0.50	12.8
Rubber tires	200	1.62	49.1
	420	1.57	135.2
	500	4.57	172.4
	577	4.10	193.1
Municipal refuse	420	1.15	20.6
	470	1.45	32.9
	500	5.18	35.8
	575	7.38	59.0

Source: Ref. 32.

of waste. The unit normally operates at some 425 to 535°C (about 800 to 1000°F). Charging of the pit occurs on the side opposite the air blower, which has an adjustable output that creates the air blanket. The air pattern developed by the discharge from the blower creates a sheet of flame across the top of the pit under the manifold. The flames, as they roll across the pit, contain most of the particulates and unburned gases, which are then returned to the burning zone, thus eliminating most, if not all, of the smoke [32–34]. Table 8-4 illustrates typical particulate emissions from such an operation.

Process applications have included use for on-site combustion of construction debris and other field-generated material, which are often burned in an unlined pit. Certain chemical and industrial wastes, which cannot be burned in normal furnaces, have also been burned in such incinerators. Application of this unit will be limited by potential air pollution problems.

WASTE HEAT RECOVERY

Based on a Department of Energy a study, the potential for waste energy recovery from wastes by 1985 is estimated at 2.5 quadrillion Btu (or 2.5 quads) [36]. This potential may be achieved by mass burning of unprocessed solid waste or by other means of recovery such as illustrated in Table 8-5.

There are two basic designs that have been used to recover waste heat.

A refractory combustion chamber followed by a waste heat boiler system with the tubes following the secondary combustion area of the furnace.

Waterwall combustion chambers followed by a boiler convection section.

Waste heat boiler systems in which the heat-transfer tubes are located beyond a conventionally constructed refractory combusion chamber have been used. In a waste heat boiler recovering heat, steam production may range from 1 to 2 kg/kg of refuse incinerated and 60 to 70% of the heat may be recovered. Waterwall construction in a furnace for heat recovery has many advantages, but there are a number of limitations that must be understood if the system is to work properly. A waterwall furnace consists basically of steel tubes closely spaced together and welded to form a continuous wall. Water or steam circulates through these tubes. Integrally constructed waste heat recovery boilers reduce the temperature of the exhaust gases, and thus their volume, following the waterwall combustion zone. Figure 8-19 shows a cutaway section through a typical waterwall incinerator.

TABLE 8-5
Energy Recovery from Solid Waste

Method of Energy Recovery	Description
Burning refuse in steam-generating incinerators	In this process, the heat generated during incineration produces steam that can be used for a variety of purposes, including production of heating and cooling and to drive turbine drives to make electricity.
Burning refuse in existing heat exchangers	Refuse can be substituted, or serve as an adjunct fuel for fossil energy in existing power boilers.
Pyrolyzing refuse	Pyrolysis, described in Chapter 8, can produce a transportable fuel and/or gases and can also be utilized to produce steam.
Hydrogenation	In this process, the conversion of refuse into a heavy oil in the presence of carbon monoxide and steam under pressure creates a transportable fuel.
Anaerobic digestion	This is the process of decomposition of organic material in the absence of oxygen for the production of gas such as methane which can be used as a substitute for natural gas.
Refuse derived (or prepared) fuels	Many techniques are available for processing refuse into fuels which can be stored and/or transported more easily than raw refuse.

Source: Ref. 37.

Excess air is introduced into the furnace from beneath the grates in this system as well as from above the fuel bed to provide oxygen for combustion and turbulence. Waterwall furnaces require anywhere from 80 to 100% excess air [2, 16, 37–40].

There are a number of problems associated with the direct generation of energy from such waterwall incinerators. The capital and operating costs of these systems are substantial and quantities of waste in excess of 500 tons/day may be needed to make such systems economical. The steam produced by these waste heat systems can present certain operational problems. For example, electric generators in the United States are driven

Waste Heat Recovery

Figure 8-19. Waterwall incinerator. (Courtesy of Wheelebrator-Frye, Inc.)

by high efficiency steam turbines which require superheated steam at high pressures. The corrosion of the boilers when refuse is used to generate such superheated steam appears to be a problem. The corrosion of the waterwall tubes and superheated tubes has been reported at critical points in nearly all of the high-temperature steam-producing boilers burning refuse [38]. It is suspected that this corrosion is caused by the chloride content in the waste or by high gas velocities [41].

Another problem identified with the direct mass burning of refuse is the variations in the waste composition. The fluctuation of the moisture content from 15 to 50% by weight appears to be troublesome [41].

Prepared Fuels

An alternative to the use of mass-burning waterwall incinerators for steam generation is the use of the prepared fuels in existing heat exchangers such as power generation facilities. The refuse derived fuel can be used as the primary fuel or as a supplementary fuel and fed in combination with a primary fuel such as coal into the main combustion chamber.

Refuse derived fuel (RDF) is a name given to the organic-rich fraction of processed refuse. As discussed in previous chapters in this text, refuse can be processed by means of size reduction and materials separation so as to obtain a product which has a substantial heat value. Physically, shredded and air classified RDF looks like fluffy confetti. It can be burned along with a fossil fuel such as coal because many boilers need only minor modifications to accept the RDF, and that they are already equipped with residue and fly ash handling facilities.

As might be expected, a number of important comparisons must be made between coal and RDF if proper design of joint combustion facilities is to be accomplished. Properties of concern include (1) the heat value; (2) the chemical analysis, including C, N, S, P, and Cl; (3) the ash composition; and (4) the ash fusion temperature. The values characterizing RDF obviously vary considerably, but the analytical data summarized for 223 samples indicate that 435 tonnes (479 tons) of RDF produced had an average energy content of from 5300 to 17,700 J/g (2300 to 7600 Btu/lb). These data are shown in Table 8-7 [42].

TABLE 8-7

Fuel Values of RDF

Property	Average	Minimum	Maximum
Moisture, wt%, as rec'd.	30.1	11.1	66.3
Ash, wt%, dry weight	24.9	14.3	40.5
Heat value (J/g)			
As received	11,500	5,300	17,700
Dry weight	16,500	13,600	30,200

Note: to obtain Btu/lb, multiply J/g by 0.431
Source: Private communication from D. Klumb, Union Electric Co., 1974, as noted by Alter [42].

There are a number of factors that affect the potential heat value which can be derived from RDF. One of the most significant is the effect of moisture on the material. When the RDF is burned, a portion of the energy of combustion must be used to vaporize the moisture. The "available energy" in the RDF may thus be defined as the gross energy consumption (on a dry weight basis) less the latent heat of vaporization of water. Figure 8-20 illustrates this available energy as a function of moisture content.

Another factor that has a substantial effect on the available energy from RDF is the ash content. Ash in a waste obviously increases the operating and capital costs, since one must handle both the bottom and fly ashes, it

Waste Heat Recovery

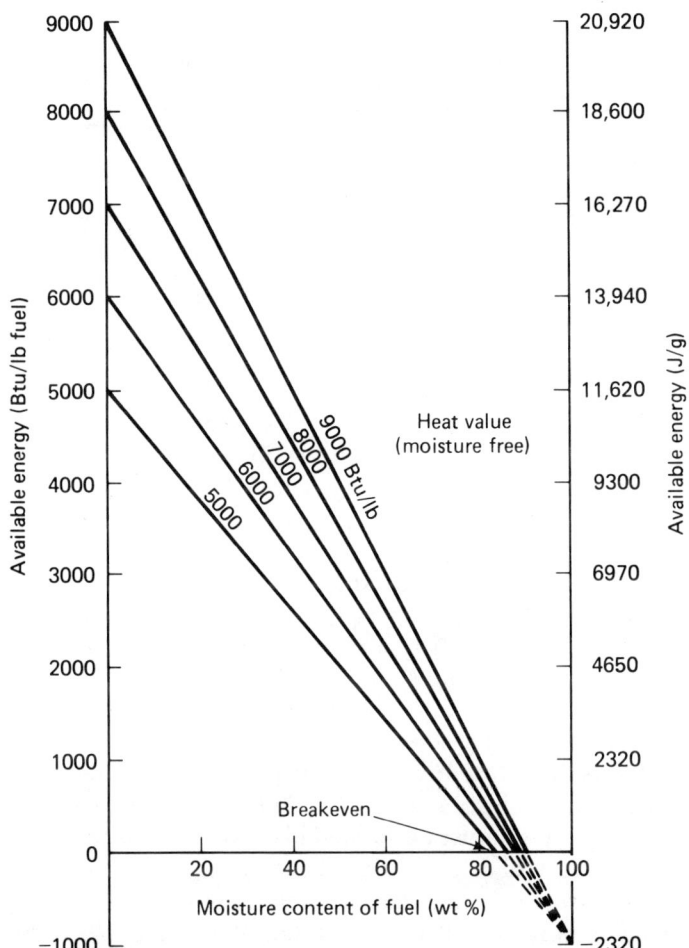

Figure 8-20. Available energy of RDF as a function of moisture content. (From Ref. 42)

is erosive on pneumatic conveying devices, and it tends to dilute the fuel delivered to the boiler.

The first full-scale firing of processed municipal waste in a suspension boiler in the United States was conducted by the Union Electric Company in St. Louis. The objective of the test was to ascertain what potential problems in the boiler might occur as a result of such an operation. A number of potential effects were considered, as noted in Table 8-8. Based on the work done in St. Louis and in Wisconsin, it appears that there is no significant indication of corrosion due to the firing of fluff refuse derived

TABLE 8-8
**Potential Effects of MSW Suspension
Firing on Power Generation Boilers**

Particle size	Effects on bottom ash handling
Non-burnable materials	Corrosion effects
Variations in heating value	Erosion effects
Moisture content	Combustion air requirements
Carryover of dust	Interference with normal boiler operation
Carryover of unburned matter	Air pollution control
Slagging effects	

Source: Ref. 43.

fuel [42]. The most important chemical that might present a problem is chlorine, since the chloride content in the RDF could form hydrochloric acid. Another element that is often associated with corrosion in power plants is sulfur. Because normal municipal waste is essentially sulfur-free, it is unlikely that sulfur will present any type of corrosion potential.

Slagging has not been a significant problem at St. Louis and it should not be a problem if the MSW is properly processed. There are indications that slagging may be a problem in the Wisconsin facility, however, where the material is much more highly refined than at St. Louis [43].

An examination of the heat-transfer surfaces at St. Louis confirmed the suspicion that little or no additional erosion should occur on the heat-transfer surfaces due to the firing of municipal solid waste over that which might be encountered with a solid fuel. This is not true, however, for the tubes conveying the solid waste to the boiler. In these pneumatic conveying lines, abrasion was evident at the bends in the lines, due to the high velocities in these lines, and the abrasive nature of shredded and air classified RDF, which contains substantial amounts of glass dust. This problem can be alleviated by lining the bends, or by removing the glass dust by screening.

The air required to pneumatically convey the RDF is relatively small and does not impact directly on the combustion process. In fact, the quantities of air required for such an operation amount to only 22 to 25% of the total excess air requirement.

There is little evidence at the Union Electric operation that any unburned materials have carried back into the backpasses of the boiler by the gas stream, but a normal amount of fly ash was carried through.

The total ash content of the solid waste delivered to the Union Electric facilities averaged about 17.6% after air classification [42]. Bottom ash seemed to be relatively free from large particles of wood, leather, rubber, and other nonmagnetic materials and did not seem to have any significant impact on the bottom ash handling operation [52].

The only other item of particular concern is the particle size of the material that is fired. Particles in the range 1 to 1.5 cm (0.4 to 0.6 in.) are necessary to ensure reasonably good firing of the municipal solid waste under suspension conditions. Particles much larger in diameter than that simply are not capable of being adequately fired.

These items point to the fact that RDF can be fired in combustion processes such as the suspension-type boiler with a bottom grate. Such oxidation is not without its problems, as noted, but appears to be an efficient means of converting processed or prepared MSW to usable energy form.

Processed shredded fuel

Some of the problems that may occur with a refuse-derived fuel are:
Inefficient material classification, resulting in low recovery efficiencies and a fuel with undesirable specifications.
Poor process control as a result of variable moisture contents of the MSW.
Relatively low bulk densities, which require expensive transportation and storage facilities.
Low bulk densities, which result in poor handling characteristics.
Relatively high ash contents, which increase particulate emissions.
Chemical and biological instability.

Considerable research has been directed at solving some of these problems by stabilizing the RDF either chemically or physically and forming it into a more dense transportable fuel. Two such processes have been developed that may be typical of systems to be developed in the future. These include the development of the use of inorganic chemicals to embrittle the cellulose in the solid waste and make a powdered product [44] or making briquets or pellets from municipal solid waste, called *densified RDF* or *d-RDF* [42,44].

In the production of the first version of the powdered RDF, it was determined that the refuse-derived fuel could be stabilized (the microbiological activity could be stopped), if the moisture content was reduced to less than 15%. After further air classification, the RDF also had a lower ash content. This material, however, had some substantial disadvantages, in that the size of the particles was still relatively large—between 1 and 1.5 cm (0.5 to 0.75 in.) and the bulk density was low—on the order of 66 to 112 kg/m^3 (4 to 7 lb/ft^3) [44].

The major modification to this process was the addition of an inorganic chemical such as sulfuric acid and heat, which embrittled the cellulose fraction of the MSW. This was particularly important, as it enabled the

MSW to be ground easily to a fine powder. In the process, the chemical is added in small quantities to the screened material.

After embrittlement of the cellulosic material, the chemically treated waste is mixed with a hot grinding media (steel balls) in a ball mill. The violent agitation and mixing in this mill results in the almost immediate evaporation of the remaining moisture in the waste. At the somewhat elevated temperatures at which the ball mill operates, approximately 95 to 205°C (about 200 to 400°F), the combustibles are ground to a fine particle size. This occurs with a relatively low energy input, since the chemical embrittlement helps with the size reduction of the material.

The properties of one type of powdered RDF are noted in Table 8-9. The particle size can be varied widely to meet the specific needs of the particular unit in which the material will be fired. It is claimed that this material can be fired in a direct-firing boiler, a slurry in an oil or waste fuel, or in a pyrolysis process [44].

While the economics of the system are not well established, it appears that such heat treatment processes may have a potential application. The

TABLE 8-9

Properties of a Powdered RDF

	Percent by Weight (as fired)
Estimated Chemical Analysis	
Carbon	41.6–47.3
Hydrogen	5.5–6.3
Oxygen	33.9–38.6
Nitrogen	0.6–1.5
Ash	5.0–12.0
Sulfur	0.1–0.6
Chloride	0.1–0.7
Water	1.0–5.0
Generalized Properties	
Combustible	88.6
Ash	9.4
Moisture	2.0
	100.0
Higher heating value	7800 Btu/lb
Particle size	< 0.015 in.
Bulk density	30–35 lb/ft^3
Storage life	Indefinite

Source: Ref. 44.

use of an easily transportable RDF could be useful even to small communities considering recovery of the paper fraction of their wastes.

One of the more common ways of combusting powered RDF is by suspension firing. Suspension firing is not incineration, but is a means of firing fuels, primarily in utility boilers, and has been used for decades in the burning of pulverized coal oil or gas in power generation facilities. In such a process, the fuel is burned while in suspension in the furnaces, which requires that the fuel be fairly finely shredded.

One system for burning such highly pulverized fuels consists of tangentially firing the material into the boiler. In this system, the material is fed into the combustion chamber at four points, one at each corner of the unit. The refuse and the preheated combustion air are directed tangentially toward the middle of the combustion chamber. The fuel and air meet in the center of the unit, where oxidation takes place. A tangentially fired burner is illustrated schematically in Fig. 8-21. Such units have been in operation for many years and the technology dealing with waste burning dates back to use of bark and bagasse in such boilers in the 1940s.

One other potential method of preparing refuse derived fuels is the briquetting or pelletizing of the waste. Such systems basically take a shredded municipal waste and cause the particles to stick together under pressure in the form of a typical charcoal briquet or as a pellet from a machine which was originally developed for making animal feeds. While the process has much to recommend it, the increased energy required to briquet or pelletize the waste can only be justified if the waste must be

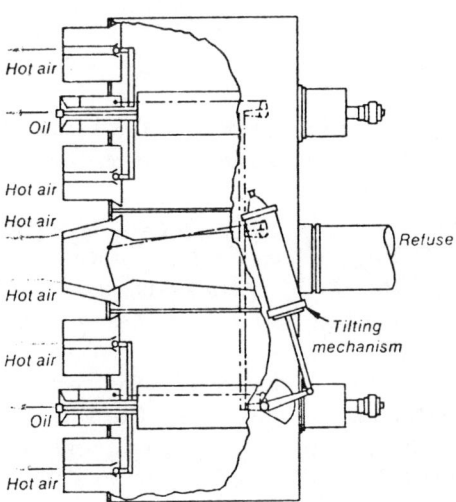

Figure 8-21. Burner for tangentially fired combusion. (From Ref. 45)

hauled or transported some distance from the location where it is to be used, or if storage of the pellets is necessary. Preliminary experiments indicate that pelletized RDF may not have as serious a problem with spontaneous combustion as does shredded refuse without further processing. Pelletized RDF can also be fed with coal in many boilers without the need for any boiler modification. The superior storage and handling characteristics of pelletized RDF hold much promise for this type of fuel.

AIR POLLUTION CONTROL EQUIPMENT

The control of the emissions from incinerator and pyrolysis units is of increasing concern to designers as air quality standards become more stringent. The major constituents of the gases from the combustion processes having air pollution significance are particulates, odors, sulfur, and nitrogen oxides. There are two practical means of controlling such emissions: (1) better design and/or operation of the combustion and (2) the use of add-on equipment which can serve as collection or removal devices to decrease the emission of a specific substance.

The primary air pollutant of concern in the combustion of solid fuels is particulate matter. The emission of fully oxidized gases such as SO_2 and NO_x is usually not a problem, while unburned combustion products, such as hydrocarbons, which can cause odor problems, can be dealt with by better control of the oxidation process.

Particulate Emissions

Particulate matter, the major air pollutant, is basically any material that exists as a solid at standard conditions. We can associate that definition with such materials as smoke, dust, fumes, or the like. A key to understanding the characterization of particulate matter is that the classification is, at best, empirical. For a given refuse, the amount of particulate matter in the flue gases is a function of the material being burned. The combustion process is directly related to the quantities of air used in the burning process.

There are at least 12 methods for reporting particulate emissions [46, 47]. One paralegal standard is the kg/tonne (lb/ton) of fuel charged to the incinerator, although the predominant conventional unit is g/m^3 at 12 percent CO_2. Table 8-10 lists the several methods of expressing emission levels, in descending order of popularity.

TABLE 8-10

Methods of Reporting Particulate Emissions

	English equivalent
1. g/m³ at 12% CO_2	0.437 grain/scf at 12% CO_2
2. kg/100 kg dry refuse charged	lb/100 lb
3. kg/h	2.2 lb/h
4. g/m³	0.437 grain/scf
5. g/kg gas at 50% excess air	lb/1000 lb
6. kg/h per kg/h charge	2.2 lb/h per 2.2 lb/h
7. g/kg gas	lb/1000 lb
8. g/m³ at 75% O_2	0.437 grain/scf
9. g/m³ at 50% excess air	0.437 grain/scf
10. g/10⁶ cal	0.556 lb/10⁶ Btu
11. g/kg at 12% CO_2	lb/1000 lb
12. kg/metric tons of fuel charged (this is not legal but is widely used)	lb/1000 lb

One reason there are so many ways of reporting particulate emissions is that a simple concentration parameter (kg/m³) is inadequate. An operator can achieve compliance by simply diluting the flue gas with air in terms of some standard excess air or a given amount of carbon dioxide.

The particle-size distribution of incinerator fly ash has been reported as in Fig. 8-22. Significantly, almost 35% of the particulate matter leaving a furnace is below 10/μm in diameter. Because it is so small, it is relatively difficult to collect and the types of equipment capable of removing particles in this size range have high capital and operating costs.

Other Emissions

Other components in incinerator emissions having potential air pollution significance include nitrogen and sulfur oxides. Nitrous oxide (NO) is produced in high-temperature combustion processes where air is used as the supply of oxygen. Nitrous oxide is oxidized to nitrogen dioxide in the atmosphere, which plays an important part in the formation of photochemical smog. Typically, incinerator emissions amount to about 1 to 4 lb of equivalent nitrogen dioxide per ton of refuse [9].

Since refuse and RDF both contain little sulfur, the emission of sulfur dioxide and sulfur trioxide is insignificant. This is one reason why RDF is an attractive substitute for coal.

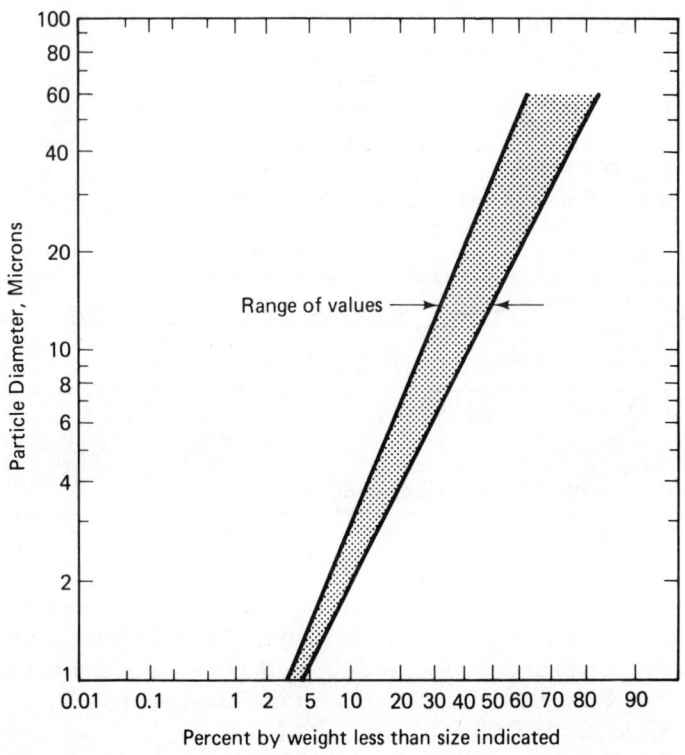

Figure 8-22. Particle size distribution of incinerator fly ash. (From Ref. 46)

The last major class of emissions from incinerators, RDF facilities, and pyrolytic reactors (see following chapter) is incomplete combustion products, which include smoke and organic compounds resulting from thermal cracking and condensation reactions. Carbon monoxide is also produced as a result of incomplete combustion and is formed when insufficient oxygen is available to carry the combustion reaction to CO_2. These emissions can be largely controlled by proper furnace design and operation.

Emission Standards

Emission standards for particulates and gases such as nitrogen and sulfur oxides are based on provisions of the Clean Air Act Amendments of 1970 and as subsequently amended. Under provisions of this act, the

Air Pollution Control Equipment

Administrator of the Environmental Protection Agency has the authority to require monitoring of the emissions from stationary sources such as incinerators or pyrolytic reactors. Compliance with the Clean Air Act requires the use of standards to determine whether or not a stationary source has met the qualifications stipulated, which must be obtained by actual measurements of the emissions rather than by calculated figures. Sampling of the emissions must be completed according to standard conditions stipulated by the Environmental Protection Agency.

Sampling techniques required for compliance with air pollution control standards are relatively complex and are based on a concept of isokinetic sampling techniques. This method of stack gas sampling is done in such a way that the flow conditions entering the sampling or collecting device are essentially the same as those in the stack. A reasonably representative sample should therefore result. A detailed discussion of sampling techniques is not warranted here, and is covered thoroughly in several references [9, 46, 48].

The pollutant that requires the most control to attain emission standards is particulate matter. Most incinerators operate well below the limits for CO, SO_x, NO_x, hydrocarbons, and oxidants; therefore, control measures are unnecessary for these pollutants.

Historically, standards for the quantities of particulates legally discharged to the atmosphere have dropped from as high as 1.8 g/kcal feed (1 lb/1 × 10^6 Btu) to a new low of approximately 0.35 g/kcal (0.2 lb/1 × 10^6 Btu). Figure 8-23 illustrates the range of changes that have occurred in standards under which facilities have been operating.

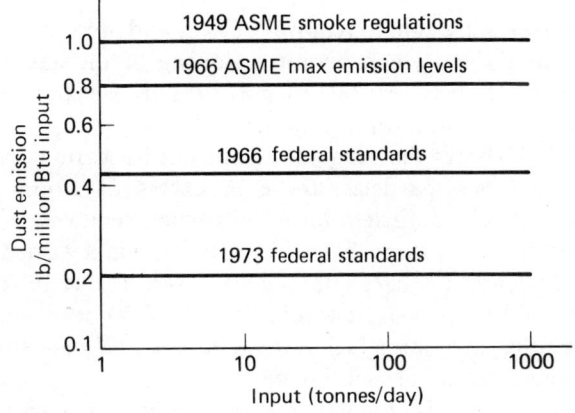

Figure 8-23. Emission control standards. (Adopted from Ref. 46.)

Control of Particulates

Particulate matter emitted from incinerators and pyrolytic reactors are suspended in a gaseous stream and are subject to most of the laws that describe the conditions of a gas as well as the actions of particulate matter in compressible fluid flow. The particles in the gas stream are subject to a variety of forces, including gravity and attractive forces of the particles to each other. Some of the earlier attempts to control particulate emissions took advantage of certain of those forces, such as a settling chamber, where by gravity the particulate matter was allowed to settle out. But as standards become more stringent, more sophisticated control devices using electrostatic and centrifugal forces, must be used to attain more efficient particulate removal.

The free fall of a particle, whether in a gaseous or a liquid phase, is most often analyzed by the Stokes law (see page 165).

The problem with using Stokes's equation explicitly for small particles, that is, those less that $10\mu m$ in diameter, is that the interparticle forces come into play and particles tend to have a lower terminal velocity than predicted. This fact is important, since particles of this size make up at least one third of the total often encountered in incinerator and pyrolysis reactor emissions.

Another important consideration in the evaluation of equipment for the removal of particulate matter is the agglomeration effect of smaller particles into larger bodies as a result of interparticle collisions and adhesion. Generally, particle agglomeration is to be encouraged, since the larger the particle, the more easily it can be removed from the gaseous phase.

Settling chambers

Settling chambers are simple removal devices and are nothing more than enlargements in a duct which permit a slowing of the gas velocity to a point where particles may separate by gravity. This is exactly analogous to a sedimentation basin in a water treatment plant.

Figure 8-24 illustrates the range of equipment for various particle sizes. Except for the larger particles, those in excess of $10\mu m$ the settling chamber is a relatively inefficient means of particle removal.

If we assume for a moment that a $10\text{-}\mu m$ particle in a gas is flowing with a horizontal velocity of 3 m/s (10 ft/s) in a 3-m (about 10-ft)-high duct, the particle would be deposited within 45 m (150 ft) after entry. This, of course, implies that very large facilities are necessary if such small particles are to be removed by simple settling chambers.

The efficiency of gravity-type separators may be increased by the use of baffles on which the particles may impinge, or by the use of wet scrubbers,

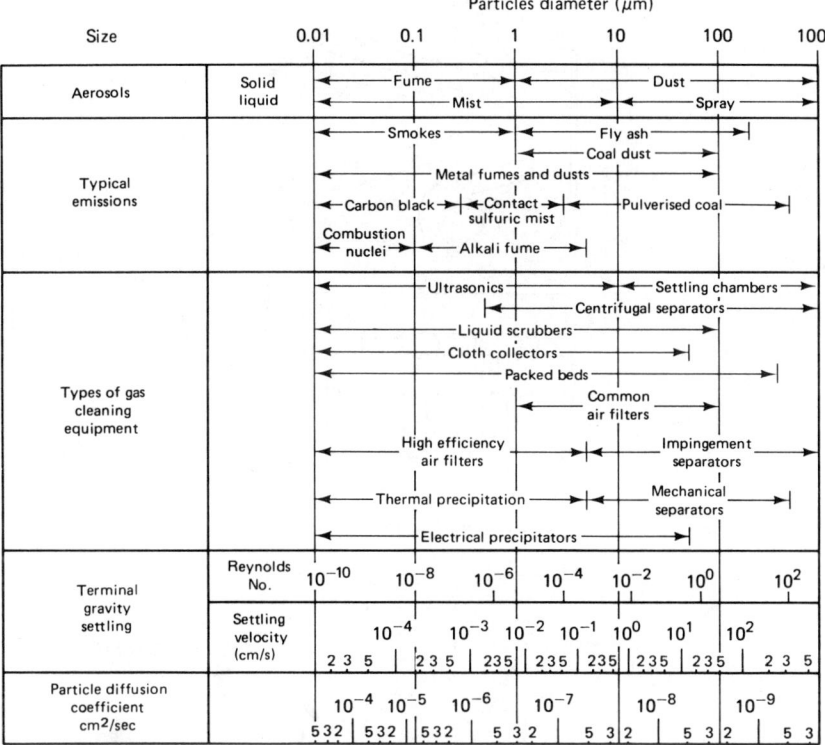

Figure 8-24. Equipment utilized to remove varying sizes of particulates. (From Ref. 9)

which are described subsequently. The efficiencies of settling chambers are reputed to be about 10%, although some have reported removals of up to 60% [2, 4, 9, 46].

Cyclones

A mechanical *cyclone collector* has a relatively low capital cost and is relatively inexpensive to operate. A cyclone consists of a vertical cylinder with an entry port at the top tangential to the section. A reducing conical section is located at the bottom of the cyclone and is used to collect the materials separated from the gas stream. The particulate matter flows into the unit in a vortex path, and the larger particles are removed by centrifugal force. The gas, cleaned of the particulate matter, leaves the top of the cyclone. The hopper into which the larger particles fall is emptied as required. A bank of such cyclones is shown in Fig. 8-25. The operation and theory of cyclones is covered further in Chapter 4.

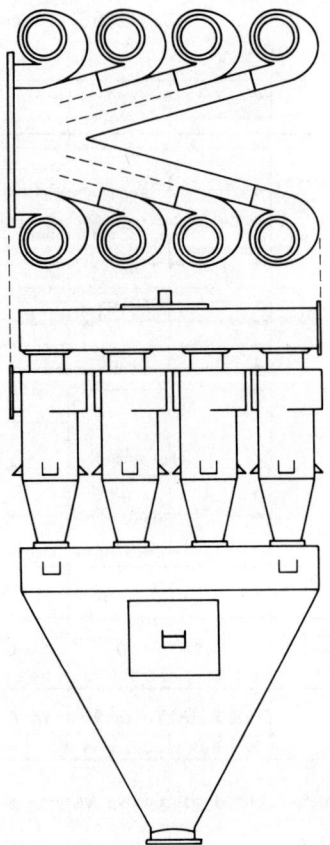

Figure 8-25. Large involute type cyclone (Courtesy of Research-Cottrell, Inc.)

Wet gas scrubbers

Wet scrubbers are used for removing particulate matter and are often confused with spray chambers and wet baffled collectors, which are normally used in conjunction with simple settling chambers. There are many configurations of scrubbers, including cyclonic scrubbers, Fig. 8-26, packed bed scrubbers, Fig. 8-27; a Venturi scrubber, Fig. 8-28; and a spray tower, Fig. 8-29. These scrubbers can be classified into either low-energy units, such as the spray tower, or high-energy units, where the water spray is much finer and more evenly distributed, and where more resistance to the air flow and much smaller water droplets are formed such as the Venturi scrubber.

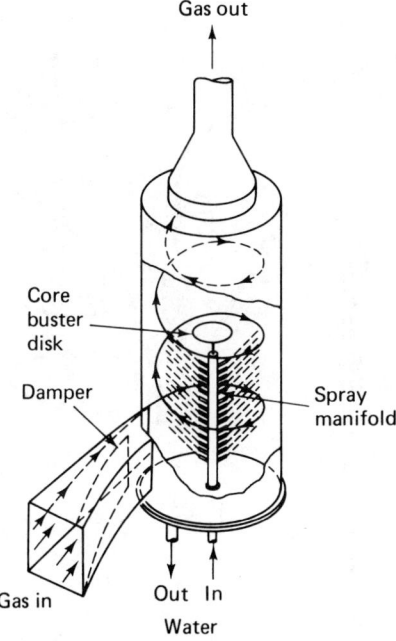

(a) Cyclonic spray wet scrubber

Figure 8-26. Cyclonic spray wet scrubber. (From Ref. 48)

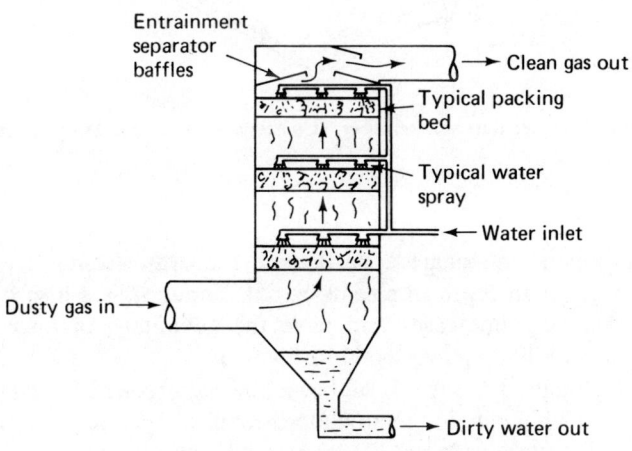

(b) Packed bed wet scrubber

Figure 8-27. Packed bed wet scrubber. (From Ref. 48)

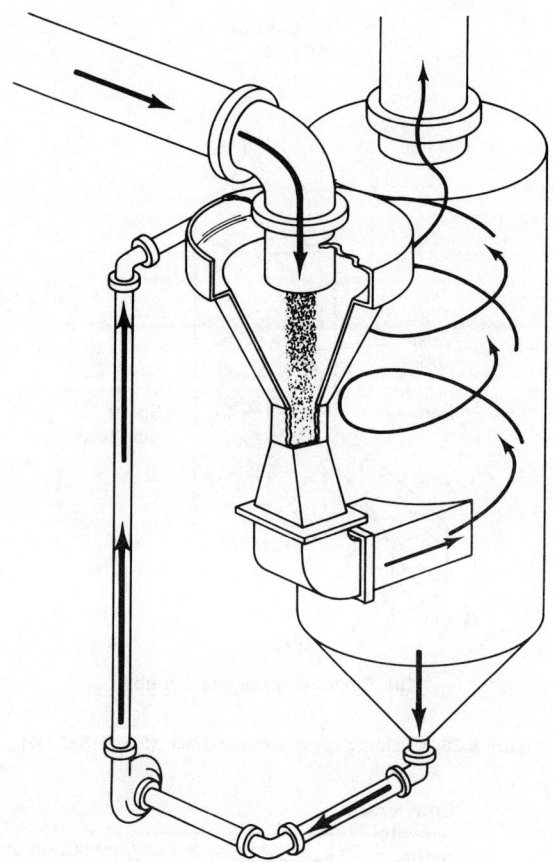

Figure 8-28. Venturi wet scrubber. (Courtesy of American Lung Association)

Wet scrubbers can achieve particulate removals from 90 to 97% of particles larger than 5 μm in size [2, 8, 65]. The units are also capable of handling relatively hot gases and have the advantage of also removing certain water-soluble gaseous pollutants.

A disadvantage of the wet scrubber is the requirement for large quantities of water, which if not recirculated, require treatment to meet strict water pollution standards or, if recirculated, require extensive treatment for the removal of the solubilized gases and suspended particulate matter. The units require substantial power to attain high collection efficiencies, are subject to corrosion and have not achieved desired collection efficiencies [51].

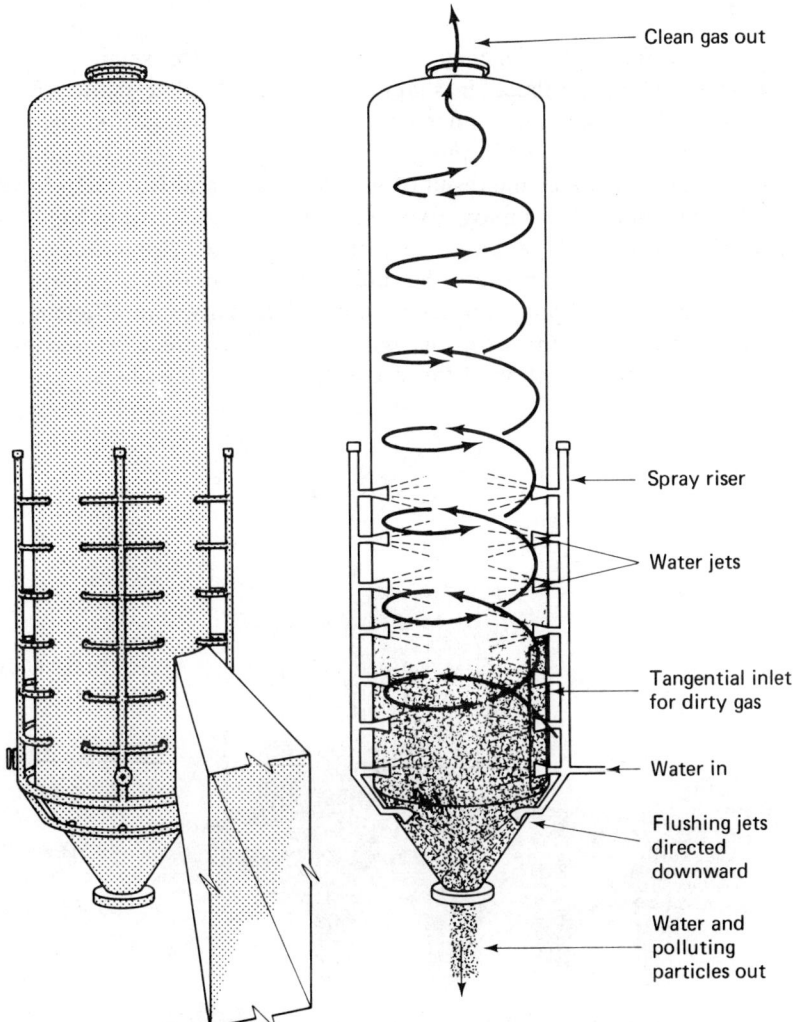

Figure 8-29. Spray tower wet scrubber. (Courtesy of American Lung Association)

Wet scrubbers almost always produce a plume which is mostly steam. Although the water vapor is not a pollutant, the visible plume can cause public relations problems. There are two possible means to eliminate a vapor plume from scrubber applications; by heating the stack gases to increase their temperature, which will permit a greater dispersion of the gases before condensation takes place (and therefore will reduce the plume) and cooling the gas stream to reduce its original water vapor concentration.

Filters

Filters can be high-efficiency collection systems for small particles. Filter bags, generally tubular in shape, trap the particulate matter. Much as in a household vacuum cleaner small particles are caught on the filter fabric by interception and impingement, and the particles continue to build and agglomerate as the result of van der Waals and Brownian forces.

Such filters normally employ a woven fabric, as illustrated in Fig. 8-30. The choice of material is based on the specifications for efficiency of removal of particulate matter, the pressure drop across the unit, the allowable gas throughput, and temperature. Normally, the efficiency of removal and the pressure drop across the bag filter are closely related, because it is a direct function of the size of the pores in the filters.

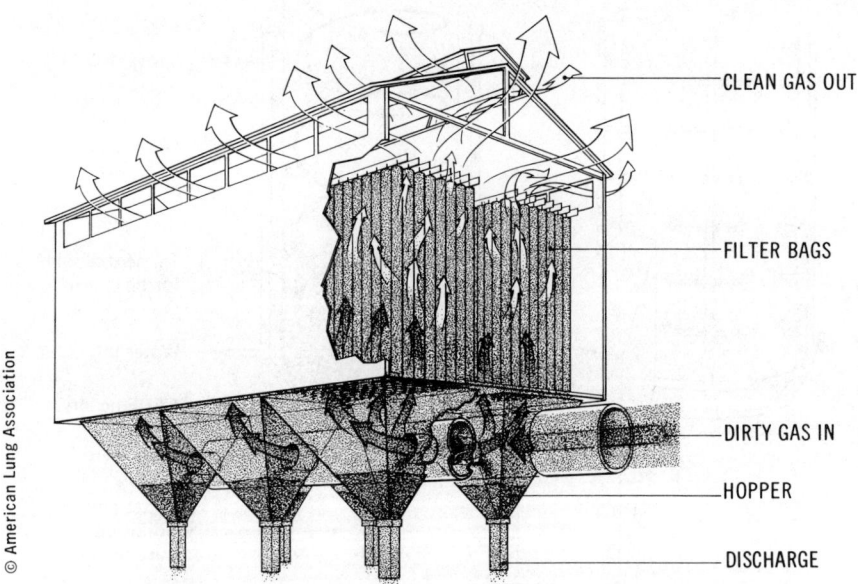

Figure 8-30. Bag house filters (Courtesy of American Lung Association)

Design velocity for fibrous filters ranges from 6 to 23 m/min (20 to 75 ft/min) for those that remove relatively small particles. Velocities of 75 to 150 m/min (about 250 to 500 ft/min) are not uncommon for the core filters removing relatively large dust particles in very low concentrations [49].

A filter house, sometimes known as a bag house, has generally not been used in cleaning incinerator gases for a number of reasons, including the high initial cost, the problems associated with the replacement of the filters, and the gas cooling that is necessary before the gases can be passed through the bag house. It has recently been demonstrated that some of the new fabrics used in bag house applications can withstand temperatures in excess of 400°C (750°F), which is about the normal exit temperature of an incinerator.

Electrostatic precipitator

Figure 8-30 illustrates a basic *electrostatic precipitator*, which consists of a discharge electrode that provides the electrons for charging the particulate matter, and a series of collecting electrodes that provide a number of surfaces which collect and hold the charged particulate matter.

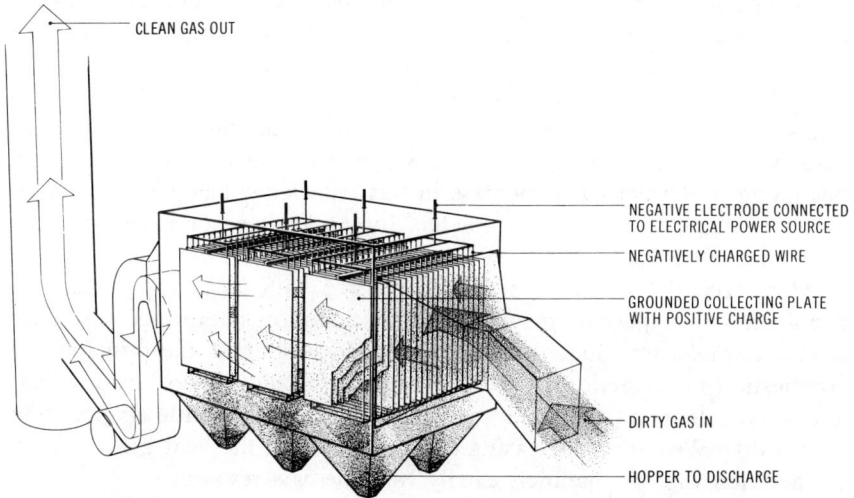

Figure 8-31. Electrostatic Precipitator (Courtesy of American Lung Association)

The strength of the electric field normally ranges from 40,000 to 80,000 V and obviously affects the efficiency of removal. Other factors affecting the efficiency include the gas temperature, the moisture content of the gas, and the resistivity of the dust itself. Proper collection efficiency requires that the particulate matter should have a resistivity of 1×10^5 to 2×10^{10} ohms/cm. Dust with a resistivity of less than 1×10^5 ohms/cm cannot be precipitated [46].

Electrostatic precipitators are generally cleaned by either shaking the particles loose from the unit or washing the particles off the units at intervals.

There are obvious drawbacks to the use of precipitators, which include their high capital cost, the need to maintain optimum operating temperatures [which have been reported to be between 243 and 260°C (470 and 500°F) [4], and the fact that most incinerators have relatively nonuniform emissions. This is particularly important, because the particles have a range of electrical resistivities which will obviously affect the overall performance of the unit.

PARTING SHOTS

The process of incineration is not an exact science. In fact, it is really a combination of science and "art." The art involves the evolutionary development of new designs based on an analyses of the data collected from existing facilities. We have seen incinerators progress from the simple batch fed "destructors" to the sophisticated high technology incinerators of today. While we have a good understanding of combustion, the science associated with the burning of solid wastes in an incinerator is just now beginning to develop. The increase in this knowledge has permitted us to better evaluate the design parameters for an incinerator and apply the principles to good design.

More recently we have become concerned with air-pollution and the problems an incinerator may create. Air pollution control devices have added a whole new dimension to our understanding of the principles of combustion in incinerators. The challange is to continue to blend the art and science together to assure an efficient burn with as little air pollution as possible. With ever decreasing energy resources, the heat generated by incineration will become increasingly valuable as a resource.

REFERENCES

[1] HERING, R. and S. A. GREELEY, *Collection and Disposal of Municipal Refuse*, McGraw-Hill, New York (1921).

[2] BAUM, B., and C. H. PARKER, *Solid Waste Disposal*, Ann Arbor Science, Ann Arbor, Michigan (1973).

[3] STEPHENSON, J. W., "Incineration Today and Tomorrow," *Waste Age*, 2:5 (1970).

[4] National Center for Resource Recovery, *Incineration*, D. C. Heath and Company, Lexington, Massachusetts, 1974.

[5] DEMARCO, J., D. J., KELLER, J. LECKMAN, and J. L. NEWTON, Municipal-Scale Incinerator Design and Operation, U.S. Public Health Service, Washington (1969).

[6] National Center for Resource Recovery, "Incineration Fact Sheet," Washington, D.C., undated.

[7] Incinerator Institute of America, *Incinerator Standards*, November 1968.

[8] SKITT, J., *Disposal of Refuse and Other Waste*, Charles Knight & Co. Ltd., London, 1972.

[9] COREY, R. C., *Principles and Practices of Incineration*, Wiley-Interscience, New York (1969).

[10] STENBURG, R. L., R. P. HANGEBRAUCK, D. J. VON LEHMDEN, and A. H. ROSE, JR., "Field Evaluation of Combustion Air Effects on Atmospheric Emissions From Municipal Incinerators," *Journal of the Air Pollution Control Association*, February, 12 (2): 1962.

[11] EBERHARDT, H., "European Practices in Refuse and Sewage Sludge Disposal by Incineration," ASME National Incinerator Conference, New York, 1966.

[12] Ametek Corporation, Durham, North Carolina, Personal Communication.

[13] ROGUS, C. A., "An Appraisal of Refuse Incineration in Western Europe," ASME National Incinerator Conference, New York, 1966.

[14] VELZY, C. O., "The Enigma of Incinerator Design," ASME Winter Annual Meeting, New York, December, 1968.

[15] ROSE, A. M., Air Pollution Effects of Incinerator Firing Practices and Combustion Air Distribution, *Journal of the Air Pollution Control Association*, No. 2, February, 1959.

[16] MANTEL, C. L., *Solid Wastes: Origin, Collection, Processing and Disposal*, John Wiley and Sons, New York, 1975.

[17] American Public Works Association, *Municipal Refuse Disposal*, Public Administration Service, Chicago, 1970.

[18] CRISS, G. H., and R. A. OLSEN, "The Chemistry of Incinerator Slags and Their Compatibility with Fireclay and High Alumina Refractories," ASME National Incinerator Conference, New York, 1968.

[19] CRISS, G. H., and R. A. OLSEN, "Further Investigation of Refractory Compatibilities with Selected Incinerator Slags," ASME National Incinerator Conference, New York, 1968.

[20] KENT, R. T., *Mechanical Engineer's Handbook*, Wiley, New York, 1968.

[21] SCHOENBERGER, R. J., and P. W. PURDOM, "Classification of Incinerator Residue," ASME National Incinerator Conference, New York, 1968.

[22] SCHOENBERGER, R. J., and P. W. PURDOM, "Residue Characterization According to Furnace Design," 1968 ASCE Environmental Engineering Conference, Chattanooga, Tennessee, 1968.

[23] SCHOENBERGER, R. J., N. M. TRIEFF, and P. W. PURDOM, "Special Techniques for Analyzing Solid Waste on Incinerated Residue," ASME 1968 National Incinerator Conference, New York, 1968.

[24] STICKLEY, J. D., "Instrumentation Maintenance-A Major Problem," 1966 ASME Winter Annual Meeting, New York, 1966.

[25] STICKLEY, S. D., "Instrumentation Systems for Municipal Refuse Incinerators," ASME 1968 National Incineration Conference, New York, 1968.

[26] STEPHENSON, J. W., "Incineration Design with the Operator in Mind," ASME 1968 National Incinerator Conference, New York, 1968.

[27] "Evaluation of Small Modular Incinerators in Municipal Plants," U.S. Environmental Protection Agency, Report SW-113c, Washington, 1976.

[28] SCURLOCK, A. C., A. W. LINDSEY, T. FIELDS, and D. R. HUBER, Incineration in Hazardous Waste Management, U.S. Environmental Protection Agency, Washington, 1975.

[29] OTTINGER, R. S., and J. L. BLUMENTHAL, "Recommended Methods of Reduction, Neutralization, Recovery or Disposal of Hazardous Wastes," Publication No. PB 224-579, National Technical Information Service, Springfield, Virginia.

[30] BALLIE, R. C., "Solid Waste Incineration in Fluidized Beds," *Industrial Water Engineering*, November, 1970.

[31] ZINN, R. E., C. R. LAMANTIA, and W. R. NIESSEN, "Total Incineration," ASME 1970 National Incinerator Conference, New York, 1970.

[32] BURCKLE, J. O., J. A. DORSEY and B. T. RILEY, "The Effects of the Operating Variables and Refuse Types on the Emissions from a Pilot Scale Trench Incinerator," ASME 1968 National Incinerator Conference, New York, 1968.

[33] BELCHER, R., "Curtain Destructor," *Washington Highway News*, June, 1971.

[34] "Dulles Corporation Uses Smokeless Burning Process," *Construction*, October 4, 1971.

[35] WITT, P. A., "Disposal of Solid Wastes," *Chem. Eng.*, Oct. 1971.

[36] National Solid Wastes Management Association, "Energy Recovery From Solid Wastes-Consideration for Determining National Potential," Technical Bulletin, Volume 6, No. 10, November, 1975.

[37] National Center for Resource Recovery, "Municipal Solid Waste-A Source of Energy," *NCRR Bulletin*, Volume III, No. 3, Summer, 1973.

[38] ENGDAHL, R. B., "Water Tube Wall Incineration," Proceedings of the 1975 International Symposium on Energy Recovery from Refuse, Kentucky Center for Energy Research, Louisville, September, 1975.

[39] Mitre Corporation, "Technical and Economic Information Relating to the Resco Refuse to Energy Facility and the Refuse Disposal Requirements of Nine Communities," Mitre Corporation, May 1974.

[40] ROFE, R., "Energy Conservation Waste Utilization Research and Development Plan," Mitre Technical Report MTR-3063, Bedford, Massachusetts, July, 1975.

[41] CORDIANO, J. J., "Refuse as a Supplement to Coal Firing," Industrial Fuel Conference, Purdue University, October, 1974.

[42] ALTER, H., and H. P. SHENG, "Energy Recovery From Municipal Solid Waste and Methods of Comparing Refuse-Derived Fuels," *Resource Recovery and Conservation*, Vol. 1, 1975.

[43] MULLEN, J. F., "Report on Refuse Burning for Western Massachusetts Electric Company," Combustion Engineering, Inc., October, 1971.

[44] ARTHUR D. LITTLE, INC., "Eco-Fuel II—The Technology and Economics," A report to Combustion Equipment Associates, Inc., Boston, January, 1976.

[45] Preliminary Draft, "Appraisal of the Use of Prepared Waste as a Supplementary Power Plant Fuel in Certain Units of Louisville Gas and Electric Company in Conjunction with the City of Louisville, Kentucky," Horner & Shifren, Inc., November, 1974.

[46] FERNANDES, J. H., "Incinerator Air Pollution Control," *Proc. ASME Nat. Incin. Conf.*, Cincinnati, 1970.

[47] FINEHART, R. D., "Complete Conversion Among the Regulatory Incineration Particulate Emission Definitions," *Proc. ASME Nat. Incin. Conf.*, Cincinnati, 1970.

[48] NIESSEN, W. R. and A. F. SAROFIM, "Incinerator Air Pollution: Facts and Speculation," *Proc. ASME Nat. Incin. Conf.*, Cincinnati, 1970.

[49] SELL, W., "Dust Separation on Air Filters," *Forsch. Geb. Ingenieures Publ. 347*, Aug. 1931.

[50] "Utilities Industry Burns Refuse to Generate Power," *Power*, February, 1970.

[51] VELZY, C. O. "Materials of Construction for Wet Scrubbers for Incinerator Application" and "Resolving Corrosion Problems in Air Pollution Control Equipment," National Association of Corrosion Engineers, Houston, 1970.

PROBLEMS

8-1. A community wants to build a combination incinerator/steam generation facility. How should they go about siting the facility?

8-2. Would it be more cost effective to supply more than one days operating storage at the incinerator (Problem 8-1) or at outlying transfer stations which serve the facility?

8-3. a) What amount of grate area is required to burn 2000 kg/hour of solid waste. b) If the waste had a heat value of 3000 kcal/kg would the required amount of grate area be the same as (a)?

8-4. Bases on Figure 8-11 if gas exit temperatures were at 1000°C by what percentage could the excess air be cut back if the moisture content were dropped from 50% to 20%?

8-5. What stack height should be selected to provide a natural draft if the draft is 7.4×10^{-4} atm, the ambient temperature is 30°C and stack temperatures are approaching 600°C?

8-6. What potential air pollution impacts might be created by a rotary kiln incinerator and why might these be more difficult to handle than those from a typical incinerator?

8-7. How would you design a facility to store and feed RDF to a boiler pneumatically? What equipment would you provide in the facility?

8-8. A wet scrubber is to be placed on the municipal incinerator. What provisions would you make for control of the resulting water pollution?

9 PYROLYSIS

INTRODUCTION

Pyrolysis is an irreversible chemical change brought about by the action of heat in an atmosphere that lacks oxygen. While we are able to describe a long and generally constant development of incinerator technology from the late 1800s, pyrolysis has a much shorter history in the field of solid waste management. Although the pyrolysis process was employed successfully for many years in the production of coke from soft coal, it was not until the 1960s that research on the pyrolysis of solid waste matter was pursued seriously. Since then, the development of pyrolytic systems for solid waste has been spurred by the need to find more environmentally sound means of waste reduction than incineration. A second important benefit of the pyrolysis process is that the conversion of the organic portion of the solid waste into a usable energy form may be economical.

Many types of pyrolysis operations are associated with the processing of municipal solid wastes [1]. The form of energy developed from the selected process will vary according to the type of reactor employed, and how the reactor is operated. The products of pyrolysis are either a solid char, a liquid containing high molecular weight hydrocarbons and water and gas which has various amounts of methane, hydrogen, carbon monoxide, and other constituents. A variation of the pyrolysis process, more accurately called gasification, is where air, oxygen or steam is injected into the reactor. The availability of oxygen in the reactor allows for partial combustion to occur, and the process is thus autogenous (that is, it doesn't require an outside heat source). Gasification has also been called starved air incineration, because of the less than stochiometric quantities of oxygen available.

In the gasification reactor, the process takes place in two steps, the first one being true pyrolysis as previously described. The reaction of the hot char with the injected air or steam is the gasification step. As one would suspect, the end products of gasification differ from true pyrolysis in that much of the char is converted to gaseous products.

PROCESS HISTORY

As early as 1929, research into the pyrolysis of solid waste materials other than municipal refuse was conducted by the U.S. Bureau of Mines [2]. In 1967, Kaiser and Friedman conducted experiments on the pyrolysis of homogeneous organic wastes, which was followed by further studies on the pyrolysis of heterogeneous wastes such as MSW [3-5]. Figure 9-1 illustrates a typical pyrolysis reactor used in these experiments, which proved that the gases produced in the process could be used as raw fuel feedstock in boilers.

Hoffman and Fitz used a retort system to pyrolyze a typical range of municipal solid wastes as received at the laboratory [5]. These studies confirmed that the products of pyrolysis include gases, pyroligneous acids, tars, and various forms of solid residue (char). The study also demonstrated that once the pyrolysis reaction was started, it could be autogeneous (self-sustaining) because the products of the process can indeed be used as the sources of energy to heat to the system. In 1970, Sanner

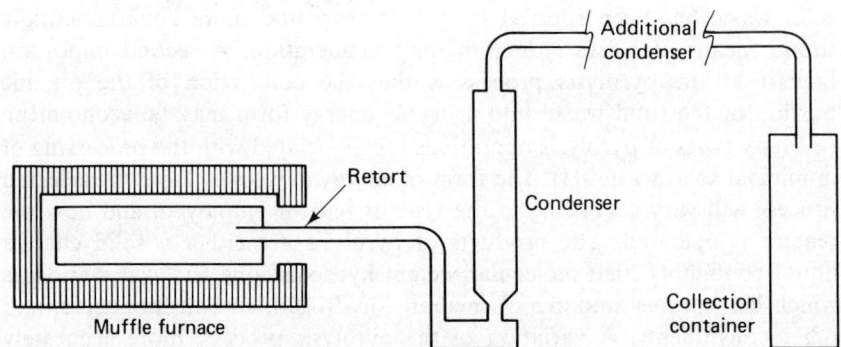

Figure 9-1. Typical experimental pyrolysis flowsheet. (Adapted from Ref. 5)

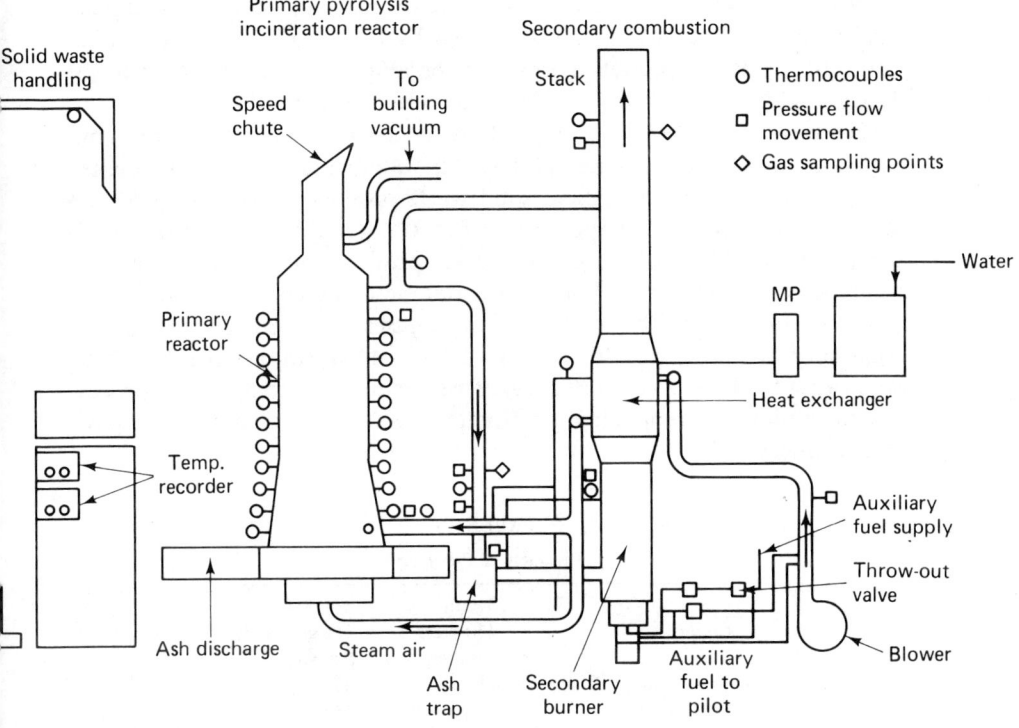

Figure 9-2. Typical gasification flowsheet. (From Ref. 7)

confirmed that the energy from MSW was clearly sufficient to provide the heat for the pyrolysis process without the need for auxiliary fuel [6].

In 1973, Battelle Northwest examined the gasification of solid wastes, where the residue from the pyrolysis of the solid waste was reduced with an air-steam mixture to produce a combustible gas [7]. Figure 9-2 illustrates a typical schematic of such a gasification facility. This unit is similar to the process that has been used by the coal industry for years to convert coal to coal gas, and it was found that the conversion of energy generally exceeded 80% and that the fuel gas produced in the process could be burned to produce steam or generate electricity as required.

PROCESS CHARACTERIZATION

The pyrolysis process, as noted earlier, may be described as a chemical change of a material which is brought about by the action of heat in an atmosphere in which there is insufficient oxygen for combustion. The process may also be defined as destructive distillation, thermal decomposition, or carbonization [8,9]. As suggested by the term destructive distillation, the volatile products (from the organic materials which can be decomposed) are distilled away from the nonvolatile residue.

Typically, during the pyrolysis of MSW, organic compounds are converted to pyroligneous acids, fuel gases, water, and char. In a detailed analysis of pyrolyzed municipal solid waste products, it was noted that a ton of MSW pyrolyzed at approximately 870°C (1600°F) yields the materials noted in Table 9-1. The gases that are produced in the process consist primarily of hydrogen, carbon monoxide, methane, and ethylene, in that order of abundance. The mixture of these gases is a good fuel with a heat content of about 6390 to 10,230 J/g (2750 to 4400 Btu/lb) of solid waste. Since about 2560 J/g (1100 Btu/lb) of heat is required to make the process self-sustaining, the remaining gas fraction becomes a valuable by-product of the process [4].

TABLE 9-1
Products of Pyrolysis[a]

Gas	510 m^3	18,000 ft^3
Liquid	430 liters	144 gal
Tar	1.9 liters	0.5 gal
Ammonium sulfate	11 kg	25 lb
Solid residue	70 kg	154 lb

[a]*Process carried out to 870°C.*
Source: Refs. 4 and 10

The reactor can also produce liquid products, oil and tars, pyroligneous acids, and water. The oils and tars could be valuable fuel stock products, while the pyroligneous acids are chemically complex mixtures.

The solid residue (char) is a lightweight carbonaceous material, and has a fuel value of from 12,800 to 21,700 J/g (5500 to 9350 Btu/lb) and a very low sulfur content [4]. This char material can be used as a fuel after briquetting.

The pyrolysis process may be characterized by both chemical equations and a physical analogy. A useful physical analogy of the pyrolysis process was suggested by Richard Bailie in which the destruction of a walnut in a simple impact tester is compared with the destruction of a particle in a thermal reactor [11].

When the weight is dropped on the walnut in the impact tester, it may be hit in one of three ways: hit hard several times, hit firmly once, or hit lightly often. A different end product results in each case (Table 9-2). The same types of reactions occur in the thermal reactor. The temperature in the reactor is what controls how hard the organic molecule is "hit," and

TABLE 9-2
Process Variables and Products—Physical Analogy of Nut Cracking to Chemical Reaction in Pyrolysis

	Physical	Chemical
Variables	a1. Weight of nut cracker a2. Distance of drop	a. Temperature of reactor (how hard it is hit)
	b. Drops per unit time (rate)	b. Rate of heating
	c. Size of walnut	c. Size of molecule
Products		
Hit hard several times	a. Powder	a. Gas
Hit firmly once	b. Chunks	b. Liquid
Hit lightly often	c. Stable core	c. Char

Source: Adapted from Ref. 11

the length of time over which the reactor is heated determines the ultimate end product.

The basic equation of pyrolysis is

$$\text{organic materials} \xrightarrow{\text{heat}} \text{gases} + \text{liquids} + \text{char}$$

A more complex and accurate equation is [12].

$$\text{carbonaceous solids} \xrightarrow{\text{heat}} \begin{cases} \text{high- and moderate-molecular-weight organic} \\ \text{liquids (tars and oils and some aromatics)} \\ + \text{char} \\ + \text{low-molecular-weight organic liquids} \\ + \text{many organic acids and aromatics} \\ + CH_4 + H_2 + H_2O + CO + CO_2 \\ + NH_3 + H_2S + HCN \end{cases}$$

At normal operating temperatures true pyrolysis (not gasification) is an endothermic reaction (one requiring heat input). Heat must be applied to the solid waste to distill the volatile compounds. During pyrolysis heat causes the carbon (C) to react with water (H_2O) as follows:

$$C + H_2O + \text{heat} \rightarrow H_2 + CO \quad (9\text{-}1)$$

$$C + 2H_2O + \text{heat} \rightarrow CO_2 + 2H_2 \quad (9\text{-}2)$$

$$C + CO_2 + \text{heat} \rightarrow 2CO \quad (9\text{-}3)$$

While these equations typify the pyrolysis process, they do not clearly identify all the reactions, because most carbon is not free in MSW. Other secondary reactions that take place have been identified [12, 8]. One such exothermic reaction (a reaction yielding heat), called the water-gas shift, occurs when carbon monoxide is reacted with the water present in the reactor to produce CO_2 and hydrogen. Another secondary reaction that often takes place in the reactor is the formation of carbon dioxide from C and O_2. The formation of methane occurs when carbon and hydrogen combine.

$$CO + H_2O = CO_2 + H_2 + \text{heat} \qquad (9\text{-}4)$$
$$C + O_2 = CO_2 + \text{heat}$$
$$C + 2H_2 = CH_4 + \text{heat}$$

Little methane is produced if the reactor is operated at atmospheric pressure and low temperature. At very high reactor temperatures, however, these reactions can produce enough heat to make the overall process exothermic.

PROCESS CONTROL VARIABLES

The critical parameters of concern in the pyrolysis process are: heating rate, temperature, time, waste composition, and the relative direction of flow of the product gases and char. Each of these parameters directly

TABLE 9-3
Effect of Rate of Heating on Gas Analysis from Shredded Newsprint[a]

	Minutes to 1500 F°						
	1	6	10	21	30	40	60	70
CO_2	15.01	19.16	23.11	25.1	24.7	25.7	22.9	21.2
CO	42.60	39.59	35.20	36.3	31.3	30.4	30.1	29.5
O_2	0.92	1.61	1.80	2.5	2.3	2.1	1.3	1.1
H_2	17.93	9.85	12.15	10.0	15.0	13.7	15.9	22.0
CH_4	17.54	21.70	19.95	20.1	20.1	19.9	21.5	20.8
N_2	6.00	8.09	7.79	6.0	6.6	8.2	8.3	5.4
Btu/ft.^{3}	372	380	355	354	354	344	367	378
Gas volume (ft^3/ton)	11,000	7,500	6,800	6,170	6,750	6,560	7,260	9,170

[a]Data are volume percent of gas, dry basis.
Source: Ref. 4.

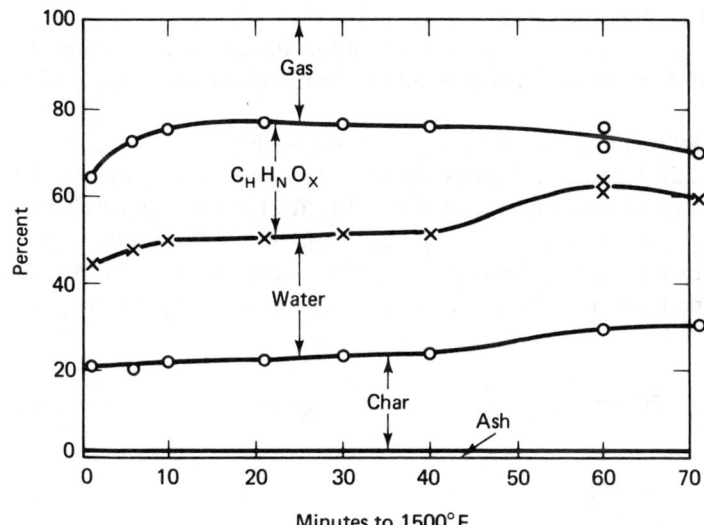

Figure 9-4. By-product yields from pyrolysis of newspaper as related to rate of heating. (From Ref. 4)

the water content and organic liquids decrease with an increase in the heating rate [13].

As the temperature of pyrolysis increases, the gas content increases proportionally while the amount of acids, tars, and char decreases [5,14]. Figure 9-5 illustrates the effect of operating temperatures on product yield. The operating temperatures influence not only the quantity of the gas produced but also the quality of that gas, as shown in Table 9-4.

The third factor which affects the products of pyrolysis is waste composition and the method of preparation of the feedstock. Municipal waste generally could be expected to produce gases, tars, and liquids but fewer solid residues than most industrial wastes [6]. As would be expected, a smaller particle size enhances heat transfer and thus makes it easier for the pyrolysis reaction to proceed [12].

The moisture content of the feedstock also has an effect on the end products. The lower the moisture content in the feedstock, the less time will be required to heat the materials to operating temperature. Pre-drying the feed however consumes energy, is thus expensive.

In the gasification process, the rate of steam injection substantially influences the yield of various products [4, 7, 12]. For example, steam reduces the percentage formation of char because the water reacts to form gaseous products. Thus the higher the operating temperature, the more

Process Control Variables

influences the mix and the yield of the products. Other factors that influence the reactor yield in gasification include the injection of steam, the use of air versus oxygen as an oxidant, and the preparation of the feed stock.

Table 9-3 illustrates the effect of pyrolysis heating rate on gas composition, while Fig. 9-3 illustrates the gas yields from newspaper as related to the rate of heating [4]. Figure 9-4 illustrates the percentage change in the yield of gas, water, organic liquids, and char during the pyrolysis of newspaper at varying heating rates. These data illustrate that the gas yield might be highest at the low and high rates of heating. On the other hand,

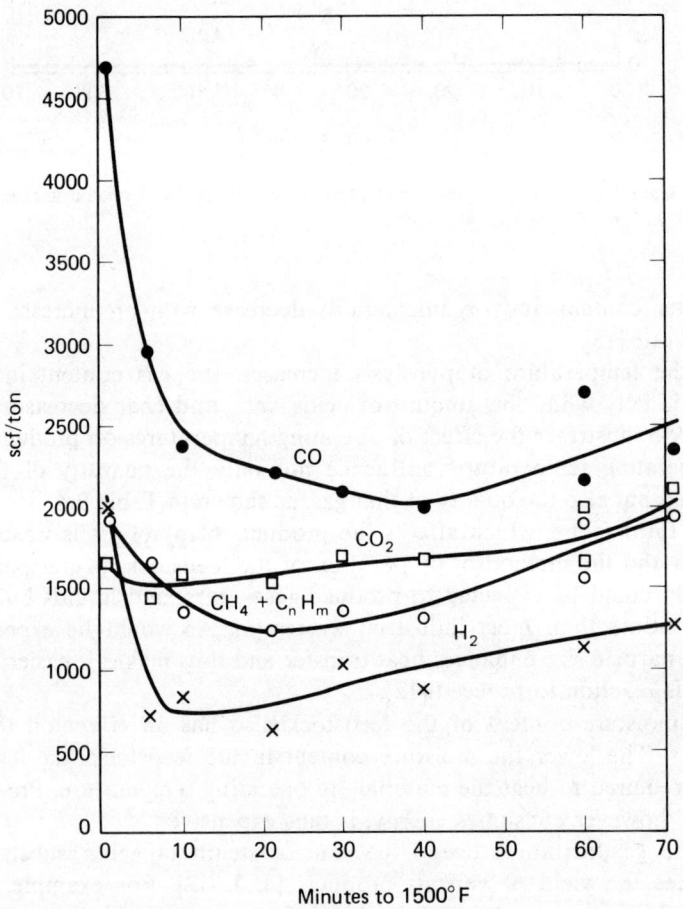

Figure 9-3. Gas yields from newspaper as related to the rate of heating. (From Ref. 4)

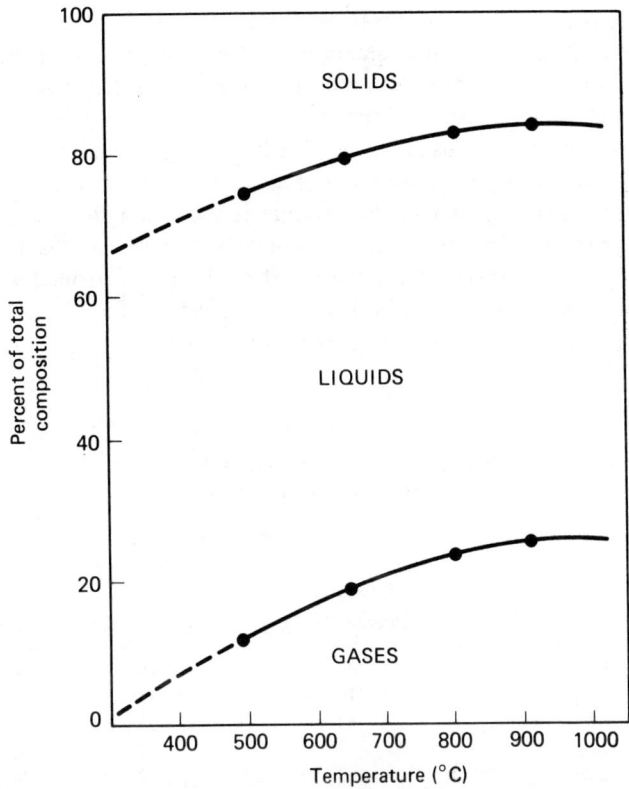

Figure 9-5. Effect of pyrolysis temperature on product yield. (From Ref. 5)

TABLE 9-4

Effect of Pyrolysis Temperature on Gas Composition

Constituent	480°C	650°C	815°C	925°C
H_2	5.56	16.58	28.55	32.48
CH_4	12.43	15.91	13.73	10.45
CO	33.50	30.49	34.12	35.25
CO_2	44.77	31.78	20.59	18.31
C_2H_4	0.45	2.18	2.24	2.43
C_2H_6	3.03	3.06	0.77	1.07
Accountability	99.74	100.00	100.00	99.99

Source: Refs. 2 and 5.

water is consumed to form gases and thus reduce the char content. At higher operating temperatures, steam may also be used to shift the ratio of hydrogen to carbon monoxide, which is important in processes where the gases will not be used for combustion.

Air or oxygen can be used as the oxidant in a gasification reactor. The product gas formed by the admission of air contains significant amounts of nitrogen; an advantage where the product is used as a synthesis gas. The use of oxygen in a pyrolysis reaction however yields a gas that has a significantly higher heat content than that of the gases formed when only air is used because the gas is basically nitrogen-free. Table 9-5 illustrates typical gaseous products of various pyrolysis processes.

TABLE 9-5

Gaseous By-Products of Typical Pyrolysis and Gasification Processes

	Percent by Volume			
	Pyrolysis in the absence of oxygen	Oxygen added	Air added A	Air added B
CO	25	49	24	7
CO_2	16	15	7	12
H_2	44	31	15	7
CH_4	7	3	2	3
C_2's	8	2	0	2
N_2	—	—	52	69
HHV (Btu/scf at 60°F)	421	324	139	111

Source: Ref. 12.

REACTORS

Various pyrolysis systems can be characterized by the type of reactor, its bed conditions, and the direction of solids flow within the reactor. Some typical units include [9]:

Directly Fired	Indirectly Fired
Fluidized bed reactor	Fluidized bed reactor with recirculating heat carrier
Moving packed bed reactor (shaft furnace) with countercurrent feed flow	Moving packed bed reactor (shaft furnace) with cocurrent feed flow

Directly Fired	Indirectly Fired
Moving stirred bed stacked reactor (multiple hearth furnace) with countercurrent feed flow	Entrained bed reactor with cocurrent feed flow
Refractory lined rotary kiln with countercurrent feed flow	Rotary retort with countercurrent or cocurrent feed flow
	Refractory lined rotary kiln with cocurrent feed flow and recirculating heat carrier.
	Tunnel reactor with agitated or static bed
	Molten metal or molten salt reactor with cocurrent feed flow

The reactors may be provided the necessary heat for operation by the utilization of the system fuel in an external heater, through partial oxidation of the products of pyrolysis in the reactors, or by the use of an external heat source. A true pyrolysis reactor is one that must be indirectly heated and into which no oxygen or steam is injected.

The gasification process, on the other hand is directly fired since partial combustion occurs within the reactor. Gasification of solid fuels generally occurs in fixed bed, fluidized bed, or suspension bed systems and refers to the manner of contacting the waste with the gases in the primary reactor. In the fixed bed unit, the fuel moves slowly down through the reactor while being contacted with the off-gas. In the fluidized bed process, the fuel is fluidized by the gasifying medium, while in the suspension fired system, the gasifying medium actually carries the material along.

The gasification process may be classified in a variety of other ways besides reactor type [13] and might include: (1) by the flow reactants (cocurrent, countercurrent, or cross current), (2) by the types of gaseous reactants (steam, hydrogen, oxygen, air, or mixtures thereof), (3) the method of providing for the endothermic reaction (internal or external), and (4) by the residue generated (dry ash or slag) [7]. Since it is not possible to separate any of these noted characteristics completely, all four of these considerations and the methods of contacting the reactants must be described in order to adequately classify a process. For the sake of clarity, they are classified by the method of fuel contact with the reactant gases.

A matrix of typical gasification systems is noted in Table 9-6. This system table illustrates the typical developmental work conducted on reactors.

TABLE 9-6
Representative Gasifiers

Description			Slagging vertical furnace O_2-blown partial oxidation Pyrolysis Gas cleanup	Rotary kiln Air-blown Partial oxidation Pyrolysis suppl. fuel	Slagging vertical furnace Air-blown partial oxidation-Pyrolysis	Vertical furnace Air-partial oxidation Pyrolysis-with grate steam added to inlet air	Horizontal retort Moving lead bath Gas-heated radiant tubes—pyrolysis	Horizontal retort Vibratory conveyor Radiant tube heater
Yields[a]	scf/ton		24,000	50,000	56,000 30% recycle	47,000 Part recycled to heat air	18,000 30% for recycle	16,000 30% used for heating
Gas composition	Dry mole		CO 49 H_2 30 CO_2 15 CH_4 3 C_2's 2 N_2^+A 1	CO 6.6 H_2 6.6 CO_2 11.4 CH_4 2.8 C_2's 1.7 N_2^+A 69. O_2 1.6	CO 15 H_2 14 CH_4 3 C_2's 1 CO_2 12 N_2 55	H_2 21 CO 21 CH_4 1.8 CO_2 12 N_2 43 Oil 1.2	H_2 16.4 CO 29.4 CO_2 18.4 CH_4 23.3 C_2 3.9 C_6H_6 7.9 C_7H_8 0.6	H_2 19 CO 26 CH_4 13 CO_2 18 N_2 18 C_2's 4
Heating value, Btu/gas	300–320 DSCF			98 (120 with sensible heat)	120 Btu/scf	152 Btu/scf	with condensible-790 Noncondensible-490	350
Conversion efficiency gas/refuse	%		75	60	55	72	60	56 40 with/recycle gas for heat

Quality of gas for methane, methanol, ammonia	good primarily CO and H_2	Poor, high in nitrogen	Poor, high in nitrogen	Poor, high in nitrogen	Poor, high in nitrogen	Poor, high in CH_4	Poor, too high CH_4, and N_2 content
Reliability	Reactor operated in research mode-need full-load 2–6 mo. test for reliability	After startup problems are overcome appears to be reasonably reliable	Appears to have operated successfully Has not been operated commercially	Research reactor operated only	Pilot unit operated as research unit needs demo. plant work	Operated with difficulty—vibrating conveyor questionable
Maintainability	Hearth refractories need replac. once per year-remainder operates every 3 yr estimated	Normal for shredder plant, but rotary kiln may require special attention	High-temp. air heater may be troublesome Has not been operated long enough to assess	Not known	Lead bath may pose problems	High maintenance expected
Residue	Slag frit	Glassy aggregate Char, ash	Slag frit	Ash	Char/ash mixture	Char/ash

[a]Based on tons of shredded refuse. If 75% by weight of MSW were to be used as input material, these yields would be reduced by 25% to determine yields per ton of MSW.
Source: Ref. 15.

Fixed Bed Reactors

In the *fixed bed reactor* illustrated in Fig. 9-6 the shredded (optional) municipal solid waste is added at the top of the reactor. At the interface with the material in the reactor, the temperature may range from 93 to 315°C (200 to 600°F). The material moves down through the bed, which is supported by a grate. At the bottom of the reactor, preheated air or oxygen is introduced where the temperature is normally about 980 to 1650°C (1800 to 3000°F).

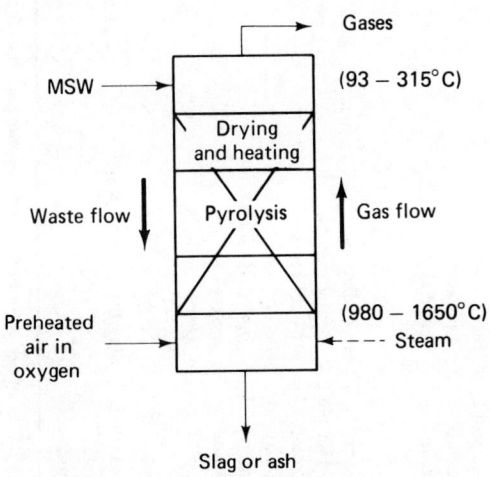

Figure 9-6. Typical fixed bed pyrolytic reactor.

The products produced by this reactor include a slag or ash discharged at the bottom, and off-gases which exit at the top of the unit. The degree to which the off-gases are cooled is dependent upon the end use to which the gases are to be put.

To achieve maximum thermal economy, the gaseous reactants are passed countercurrent to the direction of fuel flow, but some units are reported to operate efficiently with cross-current and concurrent flows [16].

The gasification of coke demonstrates the countercurrent flow concept. Here, the gas moves through the bed at a rate of 0.2 to 0.6 m/h (0.7 to 2.0 ft/h) and the fuel residence time is generally quite long [17]. The gas velocities in the reactors are also quite low—based on the cross section of the reactor, they will not exceed about 0.6 to 1.0 m/s (2 to 3 ft/s). The velocity is also a function of the fuel size: the larger the fuel, the higher the velocity.

Heat to support the reactions in a fixed bed reactor is provided by the combustion of a portion of the fuel. In a typical gas producer, for example, a substance such as coke is gasified by an air-steam mixture. The gas produced has a calorific value of 56 to 73 kJ/m^3 (130 to 170 Btu/ft^3). On the other hand, a water-gas generator can develop a gas with a calorific value of up to 215kJ/m^3 (500 Btu/ft^3 foot) [18]. Such high values are obtained in the water-gas generator by cycling the air and steam in an alternate fashion. In this process, the fixed fuel bed is first heated by combustion with air and then reacted with the steam which produces the water gas [7].

A countercurrent flow reactor produces the maximum net energy output of all the fixed bed gasifiers. Because of the long residence time in the reactor, there is an assurance that a maximum conversion of the waste material to a fuel occurs. Since the gas velocities in such a reactor are also relatively low, the particulate matter entrained in the product gases is also low. The low solids carryover and high fuel conversion minimize the loss of ungasified fuel and reduces the potential air pollution impacts.

Fixed bed reactors are not without their problems, however, and must be evaluated on the basis of potential disadvantages. For example, fuels that tend to cake, such as sludges and wet municipal solid wastes, cannot be processed without pretreatment. This condition normally includes pre-drying or further shredding of the material to ensure that it does not cake. Unshredded fuels also promote channeling through the reactor and may result in poorly gasified effluents with higher than normal solids content [19].

Tars and oils may be formed in the upper- and low-temperature zones of the reactor and thus create problems if the gas is to be used for the synthesis of chemicals and fuels. Proper control of the temperature by the use of steam, for example, can limit production of these tars and oils.

The design of fixed bed gasifiers is flexible enough to accommodate a variety of fuels. Fuel size, its caking tendency, the ash fusion temperatures, and the reactivity of the material are important design considerations. Ideally, the use of noncaking, uniformly sized fuels permits the desired uniform gas distribution throughout the reactor and results in efficient operation of the pyrolysis process.

A good example of the typical installation using the fixed bed gasification process for municipal solid wastes has been demonstrated by work at South Charleston, West Virginia [7, 20, 23]. This unit uses a fixed bed, upflow system, with partial oxidation. The unit is illustrated in Fig. 9-7. Constituents of the synthetic gas leaving the typical processor are noted in Table 9-7. Values of the heat content of this gas range from 130 to 160 J/m^3 (300 to 370 Btu/ft^3). The gas that results from this partial oxidation

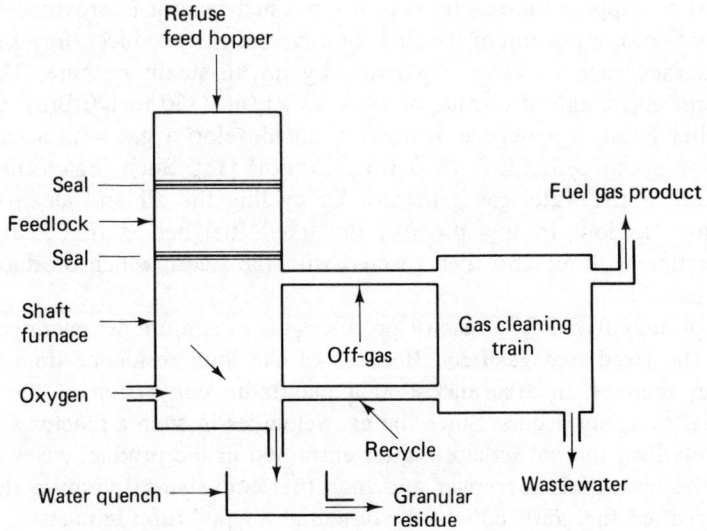

Figure 9-7. Typical fixed bed pyrolytic reactor utilizing oxygen. (Courtesy of Union Carbide)

TABLE 9-7
Typical Fuel Gas Leaving the Gasifier

Gas Constituent	Percent by Volume at 38°C (100°F)	Percent by Volume Dry
H_2	28	30
CO_2	46	49
CH_4	3	3
C_2H_4, C_2H_6, etc.	2	2
H_2 + Ar	1	1
CO_2	14	15
H_2O	6	—
Total	100	100

Source: Refs. 15 and 23.

process is a clean-burning fuel which is similar to natural gas in combustion characteristics, but has a heating value about 30% of that of the natural gas. The gas is generally free of sulfur compounds and nitrogen and burns at about the same temperature as natural gas. Since the gas must be fed at a greater volume in order to obtain the same heat value, the nozzles and the furnace must be modified accordingly so that the fuel rate can be increased to develop the same heat content. In addition, because larger volumes of the gas must be utilized to yield the same amount of energy, the compression cost per Joule may be three times greater than that for natural gas [22]. Based on these facts, it has been determined that markets for the gas must be within 2 to 5 miles of the producing facility and only short-term storage can be contemplated [22].

When oxygen is used in the partial oxidation process, approximately 0.05 to 0.06 tonne (0.038 to 0.045 ton) of oxygen is required per ton of shredded refuse, which would be the heat required to evaporate the moisture in the wastes. In the combustion zone the total heat generated can be accounted for by balancing four areas: (1) vaporization of moisture in the waste feed, (2) heat losses from the reactor vessel, (3) the pyrolytic reaction of organic fractions in the refuse, and (4) melting of the inorganic refuse residues (glass, metals, etc.).

A gas cleaning train, which might include an electrostatic precipitator and condenser, does not remove all the contaminants from the fuel gas. Contaminants most often found are noted in Table 9-8.

TABLE 9-8
Contaminants in Off-Gas from Partial Oxidation of Shredded MSW[a]

Component	Amount
Fly ash	10 ppm
Oils	150 ppm
Gasoline vapors	0.02 gal/1000 ft^3
Sulfur	15 ppm
Water	5.2–6.5 wt %
NO$_x$	<1 ppm

[a]*Gas cleaning train included electrostatic precipitator and condenser.*
Source: Ref. 15.

Fluidized Bed Gasifier

The *fluidized bed gasifier* is intermediate between the fixed bed and the suspension gasifiers. The primary difference between the fixed bed and the fluidized bed reactor is that the gas velocity in the fluidized bed is increased to the point where the particles are lifted. As they are lifted, the bed expands and the individual particles are no longer in continuous contact as they would be in the fixed bed unit. Gas and fuel are contacted in a cocurrent fashion in the fluidized bed reactor, as illustrated in Fig. 9-9.

The fluidized bed reactor can be used in a pyrolysis mode as well as in a partial oxidation mode. Operating temperatures are generally below that of the slagging temperature, which is approximately 200 to 980°C (400 to

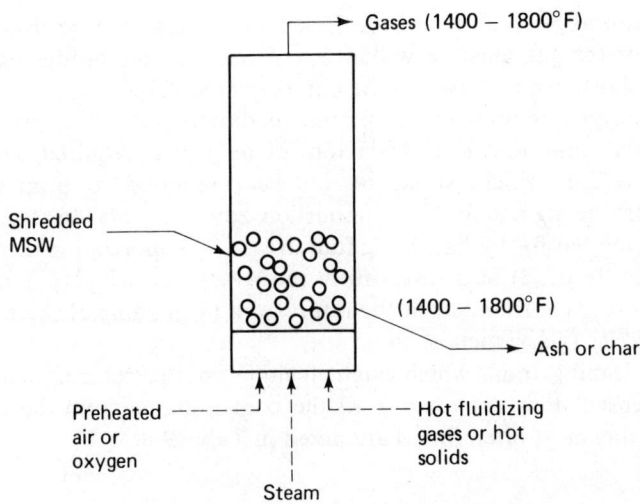

Figure 9-8. Fluidized bed reactor. (From Ref. 9)

1800°F). The effluent gases from the reactor are very hot, and heat recovery from these gases is mandatory in order to efficiently utilize the energy produced by the reactor.

Advantages of a fluidized bed system are not limited by the caking properties, such as in the fixed bed reactor. On the other hand, the operation is sensitive to the reactivity of the fuel. High reactivities are required in order to prevent the carryover of fuel that has not been adequately gasified. In fluidized processes, temperature control to avoid ash fusion is also necessary, but the reactor does have the advantage that better temperature control can probably be attained [12]. The unit also appears to have the ability to handle very finely divided fuels that have a high ash content. It also handles wastes with high moisture or with highly variable moisture content. Since the specific reaction rates in the process are quite high, the size of the units may be smaller than a typical fixed bed reactor.

Disadvantages include the high loss of ungasified fuel and high heat losses in the product gas. The heat losses occur because of the approximate thermal equilibrium of those gases with the fuel in the bed at reaction temperatures. While this heat can often be recovered in waste heat boilers, it is not as available in this system as it might be in the fixed bed reactor. Because of the heat losses, the reactor may require additional fuel (over and above that required or resulting from the process) to make the unit operate.

The use of the fluidized bed processes, based on experiences in the United States, will require more research before the application of such a process is suitable for widespread use with municipal solid wastes. On the other hands, two fluidized bed pyrolysis systems developed in Japan seem to be working well.

Suspension Systems

One of the more recent developments in gasification that may have a potential for processing of MSW is the use of the *suspension reactor*. Here the fuel is suspended in the gasifying medium and the reaction depends on the characteristics of that medium.

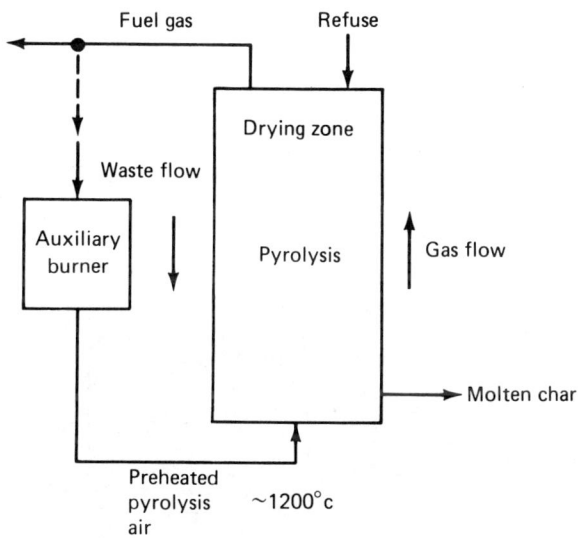

Figure 9-9. Suspension reactor. (From Ref. 9)

A typical reactor is illustrated in Fig. 9-9. In the reactor, the volatile matter is oxidized very rapidly, so the product gas generally contains no tars and oils and very little methane.

One of the more important characteristics of the suspension reactor is its capability to burn essentially any type of fuel. Since the fuel particles are discrete within the gaseous medium, they present no particular handling problems. The primary disadvantage of the suspension reactor is that

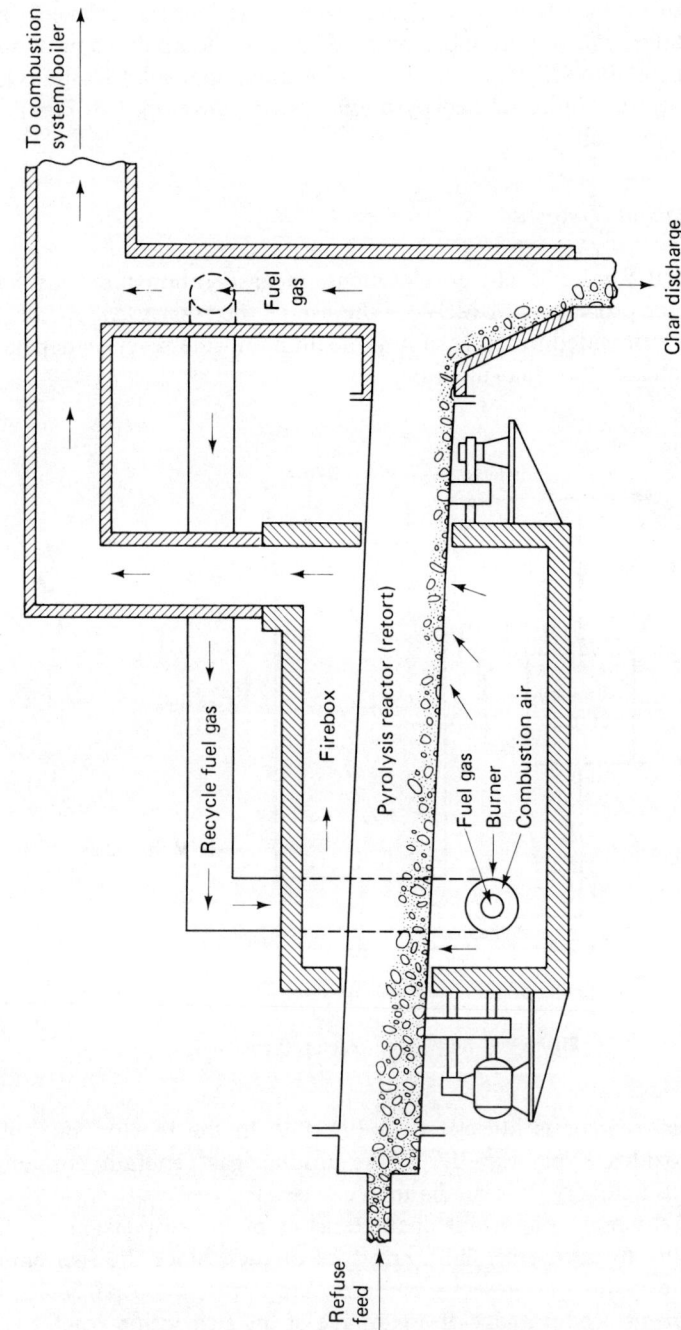

Figure 9-10. Rotary kiln reactor. (From Ref. 8)

it handles only a small amount of fuel at a time and the particles must be small (except with high gas recycle rates). The flow of the reactants is cocurrent, which is inefficient. The low fuel concentrations and cocurrent flow result in relatively low reaction rates, and fuel conversion is a problem. An increase in reactor volume does not necessarily correspond to better fuel conversion [7].

Rotary Kiln

One of the most common pyrolysis reactors is an indirectly heated *rotary kiln*, in which the retort is completely devoid of oxygen. Figure 9-10 illustrates a section of a typical indirectly heated rotary kiln. The unit is cylindrical, slightly inclined, and rotates slowly, which causes the refuse to move through the kiln to the discharge end. The retort is constructed of metal and the firebox is constructed of a refractory material. A portion of the off-gases produced by the reaction is burned in the space between the outside wall of the retort and the interior of the fire box. This heat is then used to heat the waste. Since heat transfer in this type of unit is so important, the reaction requires a relatively well shredded waste, often less than 2 in. in size, to ensure a complete reaction [24].

A relative balance of the composition of the fuel gas from a rotary kiln can be derived from an examination of reported data [2, 5, 25]. Table 9-9 represents the components of a typical pyrolyzed fuel gas (dry) from an indirectly heated rotary kiln. Based on the work of Lewis, Table 9-10 illustrates the typical pyrolysis materials balance from a rotary kiln reactor [8].

TABLE 9-9
Composition of Dry Fuel Gas

Component	Volume (%)
CO	35.0
CH_4	20.4
CO_2	19.6
H_2	16.3
C_2H_4	8.7

Source: Ref. 8.

TABLE 9-10

Indirectly Heated Rotary Kiln Pyrolysis Process and Material Balance

	Carbon (kg)	Hydrogen (kg)	Oxygen (kg)	Inerts (kg)	Total (kg)
Input					
Combustibles	235.9	31.7	186.0		453.6
Moisture		25.2	201.6		226.8
Inerts				226.8	226.8
Total	235.9	56.9	387.6	226.8	907.2
Output					
Char	77.1	1.8	11.8	226.8	317.5
Organic liquids (C_6H_8O)	97.0	10.8	21.5		129.3
Fuel Gas					
Carbon dioxide (CO_2)	13.1		35.0		48.1
Carbon monoxide (CO)	23.3		31.1		54.4
Methane (CH_4)	13.7	4.5			18.2
Hydrogen (H_2)		1.8			1.8
Ethylene (C_2H_4)	11.7	1.9			13.6
Subtotal	61.7	8.2	66.1		136.1
Water vapor					
Waste moisture		25.2	201.6		226.8
Pyrolysis		10.9	86.6		97.5
Subtotal		36.1	288.2		324.3
Total	235.8	56.9	387.6	226.8	907.2

Source: Ref. 8.

Transport Reactor

Another type of pyrolysis reactor system which has been developed is the *transport* reactor. This unit is operated at temperatures that will result in the production of a liquid fraction as the major product and generally requires the feed stock to be very finely shredded because of the short residence times of the refuse in the reactor. This process is often described as "flash pyrolysis." [26-29]

In general, the heat required for a pyrolysis reaction in the unit is supplied through the recirculation of the hot char. The hot char is removed from the reactor, passed through an external fluidized bed, in which some air is added to partially oxidize the char, and the char is then recirculated to furnish energy for the endothermic pyrolysis reaction, which yields the liquid by-products.

In the unit illustrated in Fig. 9-11, the finely shredded waste is carried into the reactor by recycled product gas. The pyrolysis takes place at

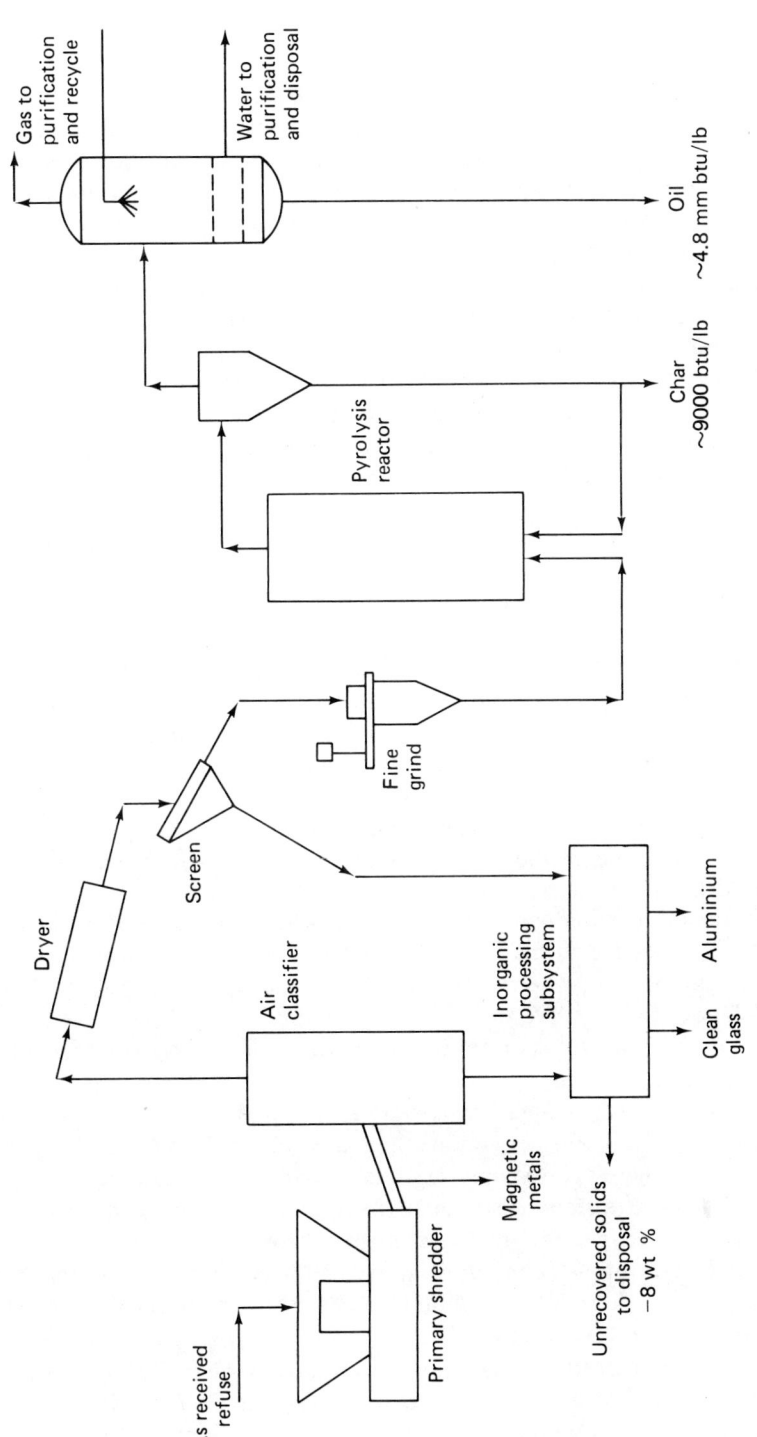

Figure 9-11. Pyrolysis transport reactor system—schematic. (From Ref. 26)

approximately 500°C (950°F) and at a pressure of about 100 KN/m² (1 gauge). In this reactor, no air, oxygen, hydrogen, or any other catalyst is used in the pyrolysis process. When the solid residue (char) leaves the reactor, the product vapors are separated with cyclone separators and the char is mixed with air and combusted. The resulting hot ash is recycled to the pyrolysis reactor entrance to supply the heat for the pyrolysis reaction. As the gas organic ash stream is very turbulent and the organic and ash particles are quite small, efficient heat transfer is achieved and the organics are rapidly pyrolyzed.

When the solids are separated from the other pyrolysis products, the pyrolytic product vapors are quenched rapidly. This prevents the large oil molecules from further cracking, which would form a less desirable product. The final products are then separated into pyrolytic oil, gas, and water.

A typical once-through distribution of product yields at an operating temperature of 500°C (932°F) is illustrated in Table 9-11. The noted yields and the product compositions, particularly for the char and gas, can be varied by changes in pyrolysis temperatures and the residence time in the reactor. These yields represent the pyrolysis reactor effluents and do not include the carrier gas or the heat-carrying ash and so do not represent the overall product yields from the process. In fact, in operation, the product gas is recycled for use as the transport medium and is eventually burned to provide heat to the kiln. The product char is also burned to provide heat for the pyrolysis reaction. The required viscosity of the pyrolytic liquid product produced in the reaction is attained by mixing it with the water formed in the reaction.

In the oil-forming pyrolysis process, approximately 38% of the incoming refuse is recovered in the form of valuable products, and 44% is consumed within the process. The remaining 18% of the mass consists of residue which goes to a landfill and effluent water [26].

A reactor that forms oil products appears to have some advantages over units that produce gas. While the gas may have a somewhat higher energy value, the oil has several important advantages, including: (1) it has a higher energy content per volume than any other energy form derived from refuse and therefore requires less storage space and can be transported more easily to distant customers; (2) it generally can be handled much more easily than solid fuels; and (3) where solid refuse-derived fuels normally have an ash content of at least 10%, synthetic liquid fuels can be produced at well under 1% ash and can, therefore, be burned at almost any location.

The oil produced by the pyrolysis process is generally intended to be sold as a substitute for No. 6 residual (Bunker C) fuel oil. Table 9-12 illustrates some properties of the pyrolytic oil as compared to No. 6 fuel

TABLE 9-11
Typical Products of Transport Pyrolysis Reactor[a]

		wt %
Char (20%)	C	48.8
Higher heating value	H	3.3
19,100 kJ/kg	S	0.4
8,200 Btu/lb	N	1.1
	Cl	0.3
	Ash	33.0
	O	13.1
		100.0
Oil (40%)	C	57.0
Higher heating value	H	7.7
24,600 kJ/kg	S	0.2
10,600 Btu/lb	N	1.1
	Cl	0.3
	Ash	0.5
	O	33.2
		100.0
		mol %
Gas (30%)	H_2	12
Higher heating value	CO	37
15.0 MJ/N-m^3	CO_2	37
380 Btu/scf.	CH_4	6
	C_2H_4	3
	C_2H_6	1
	C_3	1
	C_4 +	2
	H_2S	0.8
	HCL	0.2
		100.0
Water (10%)		

[a]Based on dry weight of feed to pyrolysis reactor.
Source: Ref. 26.

oil. Although the oils have some similarities, there are several important differences. For example, the sulfur content of pyrolytic oil is much lower than that of a typical No. 6 fuel oil based on research conducted by Occidental Research Corporation [26]. The heating value of pyrolytic oil is generally about 75 to 77% of that of a typical No. 6 fuel on a volume basis because the carbon and hydrogen contents are lower and the oxygen content higher. The viscosity of the pyrolytic oil may be a problem since it is more viscous than residual fuel oil and its viscosity is more strongly a function of temperature. It can be stored and pumped at 70°C (160°F) and

TABLE 9-12

Typical Properties of Transport Reactor Pyrolytic Oil as Compared with Number 6 Fuel Oil

	No. 6	Pyrolytic Oil
C(wt %)	85.7	57.0
H	10.5	7.7
S	0.7–3.5	0.2
Cl	—	0.3
Ash	0.05	.05
N	2.0	1.1
O		33.2
Specific gravity	0.98	1.30
Btu/lb	18,200	10,600
kj/kg	42,300	24,600
Btu/gal	148,800	114,900
kJ/kl	41,500	32,000
Pour point		
°F	65–85	90[a]
°C	18–29	32[a]
Flash point		
°F	150	133[a]
°C	66	5656[a]
Viscosity		
SSU at 190°F	340	1150[a]
N-s/m² at 88°C	0.064	0.23[a]
Pumping temperature		
°F	155	160[a]
°C	46	71[a]
Atomization temperature		
°F	220	240[a]
°C	105	116[a]

[a]*Pyrolytic oil containing 14% water, as marketed.*
Source: Ref. 26.

it can be atomized at 166°C (240°F), which is about 11°C (20°F) higher than the temperature necessary to atomize No. 6 fuel oils.

Pyrolytic oil has a high oxygen content. Three characteristics resulting from this increased oxygen content are particularly interesting. Pyrolytic oil is about 60% soluble in water, which is often added to improve its handling properties—it decreases the viscosity of the oil. The oil also tends to be acidic and is corrosive to mild steel. This acidity basically derives from the carboxylic acids formed in pyrolysis and partially from the HCl that arises from the organic chlorine compounds often found in refuse. The oxygen also seems to affect the viscosity of the pyrolytic oil. If the oil is maintained at elevated temperatures for any appreciable length of time, the viscosity of the oil increases, irreversibly degrading its handling properties. The oil must, therefore, be stored at 70°C (160°F) or below until it is atomized.

ADDITIONAL PYROLYSIS PRODUCTS

There are numerous products of the pyrolysis/gasification process, some of which have been noted previously and include pyrolytic oils, char, and certain off-gasses. The gas produced in a partial oxidation pyrolysis process using a fixed bed reactor and oxygen can also be converted to methane, methanol, or ammonia, or used as a fuel in the generation of electric power. The processes are briefly described as follows: (1) conversion to methane, where the gas is upgraded to pipeline gas for introduction into existing gas utility pipeline distribution systems; (2) generation of electric power, where the gas is used as a fuel in a conventional gas turbine system; (3) conversion to methanol, where the gas can be converted catalytically; and (4) conversion to ammonia, where the gas is mixed with a pure nitrogen base and then converted catalytically to ammonia, which is a good fertilizer. The following descriptions should serve to illustrate the versatility of using the product gas.

Methanation

The production of pyrolysis gas and its conversion to methane is illustrated schematically in Fig. 9-12. In this process, the gas is received at almost atmospheric pressure, passes through a waste heat boiler, and then through a caustic scrubber for removal of HCl, the residual sulfur compounds, and some CO_2. The scrubber gas then passes through a bulk methanator, which basically consists of a series of fixed bed, catalytic reactors. In the catalyst beds, the water shift and methanation reactions take place simultaneously [see Eqs. 9-2 and 9-4]. The reactions in these beds are exothermic, and gas temperatures reach about 540 to 650°C (1000 to 1200°F). In this process, the volume ratio of gas feed to product gas is approximately 4 : 1 [15].

The reported experience with the process indicates that it is capable of producing a good grade of methane utilizing a wide range of input gas

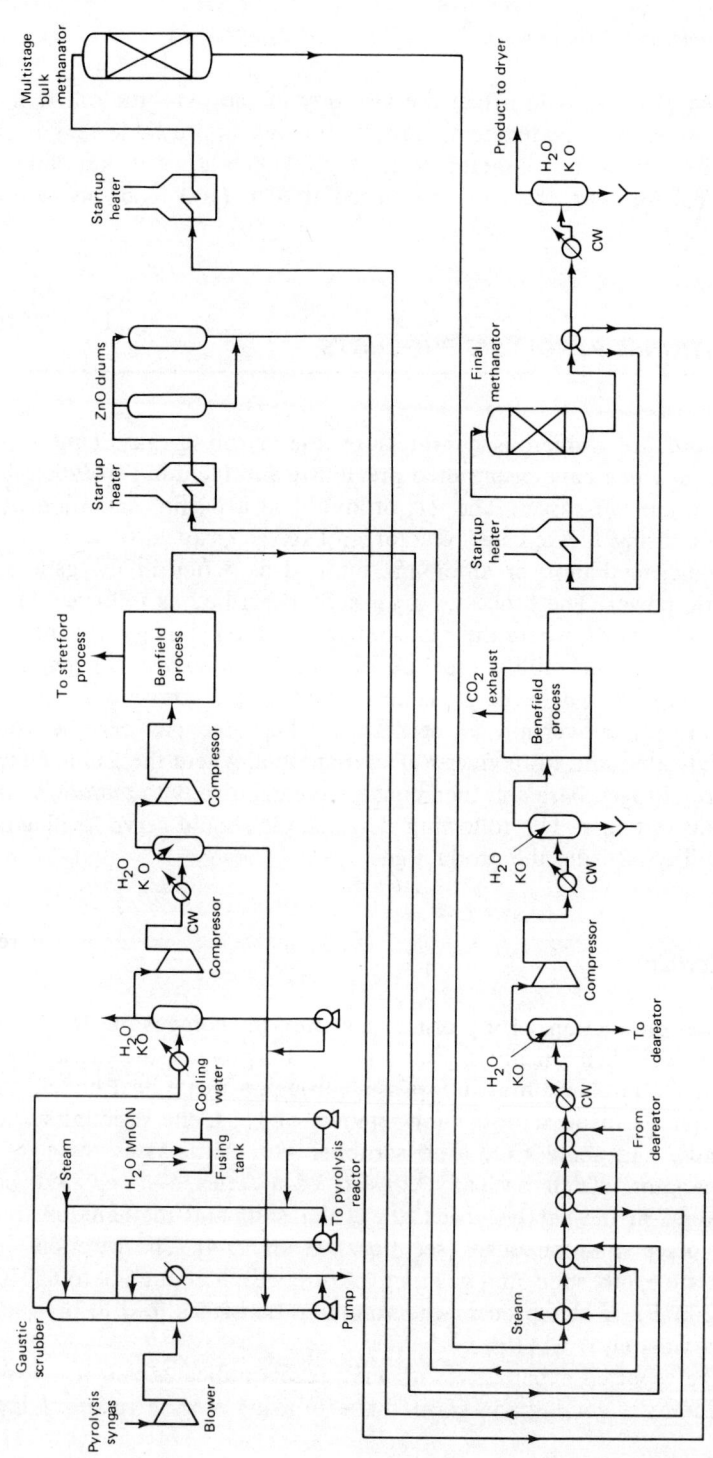

Figure 9-12. Methanation process flow diagram. (From Ref. 15)

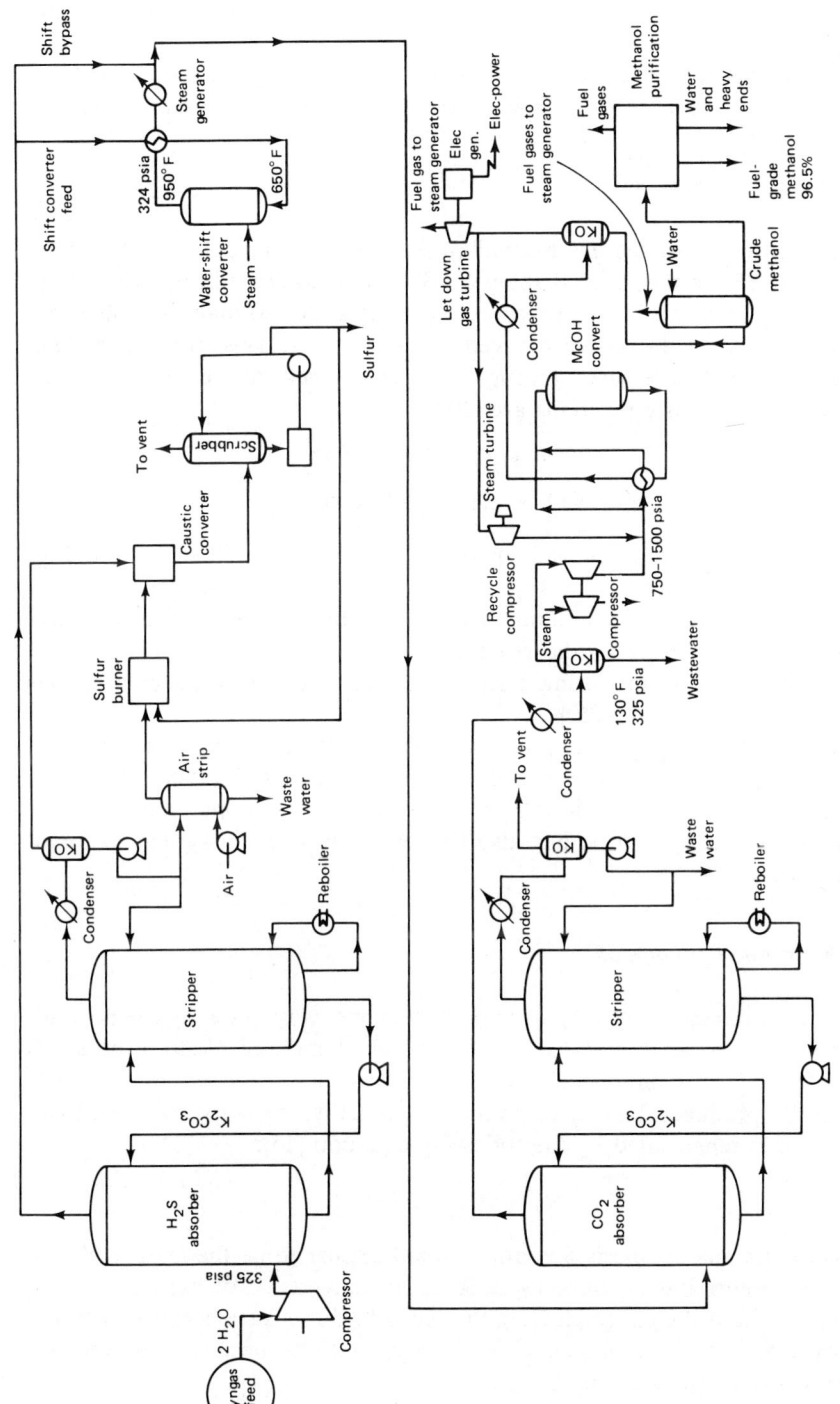

Figure 9-13. Methanol synthesis. (From Ref. 15)

compositions. The method should prove a satisfactory means of producting methane, once the economics of the system have been established.

Methanol Conversion

If conversion of the gas to methanol is desired, the CO and H_2 in the gas (generally in the ratio of about 5 : 3) can be converted to methanol by first using a water-gas shift reaction in a catalyst bed so that the mole ratio of CO to H_2 in the gas is reduced to about 1 : 2. This mixture is then passed at high pressures through a catalyst bed to be converted to methanol. The basic reactions are [20]

$$CO + H_2O \rightarrow H_2 + CO_2$$
$$CO + 2H_2 \rightarrow CH_3OH$$

The process train for such methanol production is illustrated in Fig. 9-13. In this process, the H_2S is removed from the gas and the desulfurized gas is then passed into a water-gas shift converter. The methanol synthesis takes place in the reactor at pressures of 5×10^6 to 15×10^6 N/m^2 (50 to 150 atm) using a zinc chromium oxide catalyst or a copper zinc chromium catalyst [20].

The methanol produced is an easily transportable and storable fuel. Potential problems associated with the process include the type of catalyst used, the potential air pollution problems, and the high operating pressure required in the system. It has also proven to be a costly system to build and operate.

Ammonia Conversion

The synthesis of gas to ammonia requires very pure hydrogen and nitrogen mixed at the ratio of approximately 1 mole of N_2 to 3 moles of H_2. The gas is compressed and heated, reacted in a catalyst bed, and generally produces 2 moles of ammonia. Actually, the conversion is about 94% and is represented by the following equation [15]:

$$N_2 + 3H_2 \rightarrow 2NH_3$$

An ammonia synthesis operation would appear to be the least likely of the gas conversion facilities to be used to convert MSW because of the composition of the gas (high CO), the much higher costs associated with its construction, the requirements for purity, and the necessary power required for compression of the gas.

ENVIRONMENTAL CONCERNS

There are certain environmental concerns connected with any pyrolysis process which include air, water, and solid waste problems. For example, at the present time there is a lack of sufficient data on air pollutants emitted from the pyrolysis processes described in this chapter. It is known, however, that the higher the temperature of the process, the greater tendency there will be to volatize more of the heavy metals, such as copper and lead. In a pyrolysis process where the fuel gases are combusted in a separate chamber from the solid fuel bed, it is likely that there will be significantly less particulate matter entrained in the combustion air. In fact, an indirectly heated pyrolysis reactor would be the most efficient in this respect, since there is no combustion air in the solid fuel region. The goal, then for gasification proceses when controling air pollution is to keep as much air as possible from passing through the solid fuel region.

The off-liquids, particularly in the condensate from the pyrolysis reaction, contain inorganic chemicals with very high BODs and CODs. Of particular concern are the phenols which may significantly impair the biological wastewater treatment processes which may be required in conjunction with the treatment of wastewaters from incinerators and pyrolytic reactors.

The solids formed in the pyrolysis reaction include slag, char, and some glassy aggregate. If those materials are free of the organic compounds normally associated with the gases, they probably should not present any particular problem if buried properly in a sanitary landfill.

Some water and air pollution will obviously result from pyrolysis, and gasification processes. Because of the smaller quantities of air (or other oxidants, such as oxygen) and water which are used in the process (in contrast to incineration), the adverse impacts will be less than would be experienced by incineration.

PARTING SHOTS

The possibility of using pyrolysis or gasification reactors to process MSW to produce usable end products is largely dependent on advances in the necessary technology. Economics also play a significant role in the eventual success of pyrolysis processes. Application of the technology will require a great deal of attention to details regarding the economics of the materials balances and the desired end products for specific applications.

In time, it is expected that pyrolysis will become more competitive with other energy-conversion techniques and a more widespread use of the technology will result.

REFERENCES

[1] LEVY, S. J., "Pyrolysis of Municipal Solid Waste," *Waste Age*, Oct. 1974.

[2] HOFFMAN, D. A., *Pyrolysis of Solid Municipal Wastes*, NTIS PB-222 015, U.S. Dept of Commerce, July 1973.

[3] National Center for Resource Recovery, *Pyrolysis*, Washington, D.C., Mar. 1973.

[4] KAISER, E. R., AND S. B., FRIEDMAN, "The Pyrolysis of Refuse Components," *Proc. AICHE*, 60th Annu. Meet. Nov. 1967.

[5] HOFFMAN, D. A. and R. A. FITZ, "Batch Retort Pyrolysis of Solid Municipal Wastes," *Environ. Sci. Technol.*, 2 (11) (1968).

[6] SANNER, W. S., C. ORTUGLCO, J. G. WALTERS, and D. E. WOLFON, "Conversion of Municipal and Industrial Refuse into Useful Materials by Pyrolysis," *U.S. Bur. Mines, Rep. 7428* (Aug. 1970).

[7] HAMMOND, U. L., et al., "Pyrolysis—Incineration Process for Solid Waste Disposal," Battelle Pacific Northwest Laboratories, Dec. 1972.

[8] LEWIS, F. M., "Thermodynamic Fundamentals for the Pyrolysis of Refuse," *Incinerator and Solid Waste Technology*, American Society of Mechanical Engineers, New York, 1975.

[9] JONES, J. "Converting Solid Wastes and Residues to Fuel," *Chem. Eng.*, Jan. 2, 1978.

[10] MCFARLAND, J. M., *Comprehensive Studies of Solid Waste Management*, EPA, SERL Rep. 72-3, May 1972.

[11] BAILIE, R. C., Professor of Chemical Engineering, West Virginia University, Personal communication, July 1978.

[12] LAMB, T., "Fundamentals of Pyrolysis," *Proc. 1975 Int. Sym. Energy Recovery Refuse Sept. 1975*, Kentucky Center for Energy Research, Louisville.

[13] LOWRY, H. H. (Ed.), *Chemistry of Coal Utilization*, Suppl. vol., John Wiley & Sons, Inc., New York, 1963.

[14] MULLEN, J. F., AND J. W. REGAN, "Energy Conversion of Refuse Advanced by Application of Basic Combustion Principles," *AICHE Symp. Solid Waste Disposal, Atlantic City*, 1971.

[15] *Gasification as an Alternative to Solid Fuel Concept*, Prepared for Institute for Energy Analysis, Oak Ridge Associated Universities, Ralph M. Parsons Company, Pasadena, Calif., Aug. 1975.

[16] SHIRER, G. L., *Chemi. Eng. Mining Rev.*, Aug.–Sept. 1958.

[17] GUMZ, W., *Gas Producers and Blast Furnaces*, John Wiley & Sons, Inc., New York, 1950.

[18] "Clean Fuel Gas from Coal," promotional literature, Lurgi Corporation.

[19] GRAINCER, J. W., *J. Inst. Fuel*, **28** (1955).

[20] *Gasification of Solid Waste—An Alternative to Solid Fuel*, Inst. Energy Analy. Oak Ridge Assoc. Univ. IEA (M)-75-6, Nov. 1975.

[21] PAGE, F. J., "Torrax—A System for Recovery of Energy from Solid Waste,"

[22] LEVY, S. S., *Markets and Technology for Recovering Energy from Solid Waste*, U.S. Environmental Protection Agency, Washington, D.C., 1974.

[23] GILLIES, D. M., "Union Carbide's Purox Process," *Mineral Resources and the Environment*, National Academy of Sciences, Washington, D.C., 1975.

[24] SUSSMAN, D., "Recent Applications of Pyrolysis," *Proc. 1975 Int. Symp. Energy Recovery Refuse*, Kentucky Center for Energy Research, Louisville, Ky., Sept. 1975.

[25] KNIGHT, J. A., *Pyrolytic Conversion of Agricultural Wastes to Fuels*, Rep. 74-5017, American Society of Agricultural Engineers, June 1974.

[26] PRESTON, G. T., "Resource Recovery and Flash Pyrolysis of Municipal Refuse," Presented at Inst. Gas Technol. Symp., Orlando, Fla., Jan. 1976.

[27] GARBE, Y. M., "Demonstration of Pyrolysis and Materials Recovery in San Diego, California," *Waste Age*, Dec. 1976.

[28] LANDIS, E. K., and M. D. MCKINLEY, "Urban Refuse Incinerator Design and Operation: State of the Art," Bur. Mines Grant #G0100163 (SW031), University of Alabama, Nov. 1971.

[29] National Center for Resource Recovery, *Pyrolysis*, Mar. 1973.

PROBLEMS

9-1. Describe the difference between pyrolysis and gasification utilizing a rotary kiln reactor to describe process end products.

9-2. If the size of the walnut in Table 9-2 were made four times as large as usual, how would this affect the operation of the tester to achieve the same end products described? What is the corollary argument for the pyrolysis process?

9-3. What measures might you choose to adopt to measure the heat of reactions required to bring to completion Eqs. (9-1), (9-2), and (9-3)?

9-4. A community is considering a pyrolysis process for the development of fuel gas for their fleet of vehicles. What process would you choose to gasify that community's solid waste, and what end product would you form?

9-5. Explain the substantial drop in CO_2 production when the pyrolysis is run at temperatures ranging from 480 to 925°C (see Table 9-4).

9-6. A community is considering the installation of a transport reactor to make pyrolytic oil. How many shreds of the waste should be provided? Draw a *typical* schematic of the shredding and separation facilities.

10 ULTIMATE DISPOSAL

It is highly unlikely that the processing of mixed refuse will ever be so efficient as to convert all materials to products with economic value. Some fraction of the refuse must therefore be disposed of into the environment. The total quantity of this material is obviously reduced by increasing the efficiency of the recovery process.

This chapter is a discussion of how residues that have no value can best be managed and disposed of. Although the majority of the text is devoted to the disposal of raw, untreated MSW, it is equally applicable to any fraction of MSW which remains after the recoverable materials have been removed.

OPTIONS IN ULTIMATE DISPOSAL OF RESIDUES

Discounting outer space and air (from whence it eventually comes back to the earth's surface), the two basic options for the ultimate disposal of wastes are in the oceans and other large bodies of water, or on the land.

With rare exceptions, most solid material may no longer legally be dumped into oceans. Most developed countries have enacted strong ocean dumping legislation, and the once ubiquitous refuse-loaded barges have all but disappeared. The remaining ocean disposal problems appear to result from the discharge of refuse from ships. Long-standing maritime tradition holds that all manner of refuse can be discharged from the stern, a fact well known to sea gulls, who follow ships for long distances. The total deposit of refuse on the sea bottom in some of the more frequently traveled shipping lanes such as the North Atlantic must be impressive—as well as depressing.

Once organic wastes are discharged from a ship or barge, it is either eaten by aquatic birds or fish, or it is decomposed by marine organisms on the ocean floor. The temperatures in the deeper areas are quite low, however, and the resulting slow metabolic activity hinders decomposition. This was demonstrated by the almost edible condition of the sandwiches brought up from the sunken submarine *Thresher* a few years ago.

Nonbiodegradable material, of course, remains on the ocean bottom without much change. Even the oxidation of iron and steel is slowed markedly, as witnessed by the recovery of ship anchors that were lost over 500 years ago.

In short, disposal in the ocean is more or less a storage process, not a method of treatment.

The most serious problem with ocean storage is the difficulty of retrieval. This is especially true for hazardous material and nuclear waste disposal schemes in deep oceans. It is quite probable, based on past scientific development, that what we today consider unusable and hazardous material (such as nuclear wastes) will either have some value in the future or will at least be treatable so as to make it nontoxic. Burial in deep sea canyons will, however, prevent recovery, and present potential health problems for the future.

Philosophically, therefore, if ocean disposal is indeed merely a storage process and not a method of treatment, and if it is reasonable to store (especially hazardous) materials so as to be able to retrieve them in the future, it makes little sense to use the oceans as dumping sites.

Accordingly, little is said in this chapter on ocean disposal, and only the option of land disposal (as our only remaining alternative) is discussed below.

LANDFILLS

The first covered refuse dump seems to have been in Champaign, Illinois, in 1904. Their success, as well as that of other cities, such as Dayton, Ohio (1906), and Davenport, Iowa (1916), demonstrated that many environmental problems associated with open dumps (vermin, fires, odors, etc.) can be significantly reduced by the burial of refuse underground. The term "sanitary landfill" was first used in California, where

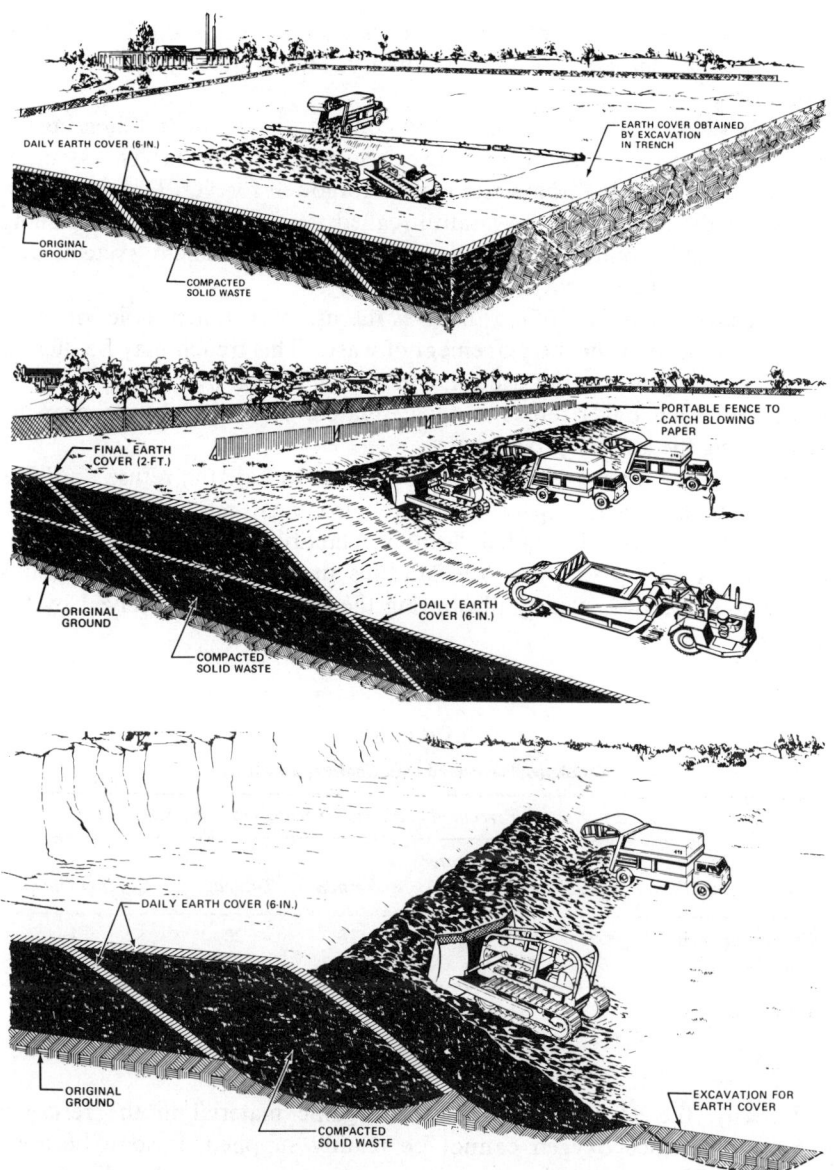

Figure 10-1. Typical sanitary landfill, showing three different methods of operation: (a) trench method, where the trucks deposit the loads into a trench, a bulldozer spreads, and compacts it, and covers it over at the end of the day; (b) area method, where a bulldozer spreads out the refuse and a scraper hauls the cover material; and (c) ramp method, where the refuse is compacted or as before but the cover is excavated from directly in front of the working face. (From Ref. 26)

waste ordinance and other military wastes were buried in the 1930s. After World War II, the concept of the sanitary landfill quickly spread across the country.

The basic objective of a landfill is to compact the waste and then bury it underground, with no subsequent fires and water pollution. The assumption is that in so doing, the waste can be forgotten forever. Unfortunately, this is not the case, since the landfill (called the "controlled tip" in many Commonwealth countries) is a biologically alive treatment system, and interfaces with the environment in many ways.

The operation of landfilling involves the use of a trench, hole, or other convenient location for the placement of waste. This trench may be dug, or an existing canyon may be used. Earth is then placed over the compacted refuse and each day's refuse is thus isolated. Figure 10-1 is a drawing representing the landfill process.

Although the objective of the earth cover is to hide the refuse and thus remove it from further impact with the environment, refuse disposal areas, while they may still be called "sanitary landfills," are seldom without problems. A survey of some 120 landfills in 1969 produced the results shown in Table 10-1, clearly illustrating the problems as perceived by the operators.

TABLE 10-1
Problems Associated with Sanitary Landfills

Type of Landfill	Percent of operators reporting problems with:				
	Fires	Groundwater Pollution	Vermin	Drainage	Gas and Odor
Canyon or ravine	55	7	18	20	
Cut and cover	59	9	17	37	12

Source: Ref. 1.

Initially, the decomposition of the organic material in the refuse is aerobic, but since oxygen cannot be readily supplied, it soon becomes anaerobic. The aerobic phase may be as long as a year in dry climates, to only a few weeks in wet areas. During the aerobic phase the temperatures within landfills may shoot up to as high as 60°C (140°F) as a result of the exothermic aerobic biological reactions taking place. As the oxygen level is depleted, however, endothermic anaerobic decomposition becomes predominant, and the temperatures drop to perhaps 40°C (104°F).

The Production and Movement of Landfill Leachate

The liquid produced during decomposition, as well as seepage of groundwater and surface water through the buried refuse, is called *leachate*. Various analyses of leachate have all shown it to be a liquid with extremely high pollutional capacity. A typical leachate composition is shown in Table 10-2.

TABLE 10-2
Composition of a Typical Leachate from a Sanitary Landfill

Component	Typical Value (mg/liter)	Range[a]
BOD_5	20,000	$0.01x$–$2x$
COD	30,000	$0.01x$–$3x$
Specific conductance	6,000	$0.5x$ – $1.5x$
Ammonia nitrogen	500	$0.01x$ – $1.5x$
Chloride	2,000	$0.05x$ – $1.5x$
Total iron	500	$0.5x$ – $5x$
Zinc	50	$0.5x$ – $5x$
Lead	2	$0.1x$ – $5x$
pH	6.0	$0.7x$ – $1.3x$

[a]For example, the range of BOD commonly reported is $0.01 \times 20{,}000 = 200$ mg/liter to $2 \times 20{,}000 = 40{,}000$ mg/liter.

Because the first biodegradation in a landfill is aerobic, the leachate should initially contain various aerobic organisms. Figure 10-2 is a plot showing the results of several investigations relating (aerobic) fecal coliform concentrations in leachate from landfills. As the character of the landfill changes from aerobic to anaerobic, these aerobic organisms would be expected to die out [2]. In one study, the total coliform count was found to be equal to fecal coliforms, strongly suggesting the presence of fecal matter in solid waste [3].

The leaching of pathogenic viruses seems to be less of a problem than pathogenic bacteria. Several investigators have found no evidence of viruses in leachate [1], and it has been estimated that polio virus, for example, can only survive 2 to 4 days in a properly constructed landfill at temperatures of 40 to 60°C (104 to 140°F) [4].

The type of leachate generated is strictly dependent on the waste involved. The available data on leachate production are all for unprocessed mixed refuse, with nothing removed. As various components are removed by recovery operations, the characteristics of leachate will

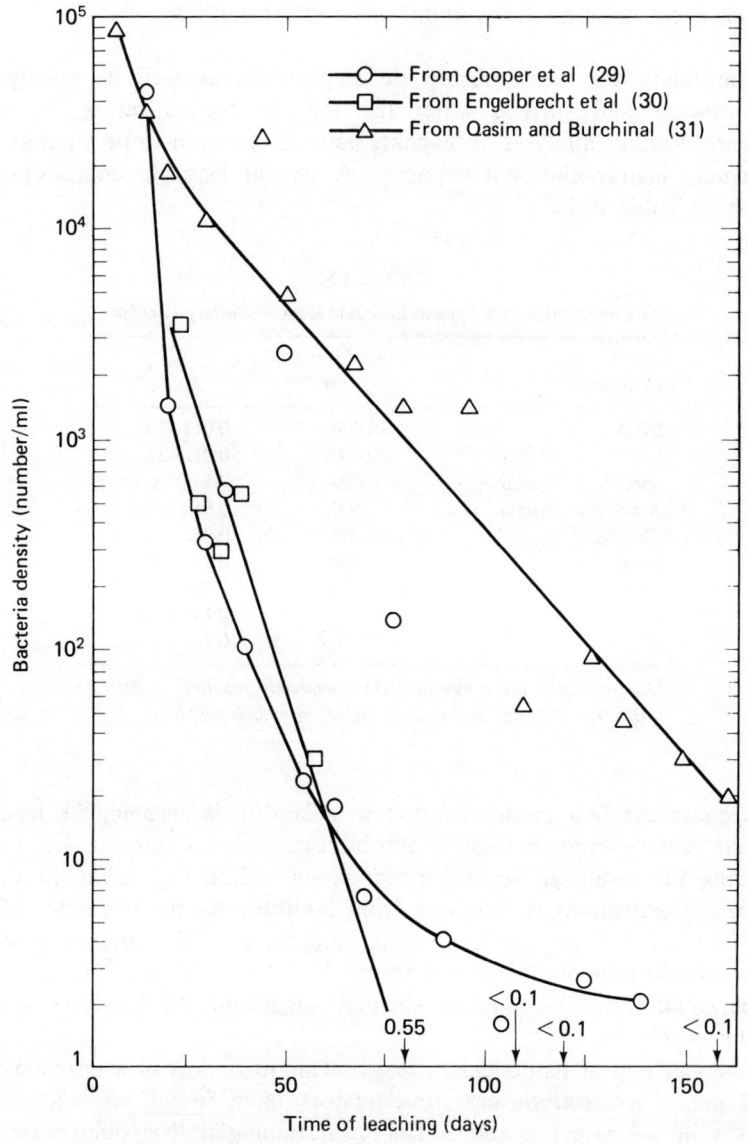

Figure 10-2. Total coliform bacteria in leachate. (From Ref. 2.)

change. For example, the leachate from incinerator residue will obviously be very different from common MSW leachate.

In one European study on incinerator residue disposal, the total amount of water extractable (dissolved) materials were found to be low for slag but extremely high for fly ash collected from electrostatic precipitators. The results are shown in Table 10-3. Sodium and potassium chloride made up about 60 to 80% of the extracted material. Considerable amounts of zinc (3 g/kg) and cadmium (0.8 g/kg) were found in the fly ash extract, indicating that fly ash would have quite a significant effect on the presence of heavy metals in leachate [5]. Recognition of this fact has prompted a number of states to require elaborate methods of incinerator residue disposal, with impervious linings, pumped leachate extraction, and monitoring wells.

TABLE 10-3
Extractable (Dissolved) Materials from Refuse Incinerator Residue

Plant	Slag (g/kg)^a	Slag and fly ash (g/kg)^a	Fly ash (g/kg)^a
A	3.7	—	103
B	—	11.4	123
C	—	7.0	115

^{a}g/kg = grams of extractable material per kilogram of dry solid sample.
Source: Ref. 5.

The effect of leachate on groundwater can range from immeasurable to severe. In dry climates, where little if any leachate is produced, there is no cause for concern, whereas in wetter climates, especially where limestone aquifiers form ready travel routes for the contaminated water, the effect can be serious. Typical results of groundwater analysis around a landfill are shown in Table 10-4. In some areas of the United States, landfills have

TABLE 10-4
Groundwater Quality in the Vicinity of a Landfill

Components	Normal groundwater	Groundwater 150 ft downstream from landfill
COD	20	71
Chloride	18	248
Sodium	30	316
Total hardness	570	820
Total dissolved solids	636	1506

Source: Ref. 4.

contaminated water supplies and made the well water unsafe for human consumption.

Prediction of the effect of leachate on groundwater is difficult, since this involves (1) the estimation of leachate character (which ranges widely, as noted in Table 10-2), (2) the quantity of leachate produced, and (3) the path of this liquid relative to groundwater movement.

The total quantity produced can be estimated by using a water balance technique. Which sets up a mass balance among precipitation, evapotranspiration, surface runoff, and soil moisture storage [6, 7, 8]. Some fraction of precipitation (dependent on runoff characteristics and soil type and conditions) will percolate into the soil and a fraction of this water is returned to the atmosphere through evapotranspiration. If the percolation exceeds evapotranspiration for a sufficiently long time, the field capacity of the soil is exceeded. The *field capacity* is the maximum moisture the soil (or buried solid waste) can retain without a continuous downward percolation due to gravity.

When a soil dries, its rate of water release is at first rapid and then levels off to a point close to its field capacity (Fig. 10-3). Without plant cover, and after a sufficiently long dry period, therefore, soil can be assumed to have a moisture content equal to the field capacity.

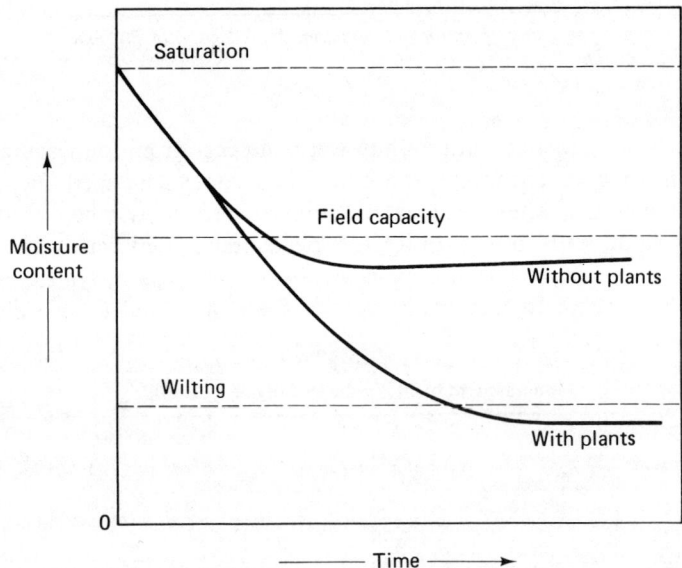

Figure 10-3. Moisture content of soil as it dries with and without plant cover.

If the soil is within the root zone, plants will extract water from the soil and release it by evapotranspiration, thus drying the soil to below field capacity. If the moisture content is reduced to the wilting points, roots can no longer obtain water from the soil and the plants die. The water consumed by various plants during one year is shown in Table 10-5.

TABLE 10-5
Approximate Annual Consumption of Water by Plants

Type of plant	Inches of Water	mm of water
Coniferous trees	4–9	100–360
Deciduous trees	7–10	180–250
Clover and alfalfa	2.5 +	60 +
Wheat	20–22	510–560
Meadow grass	22–60	560–1500
Lucern grass	26–65	660–1650

Source: Ref. 6.

When water is applied to a dry soil, it is not immediately distributed evenly throughout soil. Rather, each layer reaches field capacity before discharging to the layers below. A soil that does not attain field capacity discharges essentially no water to the deeper layers. If the field capacity of the soil covering refuse is exceeded, the water percolates through the soil and into the buried solid waste. If, in turn, the field capacity of the refuse is exceeded, leachate flows into the groundwater. The water balance method is simply a means of calculating if and by how much field capacities are exceeded, and thus is a way of calculating the production of leachate by landfills.

Some rough estimates must by used to develop the necessary calculations. Surface runoff coefficients, for example, can be estimated for different soils and slopes, as shown in Table 10-6. Precipitation data are available through the Weather Bureau. Evapotranspiration rates are also available, or can be calculated using the method of Thornthwaite [9], which takes into account the fact that evapotranspiration is reduced as the soil moisture drops. In other words, plants use less water in dry weather.

Other methods of estimating evapotranspiration are also available, and lysimeter techniques can be used to obtain accurate values when necessary [10]. In most cases, however, yearly average figures are sufficiently accurate for design. Fig 10-4 shows average yearly evapotranspiration in the United States.

The field capacity of various soils are listed in Table 10-7. The field capacity of compacted refuse has been estimated as 20 to 35% by volume,

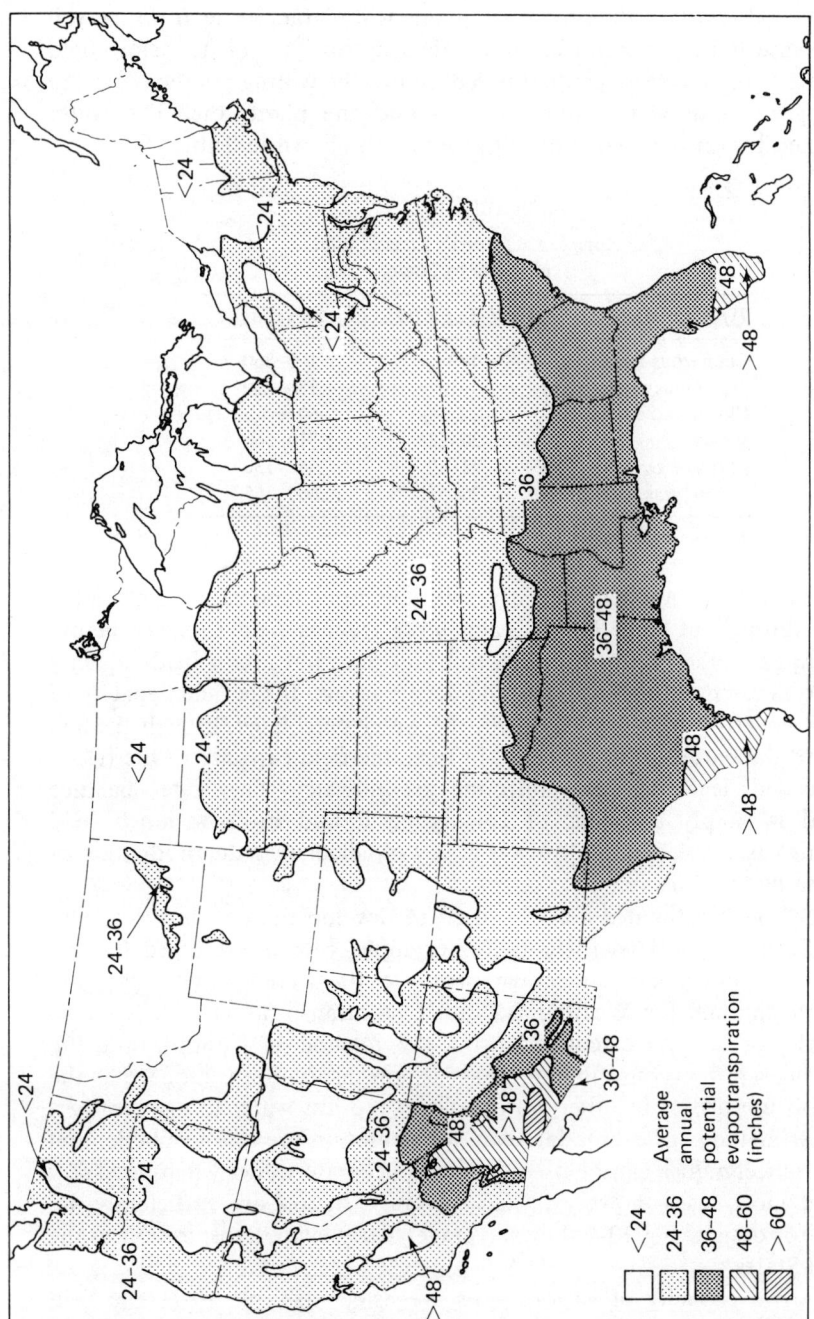

Figure 10-4. Average annual potential evapotranspiration in the continental United States. The potential evapotranspiration will occur if the soil is completely saturated; hence these figures are the highest probable values, in inches and the actual evapotranspiration is lower. (From Ref. 11.)

TABLE 10-6
Runoff Coefficients for Grass-Covered Soils

Surface	Runoff coefficient
Sandy soil, flat to 2% slope	0.05–0.10
Sandy soil, 2% to 7% slope	0.10–0.15
Sandy soil, over 7% slope	0.15–0.20
Heavy soil, flat to 2% slope	0.13–0.17
Heavy soil, 2% to 7% slope	0.18–0.22
Heavy soil, over 7% slope	0.25–0.35

Source: Ref. 6.

TABLE 10-7
Field Capacity of Various Soils

Soil	Field Capacity, as mm water/m of soil
Fine sand	120
Sandy loam	200
Silty loam	300
Clay loam	375
Clay	450

Source: Ref. 6.

or about 30%, which translates to 300 mm/m [6]. This value must be corrected for moisture already contained in the refuse, which is about 15%. A net field capacity of 150 mm/m of refuse (1.8 in./foot) is therefore not unreasonable [6].

The actual calculations of leachate production involve a one-dimensional analysis of water movement through soil and the compacted refuse, as shown in Fig. 10-5. The figure also defines the symbols used in the calculation, which are based on the following mass balance equation:

$$C = P(1 - R) - S - E$$

where C = total percolation into the top soil layer, mm
P = precipitation, mm
R = runoff coefficient
E = evapotranspiration, mm
S = storage within the soil or refuse, mm

The net percolation calculated for three areas of United States is shown in Table 10-8. Note that the net percolation for Los Angeles landfill is zero. This explains the absence of leachate in southern California landfills,

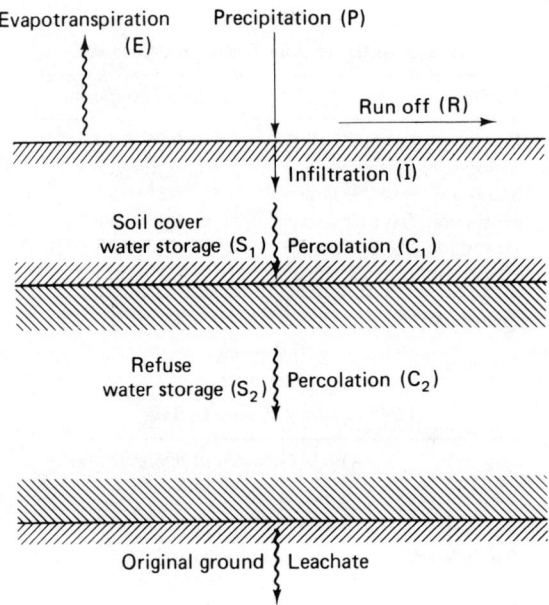

Figure 10-5. Percolation of water and leachate in a landfill.

TABLE 10-8
Percolation in Three Landfills

	Precipitation, P (mm)	Runoff, Coefficient R	Evapotranspiration, E (mm)	Percolation, C (mm)
Cincinnati, Ohio	1025	0.15	658	213
Orlando, Fla.	1342	0.07	1172	70
Los Angeles, Calif.	378	0.12	334	0

Source: Ref. 6.

a fact that contributed toward the underestimation of the impact of landfill on groundwater in other parts of the country.

Using figures in Table 10-8, it is possible to estimate the number of years before leachate is produced, since the refuse will sop up the percolated water until its capacity is reached. In Los Angeles, there will obviously never be any leachate. In Orlando, for a landfill 7.5 m deep, the first leachate would appear in 15 years. In Cincinnati, with a landfill depth of 20 m, the first leachate should appear in 11 years.

Example 10-1

Estimate the percolation of water through a landfill 10 m deep, with a 1 m cover of sandy loam soil. Assume that this landfill is in southern Ohio, and that

Precipitation, P = 1025 mm/yr
Runoff coefficient, R = 0.15
Evapotranspiration, E = 660 mm
Soil field capacity F_s = 200 mm/m
Refuse field capacity F_r = 300 mm/m, as packed

Assume further that the soil moisture is at field capacity when applied, and that the refuse field capacity is reduced by existing moisture to a net of 150 mm/m.

Percolation through the soil cover is

$$C = P(1 - R) - S - E$$
$$= 1025(1 - 0.15) - 0 - 660$$
$$= 211 \text{ mm/yr}$$

The moisture front will move

$$\frac{211 \text{ mm/yr}}{150 \text{ mm/m}} = 1.40 \text{ m/yr}$$

or it will take

$$\frac{10 \text{ m}}{1.40 \text{ m/yr}} = 7.1 \text{ years}$$

to produce a leachate that will flow into the groundwater at the rate of (211 mm × area of landfill) per year.

Many assumptions have gone into these calculations, and the accuracy of the predictions has not been tested, so considerable caution is warranted when using these figures. At the very least, this method is useful in combating arguments often heard in some health agencies that "proper landfills don't produce leachate."

Once the quantity of leachate generated has been estimated, it is useful to evaluate its movement within the soil. This is obviously an extremely difficult problem, and some quite sophisticated methods have been proposed [12, 13]. A simpler and seemingly adequate technique, originally developed for describing the movement of hazardous waste chemicals, has been reported by Elzy and Lindstrom [14]. This analysis can be modified so that it applies equally well for some components in the leachate such as chloride or COD.

Based on the Elzy–Lindstrom model, the development below allows for the consideration of the following factors: horizontal and vertical (two-dimensional) movement of the leachate, adsorption of the media, biodegradation of the leachate, variable water table, and characteristics of the

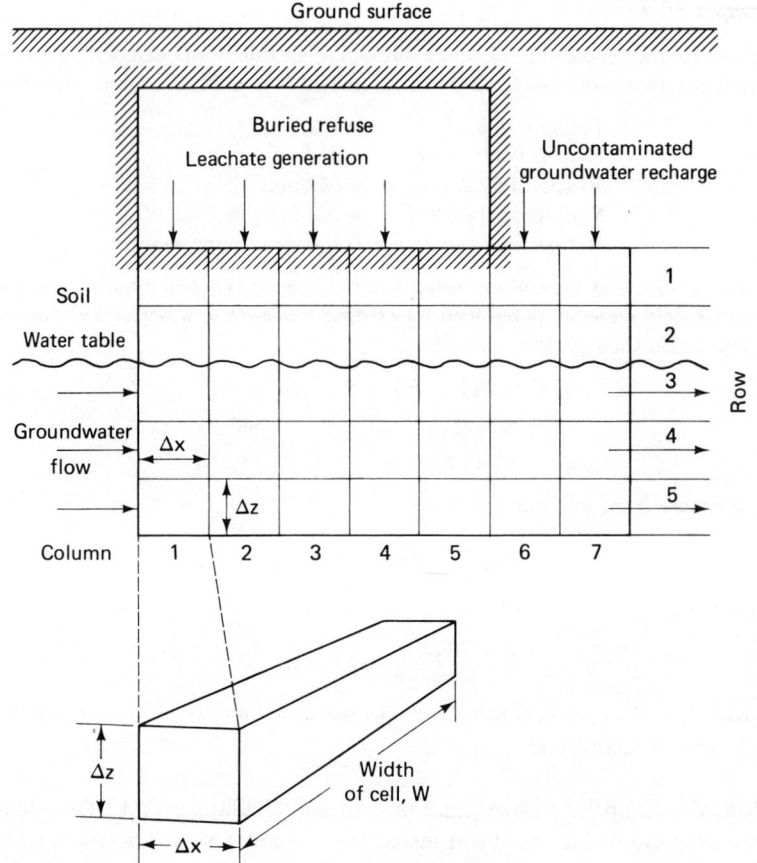

Figure 10-6. Definition of terms for the leachate movement model.

soil, such as its permeability, porosity, and hydraulic gradient. The model assumes a series of stirred tank reactors, and the soil is assumed homogeneous. Figure 10-6 is a sketch of the two-dimensional model.

The soil and buried refuse is first divided into compartments of depth ΔZ and a length ΔX, and a width (W) equal to the width of the buried refuse. The movement of water through the soil is at a uniform velocity. The amount of leachate and the chemical of interest, as well as amount of uncontaminated water entering each column, is calculated by the water balance method discussed above.

The chemical mass balances must be performed for layers above as well as below the water table. For the layers above the water table, if the field capacity of the soil or refuse above it is exceeded, or if the contaminant is percolated from the soil surface, each time period Δt results in the addition of a volume of liquid, $Q_{in}(\Delta t)$, to the liquid already in the block, V_{old}, so that

$$V = V_{old} + [Q_{in}(\Delta t)] = V_{old} + V_{in}$$

with the V terms measured in liters and Q is in liters per day. The total grams of contaminant in the soil block at the end of time Δt is

$$A_{total} = A_{old} + \Delta t Q_{in} C_{in} \times 10^{-3}$$

where A_{old} is the contaminant in the block at time zero and C_{in} is the concentration of the contaminant in the inflowing water. During the specified time period, of the total contaminant in the block, some is adsorbed, some is biograded, and some remains in free solution, such that

$$A_{total} = A_{ad} + A_{bio} + A_{free}$$

where A_{free} is the contaminant in free solution at the end of time Δt. The concentration of the chemical is

$$C = \frac{A_{free}}{V} \times 1000$$

where C is the concentration in mg/liter. The chemical adsorbed within each grid block is calculated as

$$A_{ad} = KCS \times 10^{-3}$$

where A_{ad} = chemical adsorbed, g
K = adsorption constant, liter/g solid
C = concentration of the chemical, mg chemical/liter
S = solid adsorber (soil), g

The biodegradation (if it occurs for the chemical in question) can be assumed to be a first-order reaction:

$$A_{bio} = kCV\Delta t \times 10^{-3}$$

where A_{bio} = chemical biodegraded, g
k = rate constant, h^{-1}
V = volume of solution, liters
Δt = time period, h

These equations can be combined to yield

$$A_{free} = \frac{A_{total}}{1 + \dfrac{KS}{V} + k\Delta t}$$

The chemical in free solution, A_{free}, its concentration, C, and the grams of chemical degraded, A_{bio}, can now be calculated.

The volume excess liquid that must percolate to lower layers is the difference between the actual volume in the block and its field capacity:

$$\Delta t Q_{out} = V - F$$

where F is the field capacity, in liters. On the other hand, if $F > V$, $Q_{out} = 0$.

The loss of a contaminant to the layer below is $Q_{out}\Delta t C \times 10^{-3}$ in grams, which becomes $Q_{in} \Delta t C_{in} \times 10^{-3}$ for the layer below, and

$$A_{total} = A_{old}\Delta t Q_{in} C_{in} \times 10^{-3}$$

as before. Each block in a layer collects leachate from above, mixes with the existing leachate, adsorbs some, biodegrades some, and if field capacity is exceeded, drains to layer below. The process thus proceeds downward from layer to layer. It can be assumed that the layer immediately below the water table receives all of the contaminated water from the landfill.

Below the water table, however, the situation is changed, since the groundwater adds to the flow into each block (column). Obviously, the layer below the groundwater table is saturated, or $V = V_{sat}$. The total water that flows horizontally from each block is $\Delta t Q_h$, and this has a concentration of the contaminant C_{up}. Hence the total amount of pollution transferred downstream (horizontally) from one block to another is $\Delta t Q_h C_{up} \times 10^{-3}$. Also,

$$Q_h = v\Delta Z W P$$

where v = velocity of groundwater, m/day
ΔZ = height of block, m (see Fig. 10-6)
W = width of block, m
P = porosity of soil, or the saturated volume fraction of a porous medium

The blocks immediately below the water table receive water and the contaminant from above, and from the upstream block. Unless the contaminant is naturally occurring, the block in column 1 probably does not receive any contaminant horizontally, and the first column after the landfill (column 6 in Fig. 10-6) probably receives no vertical input of contaminant. All blocks in columns 2 to 4 receive contaminant both from above and from upstream, as shown in Fig. 10-7.

The total amount of the contaminant in such a block after time Δt is thus

$$A_{total} = \Delta t Q_h C_{up} \times 10^{-3} + \Delta t Q_{in} C_{in} \times 10^{-3}$$

(note that A_{old} is not in this equation, because all of the old contaminant is washed out, i.e., there is no storage capacity.) The chemical adsorbs and

Landfills

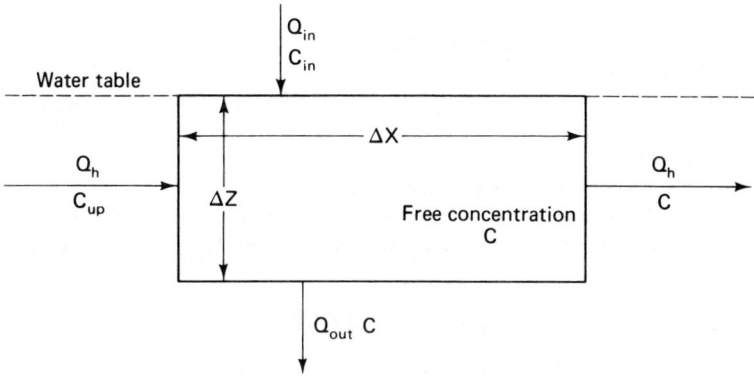

Figure 10-7. Definition sketch for the mass balance on a block immediately below the water table.

biodegrades in this block as well, or

$$A_{free} = A_{total} - A_{bio} - A_{ads}$$

The A_{free} is calculated as before, and the free concentration of the contaminant is

$$C = \frac{A_{free}}{V_{sat}} \times 10^3$$

This contaminant now seeps down into the next lower block, and is carried horizontally by the groundwater into the next column.

Obviously, this method is greatly facilitated by the use of digital computers. The final result of the repetitive calculations is a plot of concentrations over time for any distance downstream from the landfill. Because of conservative assumptions such as no lateral movement, these graphs will yield the worst-case situation.

Example 10-2

Using the sketch in Fig 10-6 and the leachate production found in Example 10-1, calculate the concentration of chloride ion in the second column, row 3 (under the water table). Assume the following:
1. The soil layers below the refuse but above the water table have moisture contents equal to the field capacity.
2. Each block is 1 m deep (ΔZ) and 5 m long (ΔX).
3. The landfill is 100 m wide (W) perpendicular to the direction of groundwater flow.
4. The concentration of chloride ion in the leachate reaching the soil below the refuse is 2000 mg/liter.

5. The concentration of chloride ion in the groundwater is 15 mg/liter.
6. Groundwater velocity is 0.5 m/day.
7. Soil porosity is 0.4.

We can select the time interval Δt as any convenient time. Although $\Delta t = 2$ days is reasonable for computer-aided solutions, we will let $\Delta t = 10$ days, to facilitate calculations.

The horizontal plane area of each grid block is $5 \text{ m} \times 100 \text{ m} = 500 \text{ m}^2$. The flow through the refuse is 211 mm/yr (see Example 10-1) or 0.578 mm/day. The total flow into the grid block is thus 0.578 mm/day $\times 10^{-3}$ m/mm $\times$ 500 m^2 = 0.289 m^3/day, or

$$Q_{in} = 0.289 \times 10^3 \text{ liters/day}$$

Since the soil is at field capacity, it can thus be assumed that this flow will enter the groundwater (row 3).

The concentration of contaminant in row 1 is calculated as

$$A_{total} = A_{old} + \Delta t C_{in} Q_{in} \times 10^{-3}$$

$$A_{old} = 15 \text{ mg/liter} \times 500 \text{ m}^3 \times 10^3 \text{ liters/m}^3 \times 10^{-3} \text{ g/mg} \times \frac{200}{1000} = 1500 \text{ g}$$

The last term is the soil field capacity, 200 m/m.

$$A_{total} = 1500 \text{ g} + 10 \times 2000 \times 0.289 \times 10^3 \times 10^{-3} = 6280 \text{ g}$$

Since the chlorine ion does not degrade biologically, and if we can assume that it is not adsorbed greatly on soil, $A_{bio} = A_{ad} = 0$, and

$$A_{total} = A_{free} = 6280 \text{ g}$$

The volume of liquid (leachate plus the original groundwater) is

$$V = V_{old} + \Delta t Q_{in}$$

$$= 200 \text{ mm/m} \times 1 \text{ m} \times 500 \text{ m}^2 \times 10^{-3} \text{ m/mm} + 10(0.289 \text{ m}^3/\text{day})$$

$$= 102.89 \text{ m}^3$$

$$= 102.89 \times 10^3 \text{ liters}$$

$$C = \frac{A_{free}}{V} \times 1000$$

$$= \frac{6280 \times 10^3}{102.89 \times 10^3} = 61 \text{ mg/liter}$$

This now becomes C_{in} for the second row, and for column 1, row 2, at $t = 10$ days after the leachate moisture front reached the soil under the refuse,

$$A_{total} = 1500 + 10 \times 61 \times 0.289 \times 10^3 \times 10^{-3} = 1676 \text{ g}$$

$$A_{free} = 1676 \text{ g}$$

$$C = \frac{1676 \times 10^3}{102.89 \times 10^3} = 16.3 \text{ mg/liter}$$

This now becomes C_{in} for the block in the first column, row 3. It is also the C_{in} for the second column, row 3, since no lateral dispersion occurred. Considering first column 1,

$$Q_h = V\Delta ZWP$$
$$= 0.5 \text{ m/day} \times 1 \text{ m} \times 100 \text{ m} \times 0.4$$
$$= 20 \text{ m}^3/\text{day} = 20 \times 10^3 \text{ liters/day}$$
$$A_{total} = \Delta t Q_h C_{up} \times 10^{-3} + t Q_{in} C_{in} \times 10^{-3}$$
$$= (10 \times 20 \times 10^3 \times 15 \times 10^{-3}) + (10 \times 0.289 \times 10^3 \times 16.3 \times 10^{-3})$$
$$= 3000 + 47 = 3047 \text{ g}$$

Since this is below the water table,

$$V = V_{sat} = 0.4 \times 500 \text{ m}^3 = 200 \text{ m}^3$$
$$C = \frac{3047 \times 10^3}{200 \times 10^3} = 15.2 \text{ mg/liter}$$

which now becomes C_{up} for column 2. As before, C_{in} is 16.3. The total contaminant in the block in row 3, column 2, is

$$A_{total} = 10 \times 20 \times 10^3 \times 15.2 \times 10^{-3}$$
$$+ 10 \times 0.289 \times 10^3 \times 16.3 \times 10^{-3}$$
$$= 3040 + 47 = 3087 \text{ g}$$
$$C = \frac{3087 \times 10^3}{200 \times 10^3} = 15.4 \text{ mg/liter}$$

In summary, 10 days after the moisture front first reached the soil below the refuse, the concentration of chloride ion immediately below the water table and 5 m from the upstream side of the landfill is estimated as 15.2 mg/liter.

Note that it is possible to continue this process for any length of time and for any part of the soil below the landfill. Concentration vs. time curves can thus be constructed. When a one-time hazardous spill is considered, it is possible to estimate the location and magnitude of the peak concentration.

As the landfill ages, the amount of pollution materials existing as leachate decreases. No long-term studies are presently available to evaluate this reduction with time. Laboratory experiments [15] on bark suggest that the decay in pollutional strength might be expressed as

$$C = C_0 e^{-kt}$$

where $C =$ concentration of pollutant in the leachate (or the strength of the leachate, e.g., COD) at time t
$C_0 =$ initial concentration
$k =$ "washout constant"

It was found that k decreases with increased contact time of water with waste.

An innovative method of landfilling has been proposed wherein the leachate is captured and recirculated. This technique is based on the early practical experiences where it was noted that the addition of water to landfill not only made it compact easier but also accelerated decomposition. It is likely that experiments will prove the value of this scheme, and thus accelerate the decomposition time from 25 years in a normal case to only a few years [15, 16].

The final alternative in leachate control is to capture and treat the leachate before discharging it to the environment. As discussed above, landfill leachate is a highly concentrated waste, with potentially high levels of toxins such as heavy metals. Accordingly, special care must be exercised in the application of existing treatment processes to leachate treatment.

In one thorough comparative study of leachate treatment [23] it was found that an anaerobic filter could remove up to 98% of the leachate COD* at organic loadings of 0.7 kg COD/m^3/day (6200 lb COD/acre/day) provided that the temperature remained higher than 22°C. Below this temperature, removel efficiencies dropped rapidly, so that at 11°C no COD removal occurred.

As shown in earlier studies [24], aerobic systems are also effective in treating leachate, provided that the organic loadings are less than 0.4 kg COD/kg MLVSS/day.[†] At organic loadings of less than 0.3 kg COD/kg MLVSS/day, a significant degree of nitrification occurs. Nitrification is the progressive oxidation of organic and ammonia nitrogen compounds, which are most prevalent in leachate, to the fully oxidized nitrate (NO_3^-), which no longer undergoes biodegradation and thus does not exert a demand for oxygen.

Because of the nature of leachate, some of the components may be toxic when subjected to biological treatment. Copper, for example, has a drastic inhibitory effect at 25 mg/liter, and zinc has a slightly adverse effect at concentrations exceeding 50 mg/liter. These concentrations are, however, an order of magnitude greater than what might be normally encountered in leachate from municipal landfills, and thus do not seem to pose a serious problem.

*Chemical oxygen demand, a measure of the strength of a material to be chemically oxidized.

†Mixed liquor volatile suspended solids, a crude measure of the total available biomass for biodegradation.

Other forms of treatment, including aerated lagoons, activated carbon, trickling filters, biodisk, and chemical precipitation all have been found successful to varying degrees.

Although the capture and treatment of leachate is usually an expensive undertaking, it is sometimes the only practical alternative. Some manufacturers of wastewater treatment equipment have, in fact, developed package systems for leachate treatment.

Production and Movement of Landfill Gases

The production and movement of gases in a landfill is a second serious environmental problem. The production of gases in landfills is discussed in Chapter 6 and is not repeated here. Only the movement of these gases through compacted refuse and soil is covered in this chapter.

The movement of the gases produced in landfills can cause serious problems outside as well as inside the immediate landfill area. As noted in Chapter 6, the composition of the gas is such that almost all of it is either methane or carbon dioxide. Methane can explode, of course, and several fatal accidents have occurred as the result of methane seepage from landfills. Methane is explosive if it is at concentrations of between 5 and 15% in air. No explosions are possible underground, under anaerobic conditions, but the gas can become deadly when vented into enclosed spaces such as basements. Carbon dioxide can cause a significant change in soil acidity if it seeps up through the ground, and the anaerobic conditions produced by the two gases have caused severe plant damage around several landfills [27].

Ventings of landfill gases is the most used method of eliminating the gas problem, although the capture and use of the gas as a fuel source is gaining acceptance, as discussed in Chapter 6.

The movement of gas underground is difficult to predict simply because of the heterogeneous nature of soil. Some analyses, based on gross assumptions, have been advanced. In most cases, however, the design engineer must resort to rules of thumb such as

1. A saturated clay soil is an excellent barrier to gas seepage.
2. Gas will seep laterally through dry and/or well-drained soils until it can vent to the atmosphere.
3. The wetter the soil, the better is the gas barrier.

Soil types are often described as Fig. 10-8, a system developed by the U.S. Department of Agriculture. An alternative means of classifying soils is the Unified Soil Classification System, and this method includes enough detail on soil properties to make it applicable to landfill design procedures, as shown in Table 10-9.

TABLE 10-9
Unified Soil Classification System and Characteristics Pertinent to Sanitary Landfills

Major Divisions		Symbol			Name	Potential Frost Action	Drainage Characteristics*
		Letter	Hatching	Color			
Coarse-Grained Soils	Gravel and Gravelly Soils	GW		Red	Well-graded gravels or gravel-sand mixtures, little or no fines	None to very slight	Excellent
		GP			Poorly graded gravels or gravel-sand mixtures, little or no fines	None to very slight	Excellent
		GM		Yellow	Silty gravels, gravel-sand-silt mixtures	Slight to medium	Fair to poor / Poor to practically impervious
		GC			Clayey gravels, gravel-sand-clay mixtures	Slight to medium	Poor to practically impervious
	Sand and Sandy Soils	SW		Red	Well-graded sands or gravelly sands little or no fines	None to very slight	Excellent
		SP			Poorly graded sands or gravelly sands, little or no fines	None to very slight	Excellent
		SM		Yellow	Silty sands, sand-silt mixtures	Slight to high	Fair to poor / Poor to practically impervious
		SC			Clayey sands, sand-clay mixtures	Slight to high	Poor to Practically impervious
Fine-grained Soils	Silts and Clays LL is Less Than 50	ML		Green	Inorganic silts and very fine sands rock flour, silty or clayey fine sands or clayey silts with slight plasticity	Medium to very high	Fair to poor
		CL			Inorganic clays of low to medium plasticity, gravelly clays, sandy clays, silty clays, lean clays	Medium to high	Practically impervious
		OL			Organic silts and organic silt-clays of low plasticity	Medium to high	Poor
	Silts and Clays LL is Greater Than 50	MH		Blue	Inorganic silts, micaceous or diatomaceous fine sandy or silty soils, elastic silts	Medium to very high	Fair to poor
		CH			Inorganic clays of high plasticity, fat clays	Medium	Practically impervious
		OH			Organic clays of medium to high plasticity, organic silts	Medium	Practically impervious
Highly Organic Soils		Pt		Orange	Peat and other highly organic soils	Not Recommended for Sanitary Landfill Construction	

*Values are for guidance only; design should be based on test results.
Source: Ref. 26.

TABLE 10-9
(Extended)

Value for Embankments	Permeability cm per s	Compaction Characteristics[†]	Std AASHO Max Unit Dry Weight lb per cu ft[‡]	Requirements for Seepage Control
Very stable, pervious shells of dikes and dams	$k > 10^{-2}$	Good, tractor, rubber-tired steel-wheeled roller	125–135	Positive cutoff
Reasonably stable, pervious shells of dikes and dams	$k > 10^{-2}$	Good, -tractor, rubber-tired steel-wheeled roller	115–125	Positive cutoff
Reasonably stable, not particularly suited to shells, but may be used for impervious cores or blankets	$k = 10^{-3}$ to 10^{-6}	Good, with close control, rubber-tired, sheepsfoot roller	120–135	Toe trench to none
Fairly stable, may be used for impervious core	$k = 10^{-6}$ to 10^{-8}	Fair, rubber-tired, sheepsfoot roller	115–130	None
Very stable, pervious sections slope protect required	$k > 10^{-3}$	Good, tractor	110–130	Upstream blanket and toe drainage or wells
Reasonably stable. may be used in dike section with flat slopes	$k > 10^{-3}$	Good, tractor	100–120	Upstream blanket and toe drainage or wells
Fairly stable, not particularly suited to shells, but may be used for impervious cores or dikes	$k = 10^{-3}$ to 10^{-6}	Good, with close control, rubber-tired, sheepsfoot roller	110–125	Upstream blanket and toe drainage or wells
Fairly stable, use for impervious core for flood control structures	$k = 10^{-6}$ to 10^{-8}	Fair, sheepsfoot roller, rubber-tired	105–125	None
Poor stability, may be used for embankments with proper control	$k = 10^{-3}$ to 10^{-6}	Good to poor, close control essential, rubber-tired roller, sheepsfoot roller	95–120	Toe trench to none
Stable, impervious cores and blankets	$k = 10^{-6}$ to 10^{-8}	Fair to good, sheepsfoot roller, rubber-tired	95–120	None
Not suitable for embankments	$k = 10^{-4}$ to 10^{-6}	Fair to poor, sheepsfoot roller	80–100	None
Poor stability, core of hydraulic dam, not desirable in rolled fill construction	$k = 10^{-4}$ to 10^{-6}	Poor to very poor, sheepsfoot roller	70–95	None
Fair stability with flat slopes. thin cores, blankets and dike sections	$k = 10^{-6}$ to 10^{-8}	Fair to poor, sheepsfoot roller	75–105	None
Not suitable for embankments	$k = 10^{-6}$ to 10^{-8}	Poor to very poor, sheepsfoot roller	65–100	None

Not Recommended for Sanitary Landfill Construction

[†]*The equipment listed will usually produce the desired densities after a reasonable number of passes when moisture conditions and thickness of lift are properly controlled.*
[‡]*Compacted soil at optimum moisture content for Standard AASHO (Standard Proctor) compactive effort.*

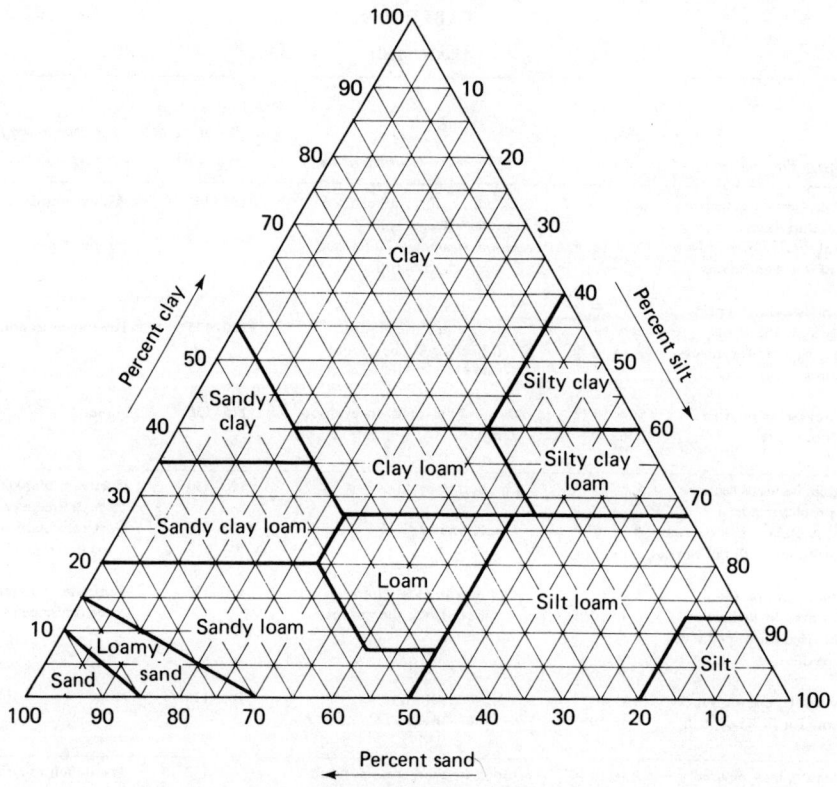

Figure 10-8. Soil classification according to the U.S. Department of Agriculture.

Table 10-10 is a listing of the soil types that might be used for cover material. Ideally, a single soil should be capable of performing all of the required functions, such as growing vegetation, preventing percolation into buried refuse, and so on. Obviously, such a soil does not exist, and the designer and planner must establish the best compromise for any particular region.

Design of Landfills

The design of landfills involves both technical and public concerns. Very simply, few people want to have landfills close to their homes, and those who might be affected by a proposed landfill will usually put up a

TABLE 10-10
Suitability of General Soil Types as Cover Material[a]

Function	Clean gravel	Clayey-silty gravel	Clean sand	Clayey-silty sand	Silt	Clay
Workability, placement of cover	G	E	E	G	F	P
Prevent rodents from burrowing or tunneling	G	F-G	G	P	P	P
Keep flies from emerging	P	F	P	G	G	E[b]
Minimize moisture entering fill	P	F-G	P	G-E	G-E	E[b]
Minimize landfill gas venting through cover	P	F-G	P	G-E	G-E	E[b]
Provide pleasing appearance and control blowing paper	E	E	E	E	E	E
Grow vegetation	P	G	P-F	E	G-E	F-G
Be permeable for venting decomposition gas[c]	E	P	G	P	P	P

[a] E, excellent; G, good; F, fair; P, poor.
[b] Except when cracks extend through the entire cover.
[c] Only if well drained.
Source: Adapted from Ref. 26.

struggle. The objective of landfill design is to locate a landfill so as to create minimum adverse environmental impact and create the minimum public outcry, and do all this for the lowest cost possible.

Landfill design is in several stages. The first phase involves the identification of all potential landfill sites. This is often facilitated by drawing transparent overlay maps. On top of a base map of the region, overlays are used to identify areas which are inappropriate for landfills (e.g., urban areas, floodplains, etc.). Using this information, the second phase involves the siting of a number of alternative locations. Once these potential sites are selected, the feasibility of establishing a landfill at these locations must be evaluated. The considerations going into this initial selection process are, among others:

Soil type, including the presence of rock.
Accessibility.
Sensitivity of surrounding area to noise.
Groundwater table and use.
Available utilities.
Future use.
Distance from town.
Cost of land.
Expected landfill life.

The adequacy of soil for landfilling is summarized in Table 10-11. Most soils can be used, with the exception of peat and rock. From the landfill operator's viewpoint, clay is the least desirable soil because of its poor workability.

Test wells should be sunk to determine groundwater level and to serve later as sampling points for monitoring leachate movement. Under no circumstances should the buried refuse be below the groundwater table. Most state standards require at least 1.5 m (5 ft) of earth between the bottom of the refuse and the groundwater table.

The future user of the landfill must consider the requirement of a minimum slope (about 2%) and the active decomposition and settlement that customarily occurs during the first years.

Volume calculations can be approximated by assuming a reasonable depth for the compacted refuse, a density of perhaps 12,000 kg/m^3 (800 lb/ft^3) and that the cover material will make up one-fifth of the volume.

When some materials are recovered from solid waste, its compaction characteristics may change markedly. It is then necessary to estimate the compaction of the waste by individual refuse components.

Bulk densities of the components can be used in such calculations, even though the actual compaction may be quite different. Table 1-9 lists some bulk densities that can be used for this purpose.

Example 10-3

For illustrative purposes only, assume that a refuse has the following components and bulk densities (from Table 1-9).

Component	Fraction by weight	Bulk density, lb/ft^3
Miscellaneous paper	50	3.81
Garden waste	25	4.45
Glass	25	18.45

Assume that the compaction in the landfill is 800 lb/ft^3. Estimate the reduction in the required landfill volume if the miscellaneous paper is removed.

Overall bulk density is $(0.50 \times 3.81) + (0.25 \times 4.45) + (0.25 \times 18.45) = 7.63$ lb/ft^3. Without the paper, the overall density is $(0.50 \times 4.45) + (0.50 \times 18.45) = 11.45$ lb/ft^3. The fraction of volume taken up by the paper is

$$\frac{\frac{50}{3.81}}{\frac{50}{3.81} + \frac{25}{4.45} + \frac{25}{18.45}} = 0.65$$

If this is removed, the new overall density is

$$800 \text{ lb/ft}^3 \times \frac{11.45}{7.63} = 1200 \text{ lb/ft}^3$$

and the reduction in volume achieved is 65%.

If preliminary studies indicate that a specific site is acceptable, and the municipality decides to use it for a landfill, the third phase is a full engineering study, which results in the development of an operational plan for the landfill. The design includes topographical maps showing the progression of filling, source of cover material, and final grading and planting plan. In certain situations, a gas and/or leachate monitoring program is also included.

HAZARDOUS SUBSTANCES

Increased technological complexity and population densities have resulted in the identification of a new type of pollutant, commonly called a *hazardous substance*.

The reported incidence of damage to the environment and to people by these materials has increased markedly in the last few years. The EPA maintains a list of such incidents, and some of the better documented ones have been published [22].

The term "hazardous substance" or "hazardous waste" is difficult to define, and yet a clear definition is necessary if different disposal standards are to be applied to such materials.

A legal definition [28], suggested by the EPA, is

> A "hazardous waste" is any waste or combination of wastes of a solid, liquid, contained gaseous, or semisolid form which because of its quantity, concentration, or physical, chemical, or infectious characteristics, may (1) cause or significantly contribute to an increase in mortality or an increase in serious irreversible or incapacitating reversible illness; or (2) pose a substantial present or potential hazard to human health or the environment when improperly treated, stored, transported or disposed of, or otherwise managed.

When deciding whether or not a specific substance is hazardous, it is useful to use a set of criteria against which the properties of the material in question can be judged. Most such rely on acute toxicity as the primary screening criterion, but there are many other factors to be considered, such as the possibility of genetic effects, carcinogenic properties, radioactivity, and so on. Further, the substance may not be toxic to human beings but be devastating to other life forms. It has thus been understandably difficult to develop one single set of toxicity criteria which are applicable in all cases [18]. Yet such a classification is necessary, since some types of wastes clearly require special handling and disposal.

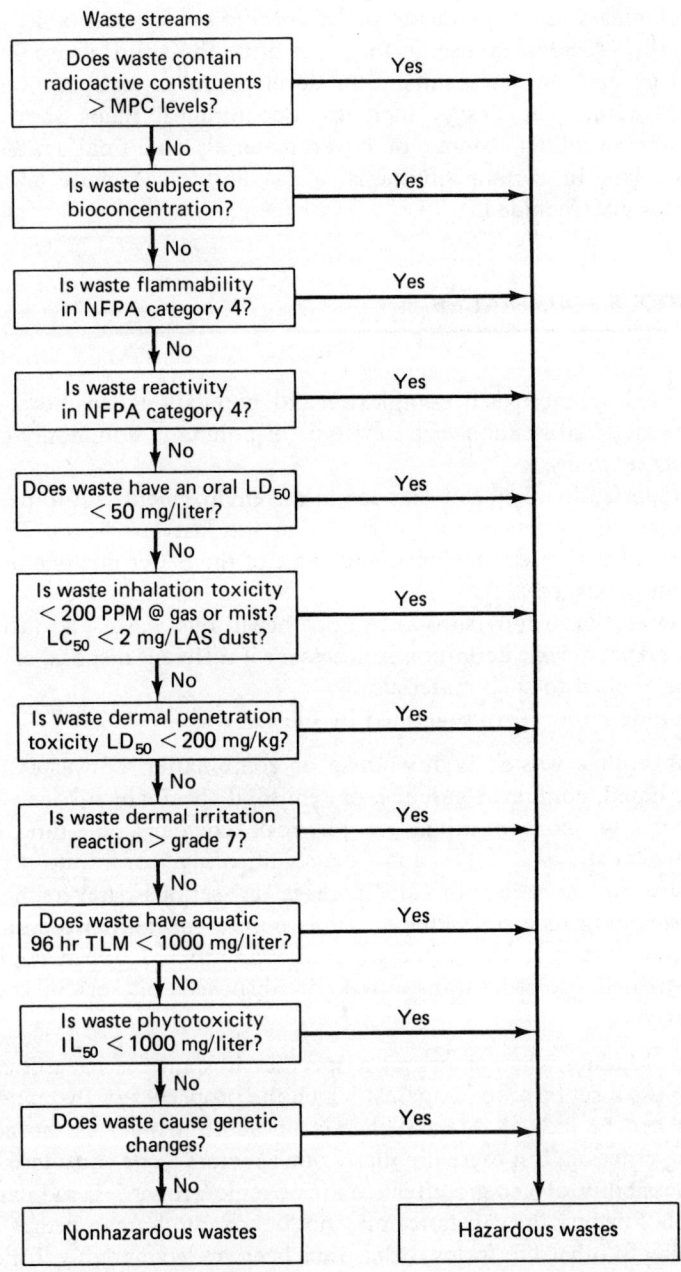

Figure 10-9. A hazardous waste decision model. (From Ref. 19.)

One example of the type of criteria developed to evaluate the hazardous nature of chemicals is the screening procedure illustrated in Fig. 10-9 [19]. The first criterion is for radioactivity, the stipulation being that the levels of radioactivity not exceed maximum permissible concentration (MPC) levels as set by the Nuclear Regulatory Commission. Secondly, the waste is classified as hazardous if it is capable of being bioconcentrated (e.g., chlorinated hydrocarbon pesticides). Third is the flammability and reactivity criteria, both based on National Fire Protection Association (NFPA) standards. Next is the oral toxic criterion, based on LD_{50}, which is the dose at which 50% of the test species (e.g., rats) die when exposed to the chemical through a route other than respiration. Inhalation and dermal toxicity is next, where LC_{50} is the lethal concentration resulting in 50% mortality during an exposure time of 4 h. Dermal irritation is measured on a Federal Drug Administration scale of 1 to 10. A grade 8 irritant causes necrosis of the skin when a 1% solution is applied. Aquatic toxicity is measured by the 96-h mean toxic limit (TLM) of less than 1000 mg/liter, although the present EPA criterion for aquatic toxicity is set at 500 ppm [16, 20]. Phytotoxicity is the ability to cause poisonous or toxic reactions in plants based on the mean inhibitory limit (ILM) of 1000 ppm or less. Since the publication of these criteria, the EPA has lowered its definition of phytotoxic poisons to 100 ppm [20]. Finally, the genetic, carcinogenic, mutagenic, and teratogenic potentials of hazardous materials are measured by tests developed by the National Cancer Institute.

Note that with this system, the actual quantity of the material is not specified. Clearly, this is a weakness when disposal schemes are to be evaluated.

The most environmentally sound disposal scheme for hazardous materials (at any quantity) is destruction and conversion to nonhazardous substances. In many cases, however, this is either extremely expensive (e.g., dilute heavy metal and pesticide wastes) or technically impossible (e.g., some radioactive waste materials). Alternative disposal schemes are thus necessary.

By far the most widely used method of disposing waste hazardous substances is the sanitary landfill. Aboveground containerized storage is also used, but this is not really a disposal scheme, but rather a means of gaining time until some better disposal method is found (or permanent disposal is forced on the owners by regulatory agencies).

The landfills used for hazardous materials are called "chemical waste landfills" and are designed to provide complete long-term protection for the quality of surface and subsurface waters from hazardous wastes deposited in the landfill, as well as prevent other public health and environmental problems.

The chemical waste landfill differs from the ordinary sanitary landfill primarily in the degree of care taken to ensure minimal environmental impact. Clay liners, monitoring wells, and groundwater barriers are some of the techniques used in such landfills. The overall philosophy is strict segregation from the environment [21].

It should be noted that just as is the case with aboveground storage, such landfills are usually not strictly disposal schemes, but rather holding operations. True disposal is still willed to future generations.

Based on a definition of a hazardous waste similar to the above, the EPA has quantified the production of hazardous wastes in the United States [25]. The 1974 figures (latest) are shown in Table 10-11. Such tabulations show that four industrial categories—primary metals; organic chemicals, pesticides, and explosives; electroplating and metal finishing; and inorganic chemicals—account for 83% of the total hazardous waste produced (wet weight).

TABLE 10-11
Potentially Hazardous Waste Quantities in United States, 1974
(million metric tons annually)

Industry	Dry Basis	Wet Basis
1. Batteries	0.005	0.010
2. Inorganic chemicals	2.000	3.400
3. Organic chemicals, pesticides, explosives	2.150	6.860
4. Electroplating	0.909	5.276
5. Paints	0.075	0.096
6. Petroleum refining	0.625	1.757
7. Pharmaceuticals	0.062	0.065
8. Primary metals	4.454	8.335
9. Leather tanning and finishing	0.045	0.146
10. Textile dyeing and finishing	0.048	1.770
11. Rubber and plastics	0.205	0.785
12. Special machinery	0.102	0.162
13. Electronic components	0.026	0.036
14. Waste oil re-refining	0.057	0.057
Totals	10.763	28.755

Source: Ref. 25.

PARTING SHOTS

Landfills are engineering projects that require an unusual mixture of technical skill and public relations acumen, in that the latter often outweighs the former. Little is mentioned here relative to the nontechnical problems associated with the planning, design, and operation of landfills, but this is not to imply that these aspects are insignificant.

A popular definition for solid waste is that it is the stuff everyone wants picked up but no one wants put down. Studies on the psychology and sociology related to the siting of landfills are both fascinating and somewhat frightening. We know entirely too little about public reaction to and the impact of such projects, and the design engineering is often placed in a somewhat uncomfortable position of being the Solomon-like judge of public reaction and public good. Nowhere does the human nature of the engineering profession become as important as in the design of the ultimate disposal of a community's residues.

REFERENCES

[1] STONE, R., and H. FRIEDLAND, "A National Survey of Sanitary Landfill Practices," *Public Works*, Aug. 1969, p. 88.

[2] POHLAND, F. G., and R. ENGLEBRECHT, *Impact of Sanitary Landfills; An Overview of Environmental Factors and Control Alternatives*, report prepared for the American Paper Institute, New York, 1976.

[3] COOPER, R. C., et al., "Virus Survival in Solid Waste Treatment Systems," in *Virus Survival in Water and Wastewater Systems*, Water Res. Symp. Ser. 7, University of Texas at Austin, 1974.

[4] HUGHES, G. M., R. A. LANDON, and R. N. FARVOLDEN, *Hydrogeology of Solid Waste Disposal Sites in Northeastern Illinois*, U.S. Government Printing Office, Washington, D.C., 1971.

[5] FICHTEL, K., "Leaching Tests of Slags and Flyashes from Municipal Waste Incinerators," *Conf. Papers, Conversion of Refuse to Energy, First Int. Conf. Convers. Refuse Energy*, Montreaux, Switzerland, 1975.

[6] TENN, D. G., K. J. HANEY, and T. V. DEGEARE, *Use of the Water Balance Method for Predicting Leachate Generation from Solid Waste Disposal Sites*, U.S. EPA OSWMP, SW-168, Washington, D.C., 1975.

[7] REMSON, I., A. A. FUNGAROLI, and A. W. LAWRENCE, "Water Movement in an Unsaturated Sanitary Landfill," *J. Sanit. Eng. Div. ASCE*, **94**(SA2), 1968.

[8] REINDL, J., "Managing Gas and Leachate Production on Landfills," *Solid Waste Manage.*, July 1977, p. 30.

[9] THORNTHWAITE, C. W., and J. R. MATHER, "Instructions and Tables for Computing Potential Evapotranspiration and Water Balance," *Publ. Climatol. Drexel Inst. Technol.*, **10**, 185 (1957).

[10] MCGUINNES, J. L., and E. F. BORDNE, *Comparisons of Lysimeter Derived Potential Evapotranspiration with Computed Values*, Agr. Res. Serv. Tech. Bull. 1452, U.S. Dept. of Agriculture, Washington, D.C. 1972.

[11] *Water Atlas*, Water Resources Information Service, U.S. Geological Service, Washington, D.C.

[12] FREEZE, A. A., "Three Dimensional Transient, Saturated-Unsaturated Flow in a Groundwater Basin," *Water Resour. Res.*, **7**, 346 (1971).

[13] PINDER, G. F., and J. D. BREDEHOEFT, "Application of the Digital Computer for Aquifer Evaluation," *Water Resour. Res.*, **4**, 1069 (1968).

[14] ELZY, E., and F. T. LINDSTROM, "Model of the Hazardous Waste Chemicals for Sanitary Landfill Sites," *Modeling of Environmental Systems*, U.S. Environmental Protection Agency, Washington, D.C., 1976.

[15] POHLAND, F. G., *Landfill Stabilization with Leachate Recycle*, U.S. EPA EP-00658, Washington, D.C., 1972.

[16] PAVONI, J. L., J. H. HEER, and D. J. HAGERTY, *Handbook of Solid Waste Disposal*, Van Nostrand Reinhold Company, New York, 1975.

[17] MERZ, R. C., and R. STONE, *Special Studies of a Sanitary Landfill*, U.S. Dept. of HEW, Washington, D.C., 1970.

[18] KOHAN, A. M., *A Summary of Hazardous Substance Classification Systems*, U.S. EPA OSWMP SW-171, Washington, D.C., 1975.

[19] Batelle Memorial Institute, *Final Report: Program for the Management of Hazardous Wastes*, Richland, Wash., 1973.

[20] *Fed. Reg.*, **39**(164), (1974).

[21] FIELDS, T., and A. W. LINDSEY, *Landfill Disposal of Hazardous Wastes: A Review of Literature and Known Approaches*, U.S. EPA OSWMP SW-165, Washington, D.C., 1975.

[22] LAZAR, E. C., "Damage Incidents from Improper Disposal," *J. Hazardous Mater.*, **1**, 157.

[23] JOHANSEN, O. J., *Treatment of Leachates from Sanitary Landfills*, Norwegian Institute for Water Research, Oslo, 1975.

[24] BOYLE, W. C., and R. K. HAM, "Biological Treatability of Landfill Leachate," *J. Water Pollut. Control Fed.*, **46**(5), 860 (1974).

[25] *State Decision-Makers Guide to Hazardous Waste Management*, U.S. EPA OSW SW-612, Washington, D.C., 1977.

[26] BRUNNER, D. R., and D. J. KELLER, *Sanitary Landfill Design and Operation*, U.S. EPA OSWMP SW-65 ts, Washington, D.C., 1972.

[27] FLOWERS, F. B., et al., "Vegetation Kills in Landfill Environs," in *Management of Gas and Leachate in Landfills*, S. K. Banerji (Ed.), U.S. EPA 600/g-77-026, Washington, D.C., 1977.

[28] NEWTON, M., *Model State Hazardous Waste Management Act*, U.S. EPA OSW SW-635, Washington, D.C., 1977.

[29] COOPER, R. C., et al., "Virus Survival in Solid Waste Treatment Systems," in *Virus Survival in Water and Wastewater Systems*, J. Malina and B. Sagik (Eds.), Water Res. Symp. Ser. 7, University of Texas at Austin, 1974.

[30] ENGLEBRECHT, R. S., et al., "Biological Properties of a Sanitary Landfill Leachate," in *Virus Survival in Water and Wastewater Systems*, J. Malina and B. Sugik (Eds.), Water Res. Symp. Ser. 7, University of Texas at Austin, 1974.

[31] QASIN, S. R., and J. C. BURCHINAL, "Leaching from Simulated Landfills," *J. Water Pollut. Control Fed.*, **42**, 371 (1970).

PROBLEMS

10-1. For your own community (assume appropriate quantities of refuse, densities, etc.), calculate how many years it would take to fill a football field 20 ft deep. Include the necessary earth cover.

10-2. Design an engineering plan for a sanitary landfill operation for the land parcel shown in Fig. 10-10. Assume that the user population is 10,000. The final surface must have a minimum slope of 2%. Specify the placement operation plan and draw cross sections to show placement. Estimate the life of the landfill. Assume conditions as local to your area.

10-3. Assume the following fraction, by weight, of material in a refuse:

Garbage	10
Yard waste	20
Mixed paper	40
Corrugated	10
Glass	10
Ferrous (cans)	10

(a) Estimate the in-place density of this refuse, using the bulk density values listed in Table 1-9.

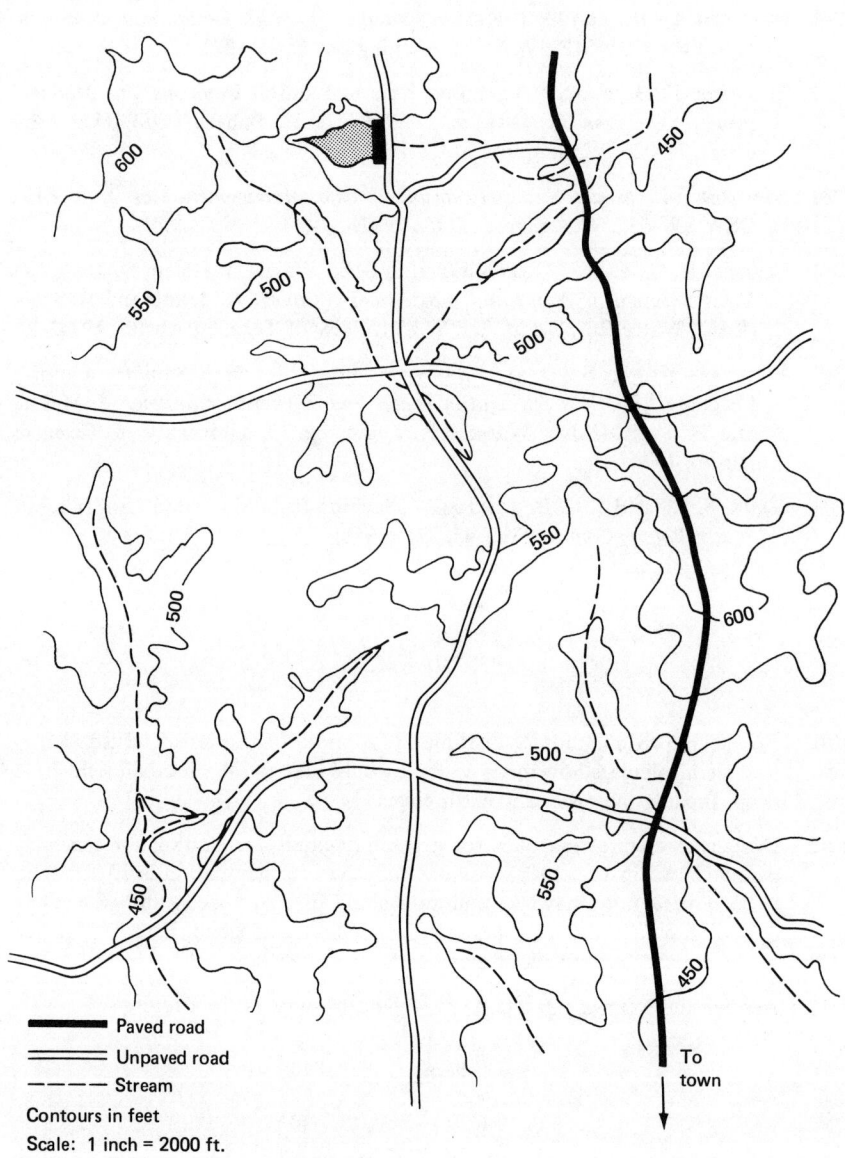

Figure 10-10. Landfill site for Problem 10-2.

Problems

(b) If the following materials were recovered in a resource recovery operation, estimate the bulk density.

	Percent recovered
Garbage	None
Yard waste	None
Mixed paper	50
Corrugated	80
Glass	60
Ferrous (cans)	95

(c) If, for this community, the life of the landfill was originally calculated as 10 years, what is the expected life if the materials listed above are recovered?

10-4. Estimate the net percolation of moisture through a 30-ft-deep landfill with a 4-ft earth cover in eastern North Carolina. If the original refuse moisture content is 20%, how many years will it take for the leachate to reach the groundwater table, which is 5 ft below the bottom of the refuse?

10-5. Consider a 5-ft-long and deep grid block (unit width) below the farthest upstream edge of the refuse. Into this block flows the leachate calculated in Problem 10-4 (area of 5 × 1 ft) and groundwater (area 5 × 1 ft) at a rate of 200 liters/day. If the leachate has a COD of 50,000 mg/liter and uncontaminated groundwater has a COD of zero, calculate the COD as the groundwater exits this 5-ft grid block. Assume no adsorption or biodegradation, and a soil porosity of 0.4.

APPENDICES

APPENDIX A

PARTICLE SIZE

Any mixture of particles of various sizes is difficult to describe analytically. If these particles are irregularly shaped, the problem is compounded. Municipal refuse is possibly the worst imaginable material for particle size analysis, and yet much of the MSW processing technology depends on an accurate description of particle size.

No single value can adequately hope to describe a mixture of particles. Probably the best effort in that direction is to describe the mixture by means of a curve showing percent of particles (by number or weight) vs the particle size. This curve can be by unit intervals, as Fig. A-1, or cumulative, as Fig. A-2. The two mixtures shown have an equal *average size* (50% of particles less than the stated size) but are obviously very different in character. Mixture A is comprised mainly of uniformly sized particles while B is quite non-uniform.

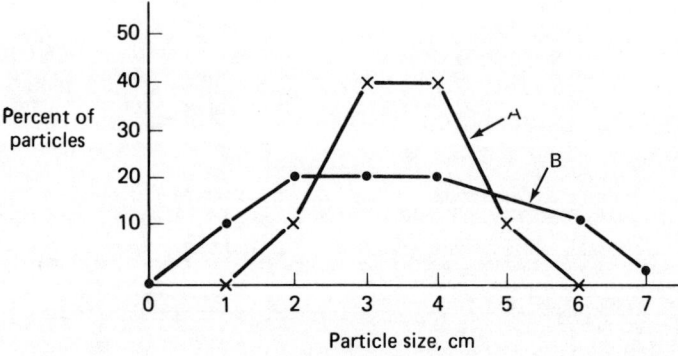

Figure A-1.

Particle Size

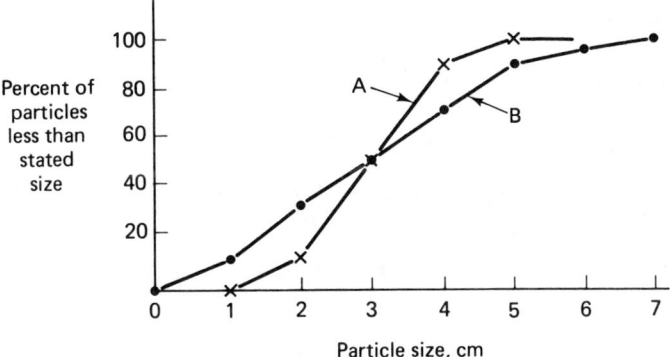

Figure A-2.

The *average particle size* can also cause difficulty. As above, "50% of particles finer than" is one way of expressing average size. Another way is by a weight basis as "50% of particles by weight finer than".

The surface area of particles can be estimated by plotting cumulative percent by weight finer than vs reciprocal of the particle diameter. The area beneath the curve is the cumulative particle surface.

A number of analysis and design equations include the particle diameter term. Invariably, the particles in resource recovery operation are nonspherical, and the material is non-uniform. Hence it becomes necessary to not only define what "diameter" is for a single particle, but also for a mixture of variable sized particles.

The diameter of a particle can be defined by a number of methods, including:

$$D = l \qquad \text{A-1}$$
$$D = \frac{w + l}{2} \qquad \text{A-2}$$
$$D = \frac{h + w + l}{3} \qquad \text{A-3}$$
$$D = \sqrt{lw} \qquad \text{A-4}$$
$$D = \sqrt[3]{lwh} \qquad \text{A-5}$$

where D = diameter
 l = length
 w = width
 h = height

Other definitions have also been advanced [2].

Which one of these equations is used to define diameters often depends on the operation. For example, Eq. A-1 is used for screening, while A-2 is often used for microscopic examinations.

When the mixture of particles is non-uniform, it is often convenient to use a single value diameter term to describe a process. Those mean diameters ($\bar{D}$) can be in terms of numbers of particles (by count) or by weight retained on a sieve. Some of the mean diameters are:

Arithmetic mean $\quad \bar{D} = \dfrac{D_1 + D_2 + \cdots D_n}{n}$

Geometrical mean $\quad \bar{D}_G = \sqrt[n]{D_1 \times D_2 \times \cdots \times D_n}$

Weight mean $\quad \bar{D}_w = \dfrac{W_1 D_1 + W_2 D_2 + \cdots W_n D_n}{W_1 + W_2 + \cdots W_n}$

Number mean $\quad \bar{D}_N = \dfrac{M_1 D_1 + M_2 D_2 + \cdots M_n D_n}{M_1 + M_2 + \cdots M_n}$

Surface area mean $\quad \bar{D}_s = \dfrac{M_1 D_1^3}{M_1 D_1^2 + M_2 D_2^2 + \cdots M_n D_n^2}$

$$+ \dfrac{M_2 D_2^3}{M_1 D_1^2 + \cdots + M_n D_n^2} \cdots$$

$$+ \cdots \dfrac{M_n D_n^3}{M_1 D_1^2 + \cdots + M_n D_n^2}$$

Volume mean $\quad \bar{D}_v = \dfrac{M_1 D_1^4}{M_1 D_1^3 + M_2 D_2^3 + \cdots M_n D_n^3}$

$$+ \dfrac{M_2 D_2^3}{M_1 D_1^3 + \cdots + M_n D_n^3} \cdots$$

$$+ \cdots \dfrac{M_n D_n^4}{M_1 D_1^3 + \cdots M_n D_n^3}$$

where $\bar{D}$ = mean particle diameter.
n = number of discrete classifications (sieves)
W = weight in each classification
M = number of particles in each classification.

The last formula is the equivalent of the weight mean, if the specific gravity is constant for all particles.

Example A-1

Given the following analysis

Particle diameter (mm)(D)	60	40	20	5
Weight of each fraction (kg)(W)	2	10	5	4
Number of particles (m)(M)	140	300	1000	2000

Particle Size

calculate the mean, geometric mean, weight mean, surface mean and volume mean diameters.

$$\overline{D} = \frac{(60 + 40 + 20 + 5)}{4} = 31.2 \text{ mm}$$

$$\overline{D}_G = \sqrt[4]{60 \times 40 \times 20 \times 5} = 22.1 \text{ mm}$$

$$D_w = \frac{(2 \times 60) + (10 \times 40) + (5 \times 20) + (4 \times 5)}{2 + 10 + 5 + 4} = 30.5 \text{ mm}$$

$$D_N = \frac{(140 \times 60) + (300 \times 40) + (1000 \times 20) + (2000 \times 5)}{140 + 300 + 1000 + 2000} = 11.7 \text{ mm}$$

$$D_s = \frac{140 \times 60^3}{140 \times 60^2 + 300 \times 40^2 + 1000 \times 20^2 + 2000 \times 5^2} + \cdots = 40.0 \text{ mm}$$

$$D_v = \frac{140 \times 60^4}{140 \times 60^3 + 300 \times 40^3 + 1000 \times 20^3 + 2000 \times 5^3} + \cdots = 47.4 \text{ mm}$$

Although, as previously noted, the most complete expression of particle diameter is graphical, showing percent of particles (either number or weight) passing a given sieve, it is difficult to use graphical information in calculations. Several suggestions have been made for a single value function which would reasonably describe the particle size distribution curve. In water engineering, the size distribution of filter sand is described by a *uniformity coefficient*:

$$\text{U.C.} = \frac{D_{60}}{D_{10}}$$

where D_{10} = particle size where 10% of the particles are smaller than that size.
D_{60} = particle size where 60% of the particles are smaller than that size (see Fig. A-3).

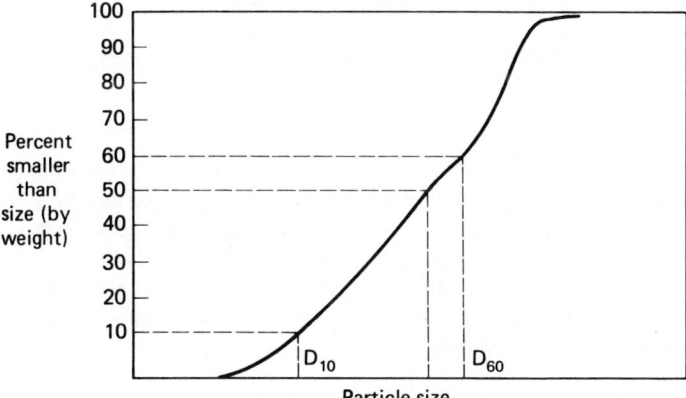

Figure A-3.

Another expression of uniformity [2] is:

$$U.C. = \sqrt{\frac{N}{2\sum_{i=1}^{k} fi^2}}$$

where N = total number of particles measured.
 f = difference in units of length between the measured diameter of a group of particles in one size category (sieve) and the average particle diameter as defined by 50% of the particles (by count) being smaller than that size.
 k = number of classifications of particle size.

Obviously, this uniformity coefficient will range from zero for perfectly uniform mixtures (perfectly vertical size distribution curve) to infinity for a theoretically perfect non-uniform distribution.

Example A-2

Using the data in Example A-1, calculate the uniformity coefficient defined as

$$U.C. = \sqrt{\frac{N}{2\sum_{i=1}^{k} fi^2}}$$

Particle Diameter D	Number of particles N	ND	$f = \frac{\Sigma ND}{N} - D$	$Nf^2 \times 10^4$
60	140	8,400	−45.34	28.70
40	300	12,000	−25.35	19.20
20	1000	20,000	−5.34	2.85
5	2000	10,000	9.65	18.62
	$N = 3440$	$\Sigma Nd = 50,000$	—	$\Sigma_{i=1}^{k} fi^2 = 69.4$

$$U.C. = \sqrt{\frac{3440}{2 \times 69400}} = 0.157$$

Particle irregularities (deviation from spherical) can be handled by redefining the particle diameter as an *effective diameter*, or by using an appropriate shape factor [2]. Some of these equivalent diameters are expressed below.

The *equivalent spherical diameter* is calculated by first knowing the volume of the particle, and then calculating the diameter as that which would exist for a perfect sphere of similar volume, or

$$D_v = \left(\frac{6V}{\pi}\right)^3$$

where V = measured particle volume.

Particle Size

The *equivalent surface diameter* is defined as the diameter of a sphere of equal surface

$$D_s = \frac{6}{S\gamma}$$

where S = measured particle surface area per unit weight.
γ = particle specific gravity.

The *equivalent diameter by projected area* is

$$D_a = \sqrt{\frac{4A}{\pi}}$$

where A = projected area of a circle with equal diameter.

Finally, it is possible to define the diameter as the *effective settling diameter* (or *Stokes diameter*) by measuring the terminal settling velocity and back-calculating the diameter using the Stokes equation; thus

$$D_k = \left[\frac{V_t 18\mu}{(\rho_s - \rho)g} \right]^{1/2}$$

where V_t = terminal settling velocity
μ = viscosity
ρ_s = density of the particle
ρ = density of the fluid
g = acceleration due to gravity

The second alternative is to leave the diameter term unchanged, but insert a shape factor into an equation to indicate non-spherical particles. The most popular shape factor is *sphericity*, defined as

$$\psi = \frac{S^1}{S}$$

where ψ = particle sphericity
S' = surface area of a sphere which has the same volume as the particle
S = actual (measured) surface area of the particle.

Often it is difficult to measure the surface area of an irregular particle, and the concept of *circularity* may be used

$$\phi = \frac{C^1}{C}$$

where ϕ = particle circularity
C' = circumference of a circle having the same cross-sectional area as the particle
C = actual (measured) circumference of the particle

Example A-3

Calculate the sphericity of a cube; w = length of a side.

$$S = 6w^2$$

$$\text{Volume of cube} = w^3 = \text{volume of sphere} = \frac{\pi D^3}{6}$$

or

$$D = \sqrt[3]{\frac{6}{\pi}}\,(w) = 1.24w$$

$$S' = \pi D^2 = \pi(1.24w)^2 = 4.83w^2$$

$$\psi = \frac{4.83w^2}{6w^2} = 0.805$$

In some instances, such as power requirements in shredding, the surface area of the material is an important parameter. This is usually expressed as *specific surface*, defined as the surface of a unit weight, or

$$S = \frac{K}{\rho}\left(\frac{\Sigma \Delta W X_m^{-1}}{\Sigma \Delta W}\right)$$

where S = specific surface resistance cm^2/g
  K = shape factor, = 6 for spheres
  ρ = material density g/cm^3
  ΔW = incremental weight, g
  Xm = mean size of the increment, cm

REFERENCES

[1] FOUST, A.S. et al, *Principles of Unit Operations*, John Wiley & Sons, 1960.

[2] TAGGART, A.F., *Handbook of Mineral Dressing*, John Wiley & Sons, 1979.

[3] VESILIND, P.A., *Treatment and Disposal of Wastewater Sludges*, Ann Arbor Science Pub., Ann Arbor, Michigan, 1979.

[4] STOCKMAN, J.D. and E.G. FOCHTMAN *Particle Size Analysis* Ann Arbor Science Pub., Ann Arbor, Mich. 1978.

APPENDIX B

SIEVE SIZES

Although the counting of individual particles would be the most accurate way of determining particle size distributions, this is quite laborious and seldom used. Most analyses are in fact made by sieving, the common ones being the Tyler series or U.S. sieves. The sizes are shown in Table B-1.

TABLE B-1
Testing-sieve

Nominal Aperture	Tyler					A.S.T.M. standard						British standard			
	Mesh $\sqrt{2}$ series	Aperture		Mesh $\sqrt[4]{2}$ series	Wire diameter, in.	Mesh	Aperture		Wire diameter, in.		Tolerance ±%a	Aperture		Wire diameter, in.	Tolerance ±a
In.		In.	Mm.				In.	Mm.				In.	Mm.		
3	—	2.97	—	—	0.207	—	3.00	76.2	0.19	to 0.32	2	—	—	—	—
2	—	2.10	—	—	0.192	—	2.00	50.8	0.16	to 0.245	—	—	—	—	—
1½	—	1.48	—	—	0.162	—	1.50	38.1	0.145	to 0.210	2	—	—	—	—
1	—	1.050	26.67	—	0.148	—	1.00	25.4	0.135	to 0.177	3	—	—	—	—
7/8	—	0.883	22.43	—	0.135	—	0.875	22.2	0.127	to 0.166	3	—	—	—	—
3/4	—	0.742	18.85	—	0.135	—	0.750	19.1	0.122	to 0.	—	—	—	—	—
5/8	—	0.624	15.85	—	0.120	—	0.625	15.9	0.108	to 0.135	3	—	—	—	—
1/2	—	0.525	13.33	—	0.105	—	0.500	12.7	0.094	to 0.122	3	—	—	—	—
7/16	—	0.441	11.20	—	0.105	—	0.438	11.1	0.088	to 0.112	3	—	—	—	—
3/8	—	0.371	9.423	—	0.092	—	0.375	9.52	0.083	to 0.102	3	—	—	—	—
5/16	—	0.312	7.925	2½	0.088	—	0.312	7.93	0.073	to 0.093	3	—	—	—	—
1/4	3	0.263	6.680	3	0.070	3½	0.250	6.35	0.063	to 0.083	3	—	—	—	—
Micron															
5,660	—	0.221	5.613	3½	0.065	3	0.223	5.66	0.050	to 0.075	3	—	—	—	—
4,760	4	0.185	4.699	4	0.065	4	0.187	4.76	0.045	to 0.066	3	—	—	—	—
4,000	—	0.156	3.962	5	0.044	5	0.157	4.00	0.039	to 0.058	3	—	—	—	—
3,360	6	0.131	3.327	6	0.036	6	0.132	3.36	0.034	to 0.052	3	5	—	—	—
2,830	—	0.110	2.794	7	0.0378	7	0.111	2.83	0.031	to 0.047	3	6	0.132	3.34	0.068	3
2,380	8	0.093	2.362	8	0.032	8	0.0937	2.38	0.0291	to 0.0433	3	7	0.1107	2.81	0.056	3
2,000	—	0.078	1.981	9	0.033	10	0.0787	2.00	0.0263	to 0.0394	3	8	0.0949	2.41	0.048	3
1,680	10	0.065	1.651	10	0.035	12	0.0661	1.68	0.0244	to 0.0354	3	10	0.0810	2.05	0.044	3
1,410	—	0.055	1.397	12	0.028	14	0.0555	1.41	0.0220	to 0.0315	3	12	0.0660	1.67	0.034	3
1,190	14	0.046	1.168	14	0.025	16	0.0469	1.19	0.0197	to 0.0276	3	14	0.0553	1.40	0.028	3
1,000	—	0.039	0.991	16	0.0235	18	0.0394	1.00	0.0169	to 0.0244	5	16	0.0474	1.20	0.024	3
840	20	0.0328	0.833	20	0.0172	20	0.0331	0.84	0.015	to 0.0217	5	18	0.0395	1.00	0.023	3
													0.0336	0.85	0.022	5

TABLE B-1 (Continued)

Nominal Aperture	Tyler					A.S.T.M. standard						British standard			
	Mesh $\sqrt{2}$ series	Aperture		Mesh $\sqrt[4]{2}$ series	Wire diameter, in.	Mesh	Aperture		Wire diameter, in.	Tolerance ±%a		Aperture		Wire diameter, in.	Tolerance ±a
In.		In.	Mm.				In.	Mm.				In.	Mm.		
710	—	0.0276	0.701	24	0.0141	25	0.0280	0.71	0.013 to 0.0189	5	22	0.0275	0.70	0.018	5
590	28	0.0232	0.589	28	0.0125	30	0.0232	0.59	0.0114 to 0.0165	5	25	0.0236	0.60	0.0164	5
500	—	0.0195	0.495	32	0.0118	35	0.0197	0.50	0.0102 to 0.0146	5	30	0.0197	0.50	0.0136	5
420	35	0.0164	0.417	35	0.0122	40	0.0165	0.42	0.0091 to 0.13	5	36	0.0166	0.42	0.0112	5
350	—	0.0138	0.351	42	0.0100	45	0.0138	0.35	0.0079 to 0.0114	5	44	0.0139	0.35	0.0088	5
297	48	0.0116	0.295	48	0.0092	50	0.0117	0.297	0.0067 to 0.0100	5	52	0.0116	0.30	0.0076	6
250	—	0.0097	0.246	60	0.0070	60	0.0098	0.25	0.0059 to 0.0087	5	60	0.0099	0.252	0.0068	6
210	65	0.0082	0.208	65	0.0072	70	0.0083	0.21	0.0051 to 0.0074	5	72	0.0083	0.211	0.0056	6
177	—	0.0069	0.175	80	0.0056	80	0.0070	0.177	0.0045 to 0.0061	6	85	0.007	0.177	0.0048	6
149	100	0.0058	0.147	100	0.0042	100	0.0059	0.149	0.0038 to 0.0049	6	100	0.006	0.152	0.004	6
125	—	0.0049	0.124	115	0.0038	120	0.0049	0.125	0.0031 to 0.0041	6	120	0.0049	0.125	0.0034	6
105	150	0.0041	0.104	150	0.0026	140	0.0041	0.105	0.0025 to 0.0034	6	150	0.0041	0.105	0.0026	8
88	—	0.0035	0.088	170	0.0024	170	0.0035	0.088	0.0021 to 0.0029	6	170	0.0035	0.088	0.0024	8
74	200	0.0029	0.074	200	0.0021	200	0.0029	0.074	0.0018 to 0.0024	7	200	0.0030	0.076	0.002	8
62	—	0.0024	0.063	250	0.0016	230	0.0024	0.062	0.0015 to 0.0020	7	240	0.0026	0.065	0.0016	8
53	270	0.0021	0.053	270	0.0016	270	0.0021	0.053	0.0014 to 0.0018	7	300	0.0022	0.053	—	8
44	—	0.0017	0.044	325	0.0014	325	0.0017	0.044	0.0012 to 0.0016	7	—	—	—	—	—
37	400	0.0015	0.037	400	0.001	400	0.0015	0.037	0.0009 to 0.0014	7	—	—	—	—	—

Base = 200 – m = 0.074 mm, $\sqrt{2}$. Base = 18 – m = 1.00 mm, $\sqrt[4]{2}$.

aPermissible variations in average openings.

Each series is developed so that the mesh, or the number of openings per linear inch, is in a geometric ratio, usually $\sqrt[n]{2}$. The standard series used in the measurement of solid waste is $\sqrt{2}$ since narrower intervals are seldom needed.

APPENDIX C

ANALYTICAL TECHNIQUES FOR SOLID WASTE*

I. Particle Size Distribution

Although sieving may provide misleading information about particle size, since a particle need be sufficiently small in only two dimensions in order to pass a sieve, it is still the most expedient means of describing particle size distribution.

Procedure

1. Weigh each of a series of soil sieves. The size is dependent on the size of the material to be screened.
2. Stack the sieves in descending order, with a weighed pan on the bottom.
3. Fill the top sieve with the sample.
4. Place the stack in a shaker and shake for 15 minutes.
5. Remove the sieves and weigh.
6. If dry weights are required, place the sieves into a drying oven at 70°C for 24 hours, cool in a desiccator, and weigh.
7. Calculate the fraction of particles sizes in any one sieve as

$$\text{Fraction of Particles} = \frac{(\text{Weight of particle} + \text{weight of sieve}) - (\text{weight of sieve})}{\text{Total sample weight}}$$

II. Starch in Compost

This is a qualitative test for estimating the progress of the composting operation by measuring the amount of starch in the compost. This test is based on the formation of a starch-iodine complex. The color of this complex changes, depending on the degree of decomposition. An unfinished compost will result in a blue color while a finished will be yellow. The progression is blue/black → light blue → gray → green → yellow.

*These methods have no official sanction, but have been found to be helpful for educational purposes and for rough analyses. The American Society of Testing and Materials (ASTM) is presently developing a series of tests on refuse through its E-38 committee. Some of these will be available in 1980.

Procedure

1. Place about 1 g compost in a 100 ml beaker, wet with a few drops of ethanol, add 20 ml perchloric acid (36%).
2. Filter through open texture filter paper (Whatman No. 90)
3. Add 2 ml iodine reagent to the filtrate and stir.
4. Place a few drops on a white background and note the color.

Reagents

1. Iodine reagent: Dissolve 2.00 g KI in 500 ml water, then add 0.08 g I_2.
2. Perchloric acid (36%).
3. Ethanol.

REFERENCE

[1] LOSSIN, R. D. "Compost Studies," *Compost Science,* Nov-Dec., 1970.

III. Kjeldahl Nitrogen

Procedure

1. Dry about 10 g of solid sample to constant weight at 70°C.
2. Weigh out about 1 to 2.5 g of dry sample (to 3 places) and transfer to Kjeldahl flask.
3. Add 16 g potassium sulfate, 0.7 g mercuric oxide, 25 ml concentrated sulfuric acid, and some glass beads.
4. Place flasks in heating apparatus and boil for at least one hour.
5. While flask is cooling, put a 500 ml calibrated flask containing 50 ml boric acid under the condenser.
6. To cooled sample, add 200 ml distilled water and mix, then add 0.5 g zinc and mix again.
7. Pour 75 ml of alkaline thiosolfate solution down the side of the flask and quickly attach to a trap-condenser apparatus.
8. Distill until volume in receiving flask is 200 ml (50 ml boric acid plus 150 ml distillate).
9. Add 4 drops of methyl purple solution to each receiving flask and titrate with 0.1 N sulfuric acid solution to light violet color. Record ml sulfuric acid.
10. Calculate per cent Kjeldahl nitrogen in sample as
$$\text{Percent } N = \frac{A \times 140}{C}$$
where A = ml 0.1 N sulfuric acid in titration
c = mg of solid waste used

Reagents

1. Mercuric oxide
2. Concentrated sulfuric acid
3. Zinc metal, granulated
4. Alkaline thiosulfate solution
    Dissolve 450 g sodium hydroxide in 700 ml distilled water, cool, add 80 g sodium thiosulfate and dilute to one liter
5. Boric acid solution
    Dissolve 40 g boric acid in distilled water and dilute to 1 liter.
6. Methyl purple indicator
    Dissolve 0.3125 g methyl red and 0.2062 g methylene blue in distilled water and dilute to 250 ml. This must be prepared weekly.
7. Sulfuric acid solution, 0.1 N
8. Potassium sulfate

REFERENCE

[1] W. H. KAYLOR, and N. S. ULMER, *Laboratory Procedures to Determine the Nitrogen Content of Solid Wastes*, Bureau of Solid Waste Management, 1970 (New printing by EPA, 1971).

IV. Total Carbon

A sophisticated technique for measuring carbon employs a dry combustion-purification-gravimetric approach where the organics are combusted in a tube at 950°C with oxygen as carrier gas. The CO is adsorbed and measured gravimetrically (1). Unfortunately, this procedure requires expensive equipment and is not readily adaptable to wide use.

Experience has shown that the carbon content of refuse is about 47% of the organic matter (2). A crude method of estimating carbon is thus to measure volatile solids and multiply this by 0.47.

Procedure

1. Grind the refuse in a laboratory mill and dry.
2. Weigh a burned, cooled and desiccated crucible.
3. Place a reasonable amount of dry refuse in the crucible and weigh.
4. Incinerate at 600°C in a muffle furnace for at least 15 min.
5. Remove, cool in a desiccator, and weigh.

6. The difference in the two weights is the volatile solids

$$\text{Volatile Solids (g/g)} = \frac{a - b}{a - c}$$

where a = weight of crucible plus refuse
 b = weight of crucible plus incinerated refuse
 c = weight of crucible

7. Estimate carbon as

$$\text{Carbon (g/g)} = 0.47 \times \text{VS}$$

REFERENCES

[1] WILSON, D. L., "Method for Macrodetermination of Carbon and Hydrogen in Solid Wastes," *Environmental Science and Technology*, July 1971.

[2] *Methods of Sampling and Analysis of Solid Wastes*, EAWAG Switzerland, 1970.

V. pH of Solids

Procedure

1. Place about 10 grams of shredded solid material prepared in a liter beaker or a flat bottomed flask.
2. Add 500 ml distilled water and stir vigorously for 3 to 5 minutes.
3. Let the mixture settle and measure pH with a pH meter.

REFERENCE

[1] CARNES, R. A. and R. D. LOSSIN, "An Investigation of the pH Characteristics of Compost," *Compost Science*, Sept.–Oct., 1970.

VI. Bio-degradability of Refuse

Refuse contains organic matter which is easily degraded (fermented) as well as organics which are quite resistant to biological action.

Through experience, it has been found that the Chemical Oxygen Demand test, run at room temperature, will give a reasonable estimate of the bio-degradability of refuse (1).

Procedure

1. Weigh out 0.500 grams of ground and dried sample in 500 ml Erlenmeyer flask.
2. Add exactly 20 ml of potassium dichromate solution from a burette and mix well.
3. From a second burette, add 20 ml sulfuric acid.
4. Let this mixture sit for one hour at room temperature. Shake it occasionally.
5. Add approximately 150 ml distilled water.
6. Add, in order
   a. 10 ml phosphoric acid
   b. 0.2 g sodium fluoride
   c. 30 drops of indicator
   Mix after each addition.
7. Titrate back with ammonium iron (II) = sulfate solution. The color change is from brownish green to green blue to blue to green. The end point of titration is a pure green color.
8. A blank is treated the same way but without sample.
9. If the green color appears when the indicator is added, the test must be repeated, using 30 ml dichromate solution.
10. Calculate the bio-degradable matter as

$$\text{BDM} = \frac{(b - a)cN(1.28)}{b}$$

where a = volume of titrant, ml
b = volume of titrant for blank, ml
c = potassium dichromate used, ml
N = normality of potassium dichromate

Note: the above calculation assumes that 1 ml $K_2Cr_2O_7$, 1N, oxidizes 3 mg C to CO_2, and that carbon content in the bio-degradable matter is about 47%.

Reagents

1. Potassium dichromate solution, 2N. Dissolve 98.08 g $K_2Cr_2O_7$ (dry) in 500 ml distilled water in a one liter volumetric flask. Carefully and slowly add 250 ml conc. H_2SO_4; keeping flask immersed in water in sink. Fill to the liter mark with distilled water.
2. Sulfuric acid, conc.
3. Phosphoric acid, conc.
4. Sodium fluoride, NaF.
5. Diphenylamine indicator
   Carefully add 100 ml conc. H_2SO_4 to 20 ml distilled water and then add 0.5 g diphenylamine.

6. Amonium iron (II)-sulfate solution, 0.5N.
   Carefully add 20 ml conc. H_2SO_4 to 780 ml distilled water in a liter volumetric flask. Dissolve 196.1 g $FeSO_4 \cdot (NH_4) SO_4 \cdot 6H_2O$ and fill to liter mark.

REFERENCE

[1] *Methods of Sampling and Analysis of Solid Wastes*, EAWAG, Switzerland, 1970.

VII. Refuse Composition

Because of the use of different materials in one product (e.g., aluminum tops on steel cans), the determination of the various materials in refuse is an imprecise effort. Nevertheless, this is often necessary when a reasonable approximation of refuse composition is required.

Procedure

1. Obtain a representative refuse sample. For fullscale studies, 200 pounds is suggested. For laboratory studies smaller samples are more reasonable.
2. Cover a large table with a plastic cloth (preferably outdoors) and dump some of the refuse on the table.
3. While wearing plastic gloves and a face mask, sort the refuse into the necessary components.
4. Using tared pans, weigh the different fractions.
5. Usually the separation procedure will end with a residual of unidentifiable fines on the plastic sheet. Collect and weigh these and report as "Fines" or "Other."

VIII. Moisture Content

The most difficult aspect of measuring moisture content in refuse is obtaining a representative sample. If the refuse is not shredded, many samples must be used and the results summed. If a homogeneous material (e.g. the glass fraction) is to be analyzed, fewer samples are necessary. The most acceptable method for obtaining a representative sample is to shred the refuse first. If this is done, care should be taken not to add moisture or dry out the refuse during the shredding.

Procedure

1. Place a reasonable amount of sample on a tared pan, weigh, and place in a 70°C oven for 24 hours.
2. Remove from oven, allow to cool in desiccator and weigh.

Analytical Techniques for Solid Waste

3. Calculate percent moisture as

$$M = \frac{(W - P) - (D - P)}{(W - P)} \times 100$$

where percent M = percent moisture
  W = weight of wet refuse
  D = weight of dry refuse
  P = weight of pan

IX. Ash Content

Ash content is defined as the material remaining after combustion at 600°C. Although it is assumed that all organics are fully oxidized at 600°C, this is not totally accurate. On the other hand, metals will form oxides at this temperature and actually gain weight. Thus the loss in weight is actually a *net* loss and is a reasonable approximation of the volatile (or organic) solids in refuse, and the non-combustibles are the ash, or inorganics.

Procedure

1. Prepare a sample for analysis by grinding the refuse in a laboratory mill.
2. Weigh several crucibles.
3. Into each crucible place a reasonable amount of dried (70° C, 24 hours) sample. Weigh.
4. Place the crucibles into a *cool* furnace.
5. Turn the furnace on and allow it to heat to 600°C.
6. Hold this temperature for 2 hours.
7. Remove the crucibles, allow to cool for a few minutes, and place in a desiccator.
8. When cool, weigh.
9. Calculate percent ash as

$$A = \frac{(R - C) \times 100}{(S - C)}$$

   where percent A = percent ash in sample
     R = weight of crucible and fired sample (after 600°C)
     S = weight of crucible and raw sample
10. Calculate weight loss (or volatiles) as

    Percent volatiles = $100 - \%A$

REFERENCE

[1] ULMER, N. "Laboratory Procedure for Determining Percent Ash and Percent Weight Loss of Solid Waste on Heating at 600°C" EPA RS-03-68-17, 1971.

APPENDIX D

CONVERSION FACTORS

Multiply	By	To Obtain
acre	0.404	ha
acre ft	1233	m^3
atmospheres	14.7	lb/in.^{2}
British thermal units	252	cal
BTU	1.054×10^3	J
BTU/ft^3	8,905	cal/m^3
BTU/lb	2.32	J/g
BTU/lb	0.555	cal/g
BTU/sec	1.05	kW
BTU/ton	278	cal/tonne
calories	4.18	joule
calories	3.9×10^{-3}	BTU
cal/g	1.80	BTU/lb
cal/m^3	1.12×10^{-4}	BTU/ft^3
cal/tonne	3.60×10^{-3}	BTU/ton
candles/ft^2	0.092	lumen/m^2
centimeters	0.393	in
cumec	1	m^3/sec
feet	0.305	m
ft/min	0.00508	m/sec
ft/sec	0.305	m/sec
ft^2	0.0929	m^2
ft^3	0.0283	m^3
ft^3	28.3	liters
ft^3/lb	0.0623	m^3/kg
ft^3/sec	0.0283	m^3/sec
ft^3/sec	449	gal/min
ft lb (force)	1.357	joule
ft lb (force)	1.357	newton meters
gallons	3.78×10^{-3}	m^3
gallons	3.78	liters
gal/person/day	3.78	liter/person/day
gal/day/ft^2	0.0407	m^3/day/m^2
gal/min	2.23×10^{-3}	ft^3/sec
gal/min	0.0631	liter/sec
gal/min	0.227	m^3/hr
gal/min	6.31×10^{-5}	m^3/sec
gal/min/ft^2	2.42	m^3/hr/m^2
grams	2.2×10^{-3}	lb

Conversion Factors

Multiply	By	To Obtain
g/cm^3	1,000	kg/m^3
million gal/day	43.8	liters/sec
million gal/day	3785	m^3/day
million gal/day	0.0438	m^3/sec
hectares	2.47	acre
horsepower	0.745	kW
inches	2.54	cm
inches of mercury	0.49	lb/in.^{2}
inches of mercury	3.38×10^3	newton/m^2
inches of water	249	newtons/m^2
joule	0.239	calorie
joule	9.48×10^{-4}	BTU
joule	0.738	ft lb
joule	2.78×10^{-7}	kWh
joule	1	newton meter
J/g	0.430	BTU/lb
J/sec	1	watt
kilograms	2.2	lb (mass)
kg	1.1×10^{-3}	tons
kg/ha	0.893	lb/acre
kg/hr	2.2	lb/hr
kg/m^3	0.0624	lb/ft^3
kg/m^3	1.68	lb/yd^3
kg/tonne	2.0	lb/ton
kilometers	0.622	mi
km/hr	0.622	mph
kilowatts	1.341	horsepower
kWh	3600	kilojoule
liters	0.0353	ft^3
liters	0.264	gal
liters/sec	15.8	gal/min
liters/sec	0.0228	mgd
meters	3.28	ft
meters	1.094	yd
m/sec	3.28	ft/sec
m/sec	196.8	ft/min
m^2	10.74	ft^2
m^2	1.196	yd^2
m^3	35.3	ft^3
m^3	264	gal
m^3	1.31	yd^3
m^3/day	264	gal/day
m^3/hr	4.4	gpm
m^3/hr	6.38×10^{-3}	mgd
m^3/sec	1	cumec
m^3/sec	35.31	ft^3/sec

Multiply	By	To Obtain
m^3/sec	15,850	gpm
m^3/sec	22.8	mgd
miles	1.61	km
mi^2	2.59	km^2
mph	0.447	m/sec
milligrams/liter	0.001	kg/m^3
million gallons	3,785	m^3
mgd	43.8	liter/sec
mgd	157	m^3/hr
mgd	0.0438	m^3/sec
newton	0.225	lb (force)
newton/m^2	2.94×10^{-4}	inches of mercury
newton/m^2	1.4×10^{-4}	lb/$in.^2$
newton meters	1	joule
newton sec/m^2	10	poise
pounds (force)	4.45	newton
pounds (force)/in^2	6895	N/m^2
pounds (force)/in^2	6.89	kPa
pounds (mass)	454	g
pounds (mass)	0.454	kg
pounds (mass)/ft^2/yr	4.89	kg/m^2/yr
pounds (mass)/yr/ft^3	16.0	kg/yr/m^3
pounds/acre	1.12	kg/ha
pounds/ft^3	16.04	kg/m^3
pounds/$in.^2$	0.068	atmospheres
pounds/$in.^2$	2.04	inches of mercury
pounds/$in.^2$	7140	newton/m^2
pounds/ton	0.5	kg/tonne
pounds/yd^3	0.593	kg/m^3
tons (2000 lb)	0.907	tonne (1000 kg)
tons	907	kg
ton/acre	2.24	tonnes/ha
tonne (1000 kg)	1.10	ton (2000 lb)
tonne/ha	0.446	tons/acre
yd	0.914	m
yd^3	0.765	m^3
watt	1	J/sec

INDEX

A

abrasiveness 84
acid hydrolysis 270–275
adiabatic flame temperature 286
aerobic decomposition 248, 254
agricultural wasters 271
air classifiers 154–177
  rotary 157
  Utah 158, 161
  zigzag 155
air pollution 270, 330, 340–352
air pollution control 340–352
air pollution control in pyrolysis 342
air/solids ratio 162
air stream cleaning 174–177
Allegheny City, PA 295
allocation models 51
Alter, H. 148
aluminum 9
ammonia conversion 386
anaerobic decomposition 248, 255
anaerobic digestion 247–253
analytical technique 435–441
angle of nip 128

angle of repose 84
aquatic toxicity 419
arithmetic equivalence 30
ash content of refuse 441
ash, residential 10
ashes from incineration 318–321, 334–336, 397
Atlanta, GA 295
Austin, L. 109
Australia 21

B

bagasse 271
bag filter 350–351
bag house 350–351
Baile, R. 360
bark 409
Battelle Northwest 359
belt-type magnet 212–213
beverage containers 8
binary separation 135–137
biodegradability of refuse 438–440
bomb calorimeter 283

Bond, F. 155, 124
Bond Work Index 115–116
bottom ash 334–336, 397
breakage function 104–111
Broadbent, S. 104, 108
bulk density 24, 92, 337, 416
Bureau of Mines 148, 196, 358, 359

C

California, U. of 104
Callcott, T. 104, 108
calorimeter 283
carbon, total 437
carbonization 360
carbon/nitrogen ratio 250, 267, 268
Carruth, D. 25
cascading in a trommel 143
cataracting in a trommel 143
cellubiase 249, 273–275
cellulase 249, 273–275
cellulose 246, 270, 337
centrifuging in a trommel 143
Champaign, IL 392
characteristic size 102, 108, 123
chemical composition of refuse 26
chlorine 336
circular grate 307
classification of solid waste for incineration 298–299
Clean Air Act 342
Club of Rome 2
coding and separation 141
coliforms in leachate 396
collection effectiveness 64–69
collection efficiency 64
collection of solid waste 43–78
color sorting 198–200
combustion 280–291
combustion chambers 303–304
Commoner, B. 2
community effects index 68
compaction 91–93
composition of solid waste 21, 26, 440

composting 263–270
compression 32, 91
compressive characteristics of refuse components 33
conductivity 217–218
conversion equivalence 30
conversion factors 442–444
conveying of solid waste 86–91
conveyors
  pneumatic 88, 335–336
  rubber belted 86
  screw 89
  vibrating 88
controlled tip 394
costs of disposal 63
costs of landfills operation 64
Coulombs Law 231
Crawford, B. 148
critical speed in a trommel 146
curb side collection 43
cyclones 174–177, 345
cyclonic spray wet scrubber 347, 349

D

Dalluvalle model 168
Davenport, IA 392
Dayton, OH 392
deadheading 47
densified refuse derived fuel 337
density 31, 84, 91–94, 257
  bulk 24, 92, 337, 416
  dry 92
  effective 94
  wet 91
dermal irritation 419
destructive distillation 360
destructor 295
DeVaney 187
digestion, anaerobic 247–253
dismal science 2
d-RDF 337
drum grate 308
drum magnet 211–213
dry density 92

Index

Dulong formula 29, 283–284
Dusseldorf grate 308
dust 119

E

Eco Fuel 161
eddy current separators 214–231
efficiency of materials separation 138–139
efficiency of solid waste collection 64–69
effective density 94
effective diameter 166, 340
effectiveness of solid waste collection 64–69
electrostatic separators 231–236
Elzy, E. 403
embrittlement 338–339
emission standards 342–343
energy equivalents 29
energy recovery 9, 294–388
enthalpy 284–285
enzymatic hydrolysis 270–275
Epstein, B. 104
Erlich, P. 2
erosion 335
ethanol 275
ethics 40
Euler, L. 47
Euler's tour 48
Evans, I. 104
evaportranspiration 399–401
excess air 332
explosions 119
extract 137

F

Fan, D. 170
Faraday's Law 206, 214
feed arrangements in incinerators 302–303
fermentation 275

ferrosilicon 188
field capacity 398, 401
filter, bag 350
fires 85, 94, 119–121
fixed bed pyrolysis reactors 370, 373
flame suppressors 120
flammability 416
flash pyrolysis 378–383
flotation 194–198
fluidized bed incinerators 327–328
fluidized bed pyrolysis reactor 373–375
fly ash 334, 341, 397
Forrester, J. 2
Franklin, OH 196
Friedman, S. 358
Fritz, R. 358
furnace design 303

G

garbage 10
garbage grinders 246
Gardner, R. 109
gas production 247–263
gasification 357, 364, 366, 383–386
gasification, products of 366, 372, 377, 378, 381, 382, 383–386
Gaudin, A. 101, 108
generation of solid waste 13, 59–62
glass 9
glass, composition 197
Glass Container Manufacturers 11
glass spectra 197
glucose 249, 270–275
Golueke, C. 267, 275
goniometer 197
Governor's Island, NY 295
grate design 310–312
grate selection 305
grates 304–312
  circular 307
  drum 308
  Dusseldorf 308
  Martin 308

grates (*cont.*)
  rocking 307
  rotary 308
  traveling 306
  von Roll 309–310
Greeley, S. 295
green can system of collection 45
groundwater 397

H

Ham, R. 93
hammermill 95
hammer wear 120–122
hand sorting 140–141
Harz jig 177
hazardous substances 417–420
hazardous wastes 417–420
health 118
heat of combustion 283, 287
heat of vaporization 284, 287
heavy fluid separator 184
heavy media separator 185
heavy metals 250
Henrikson, R. 170, 173–174
Hering, R. 295
Hickman, W. 84
high heat value 284
Hoffman, D. 358
home scrap 6
Honolulu, HI 10
hydrolysis, acid and enzymatic 270–275

I

incineration 280, 294–331
incinerator classification 297
Incinerator Institute of America 296
incinerators
  air pollution 340–352
  bottom ashes 318–321, 334–336, 397
  excess air 332
  feed design 302–303

incinerators (*cont.*)
  fluidized bed 327–328
  fly ashes 334, 341, 397
  furnace design 303
  grates 304–312
  modular 323–324
  multiple hearth 324–325
  open pit 329–331
  slagging 315
  waterwall 291, 332
inclined table 189–191
inclined table separator 224
industrial solid waste 17
injuries to collection personnel 46, 118
insects 94

J

jigs 177–183
  Harz 177
  plunger 178
Johnson City, TN 264, 265

K

Kaiser, E. 358
Keep America Beautiful 74
Kenney, R. 216
Kick's Law 114
Kjeldahl Nitrogen 436
Klee, A. 25
Klumb, D. 334
Königsberg 47
Kwan 48

L

landfill 253, 392–417
  costs 63, 64
  gases 252–263, 411, 414
  leachate 94, 395–411
latent heat of vaporization 286

Index

leachate   94, 395–411
  capture   410
  coliforms in   396
  composition   395
leaves, composting   270
Lenz's Law   215
lignin   267, 270, 274
Lindstrom, F.   403
linear induction motor   218
litter   8, 73–78
litter index   75
litter survey   73–76
Lorentz Force   216
Los Angeles, CA   401
Louisiana State Univ.   275
low heat value   284

M

magnetic fluids   236–241
magnetic separation   205–214
Marblehead, MA   71
marcorouting   47
Martin grate   308
materials balance in combustion   239
materials recovery   8
Malthus, T.   1
Meadows, D.   2
mechanical properties   34
Meloy, T.   104, 108
mesophilic digestion   249
methanation   383–386
methane   253–263, 282, 411, 414
methane generation   247–263
methane in landfills   258–263, 411
methanol   385–386
microrouting   46
modular incinerators   323–324
moisture content   19, 84, 113, 114, 229, 236, 257, 337, 364, 398, 440–441
molecular sieves   262
monosaccharides   272
motor inertia in shredders   122
multiple hearth incinerators   324–325
municipal solid waste   10

N

Newton's Law   165, 167
New York City   140, 295
New Zealand   21
nitrogen in compost   267
nitrogen oxides   341
noise   118
Nottingham, England   295

O

obstacles to resource recovery   37
ocean disposal   391
odor   94, 269
office building   18
oil, pyrolytic   380–383
open pit incinerators   329–331
operation costs for landfills   64
Oregon bottle law   7
organic analysis of refuse   247
Orlando, FL   402

P

packed bed scrubber   347
packers   45
Palos Verdes, CA   260, 261, 354
paper from offices   18
parasites   266
particle size   20–21, 84, 98–111, 217, 257, 341, 426–434, 435
particulate emissions   330
particulates   340, 343
particulates, control of   344–352
pathogens   266
pelletizing   339
percolation   402
permanent magnet separator   225
pH   438
photogrammetry   23
phytotoxicity   419
pickers   140–141
plunger jig   178
pneumatic conveying   88, 335–336

pneumatic tables 183–184
polynary separators 137–138
polysaccharides 271
Pomeroy, C. 104
population changes 59
population density 16
post-consumer solid wastes 10
powdered RDF 338–339
power requirements in shredding 111–118
prepared fuels 333–340
product 137
products of gasification 366, 372, 377, 378, 381, 382, 383–386
products of pyrolysis 360–366, 372, 377, 378, 381, 382, 383–386
projections of solid waste generation 13, 59–62
prompt industrial scrap 6
proximate analysis 26
pyrolysis 357–388
  oil 380–383
  products of 360–366, 372, 377, 378, 381, 382, 383–386
  reactors 366–383
pyrolysis reactors 366–383
  fixed bed 370–373
  fluidized bed 373–375
  suspension 375–377
  rotary kiln 376–378
  transport 378–383
pyrolytic oil 380–383
pulping 125
purity 136

R

radioactivity 419
radius of influence 256, 260
Rammler, E. 101, 103, 124
rats 94
RDF 29, 30, 334–340
reciprocating screen 150–153
recovery equations 135–139, 142
refractory lined incinerators 301, 314
refractory materials 314–316

refuse 10
refuse composition 440
refuse derived fuel 29, 30, 334–340
reject 137
residuals 318–321, 397
residues from incineration 318–321, 334–336, 397
Resnick, W. 86
reuse 8
Ricardo 1
Rietema, K. 139
Rittinger's Law 114
rocking grate 307
roll crushing 126
rotary air classifiers 157
rotary kiln 308
rotary kiln pyrolysis reactor 376–378
routing of trucks 46
Rosin, P. 101, 103, 124
rubber belted conveyors 86
rubbish 10
runoff coefficients 401

S

safety 45, 46, 118
sampling of refuse 22, 25, 440
sanitary landfill 392
Sanner, W. 358
Savage, G. 103, 112
Schur, D. 49
screens 141–154
  reciprocating 150–153
  trommel 143–150
screw conveyors 86
Senden, M. 170, 173
sensible heat 284
settling chamber 344
shaking table 191–193
Shelton 187
shredding 93–125
Shuster, K. 49
sieve sizes 432–435
sink/float separation 184–189
  heavy fluid 184
  heavy media 185
  upflow 189

Index **451**

slag 397
socioeconomic levels 15
soil classification 411–414
soil types 411–414, 415
solid waste
  characteristics 19
  chemical composition 26
  classification for incineration 298–299
  collection 43–78
  combustion 280–291
  compaction 33, 91–93
  composition 21, 26, 440
  composting 236–270
  conveying 83, 86–91
  generation 13, 59–62
  organic analysis 247
  proximate analysis 26
  sampling 22, 25, 440
  storage 83–86, 301–302
  ultimate analysis 27–28
solid waste projections 13, 59–62
Somerville, MA 71
Sommer, E. 216
source separation 69
spalling 316
specific energy in shredding 111, 123
specific heat 285
specific heat capacity 285
spontaneous combustion 85
stacks 317
starch in compost 435
steel cans 9
stokers 304–312
Stokes Law 165, 344, 178–180, 185
St. Louis, MO 335
stoners 183–184
storage of refuse for incineration 301–302
storage of solid waste 83–86
stress-strain curves 34, 35
street cleanliness 67
substitution equivalence 30
sulfur 336
sulfur oxides 341
suspended magnets 212–213
suspension pyrolysis reactor 375–377

Sweden 21
Sweeney, P. 169

T

Taggart, A. 200, 233
Tels, M. 170
thermal balance in combustion 287
thermal decomposition 360
thermophilic digestion 249
total incineration 328–329
toxic materials 250, 253, 417–420
transfer stations 43
transport pyrolysis reactor 378–383
transportation models 51
trash 10
travel times 62
traveling bridge crane 303
traveling grate 306
Trezek, G. 103, 104, 107, 112
trommel screen 143–150
tumbleback in conveyors 86

U

ultimate analysis 27–28
unicursal network 48
uniformity coefficient 429
Union Electric Company 335
upflow separator 189
user satisfaction index 66
Utah, U. of 158

V

velocity probes 86
Velzy, C. 313
Venturi scrubber 346, 348
vibrating air classifier 157
vibrating conveyors 88
visible litter 74
von Roll grate 309–310

W

Walker, W. 57
waste allocation 56
waste generation 13, 59–62
waste heat recovery 331–340
waste reduction 6–8
Waste Resources Allocation Program 57–59
water consumption by plants 399
water-gas shift 362
waterwall incineration 291, 332
wet density 91
wet oxidation 325–327
wet scrubbers 344, 356–349
windrow 264, 268, 270
Wisconsin Electric 335
Wisconsin, U. of 93
wood chips 270
Woodyard, J. 25
work index (Bond) 115–116
Worrell, W. 139
WRAP 57–59

Z

zigzag air classifier 155–156

TD
194.5
.V47
1981